"This book by two experienced researchers and interventionists successfully balances academic research with practical advice about how to best assure the health and safety of employees. It provides comprehensive coverage of the major workplace hazards, their effects on individuals and organizations, and how best to manage them to minimize harm."

—**Paul Spector**, *University of South Florida, USA*

"This book provides a thorough, yet concise introduction to the field of Occupational Health Psychology. The writing is approachable, the twelve chapters are well-organized, and the authors clearly connect OHP research and practice to the real-world complexity of the workplace."

—**Carrie A. Bulger**, *Quinnipiac University, USA*

"Essentials of OHP takes this broadly relevant area of psychology out of its typically-academic confines and makes it accessible to those who work in real-world organizational settings. This well-organized volume will serve as an excellent primer for both students and practitioners interested in promoting worker health, safety, and well-being."

—**David W. Ballard**, *Independent Consultant, UK;*
Practice Committee Chair, Society for Occupational
Health Psychology

ESSENTIALS OF OCCUPATIONAL HEALTH PSYCHOLOGY

Essentials of Occupational Health Psychology provides a thorough overview of Occupational Health Psychology (OHP) with a focus on empowering readers to take appropriate and reasoned action to address a wide variety of worker health, safety, and well-being challenges that are present in working situations all over the world.

Although relatively new as an area of specialization, OHP research and intervention efforts are already having major impacts on the way work is done around the world. Each of the twelve chapters in *Essentials of Occupational Health Psychology* addresses an essential aspect of OHP, with a consistent emphasis on putting what is known about that area into practice. Topics include essential background information regarding the history of OHP and major areas of OHP research and practice, such as work-related stress and recovery, psychological and physical demands and resources, interpersonal mistreatment, work and nonwork role dynamics, and safety. Each chapter features a discussion of why these topics are important to workers and organizations, as well as pertinent evaluation and/or intervention recommendations to help readers better understand what they can do to improve worker health, safety, and well-being, and how to convince others of the value of such efforts. Additional supplements within each chapter include a set of targeted learning objectives to help structure student reading and in-class discussion, focused discussion questions, pertinent media resources to provide current examples of these topics, and professional profiles based on interviews conducted by the authors with fourteen well-known and widely respected OHP researchers and practitioners.

Essentials of Occupational Health Psychology is valuable to graduate and advanced undergraduate students as well as working professionals who are interested in learning how to manage work environments that support worker health, safety, and well-being. The chapters in this text could also provide supplemental reading for training and development workshops for professionals in related disciplines who could benefit from a better understanding of the psychology associated with work experiences.

Christopher J. L. Cunningham, Ph.D., is a Professor of Psychology and the Industrial-Organizational Psychology Graduate Program Director at The University of Tennessee at Chattanooga. In addition to his occupational health psychology research and consulting, he is the Chief Science Officer at Logi-Serve, LLC, and the 2020–2022 President of the Society for Occupational Health Psychology.

Kristen Jennings Black, Ph.D., is an Assistant Professor of Psychology at The University of Tennessee at Chattanooga. Her current research and consulting work focus on employee health and well-being, with specific interests in high-risk work environments, workplace stress, social support, and employee engagement.

ESSENTIALS OF INDUSTRIAL AND ORGANIZATIONAL PSYCHOLOGY
Series Editor
Scott Highhouse

ESSENTIALS OF JOB ATTITUDES AND OTHER WORKPLACE
PSYCHOLOGICAL CONSTRUCTS
Edited by Valerie I. Sessa and Nathan A. Bowling

ESSENTIALS OF OCCUPATIONAL HEALTH PSYCHOLOGY
Christopher J. L. Cunningham and Kristen Jennings Black

For more information about this series, please visit: www.routledge.com/
Essentials-of-Industrial-and-Organizational-Psychology/book-series/EIOP

ESSENTIALS OF OCCUPATIONAL HEALTH PSYCHOLOGY

Christopher J. L. Cunningham
Kristen Jennings Black

Routledge
Taylor & Francis Group

NEW YORK AND LONDON

First published 2021
by Routledge
605 Third Avenue, New York, NY 10158

and by Routledge
2 Park Square, Milton Park, Abingdon, Oxon, OX14 4RN

Routledge is an imprint of the Taylor & Francis Group, an informa business

© 2021 Taylor & Francis

The right of Christopher J. L. Cunningham and Kristen
Jennings Black to be identified as authors of this work has been
asserted by them in accordance with sections 77 and 78 of the
Copyright, Designs and Patents Act 1988.

Library of Congress Cataloging-in-Publication Data
Names: Cunningham, Christopher J. L., author. | Black, Kristen
Jennings, 1991- author.
Title: Essentials of occupational health psychology / Christopher
J. L. Cunningham, Kristen Jennings Black.
Description: New York : Routledge, 2021. | Includes
bibliographical references and index
Identifiers: LCCN 2020056146 (print) | LCCN 2020056147
(ebook) | ISBN 9781138541115 (hardback) | ISBN
9781138541122 (paperback) | ISBN 9781351011938 (ebook)
Subjects: LCSH: Psychology, Industrial. | Job stress. |
Industrial safety.
Classification: LCC HF5548.8 .C86 2021 (print) | LCC
HF5548.8 (ebook) | DDC 158.7—dc23
LC record available at https://lccn.loc.gov/2020056146
LC ebook record available at https://lccn.loc.gov/2020056147

ISBN: 978-1-138-54111-5 (hbk)
ISBN: 978-1-138-54112-2 (pbk)
ISBN: 978-1-351-01193-8 (ebk)

Typeset in Galliard
by codeMantra

CONTENTS

Series Foreword   xi
SCOTT HIGHHOUSE

Preface   xii
Acknowledgements   xv
About the Authors   xvi

1 **Understanding the Psychology of Occupational Health**   1
CHRISTOPHER J. L. CUNNINGHAM AND KRISTEN JENNINGS BLACK

2 **Designing and Evaluating Occupational Health
Psychology Interventions**   24
CHRISTOPHER J. L. CUNNINGHAM AND KRISTEN JENNINGS BLACK

3 **Individual Differences That Matter in OHP**   55
CHRISTOPHER J. L. CUNNINGHAM AND KRISTEN JENNINGS BLACK

4 **Worker Psychological Health**   81
KRISTEN JENNINGS BLACK AND CHRISTOPHER J. L. CUNNINGHAM

5 **Worker Physical Health**   107
KRISTEN JENNINGS BLACK AND CHRISTOPHER J. L. CUNNINGHAM

6 **Work-Related Stress and Recovery**   129
CHRISTOPHER J. L. CUNNINGHAM AND KRISTEN JENNINGS BLACK

7 **Psychological and Social Demands and Resources**   161
CHRISTOPHER J. L. CUNNINGHAM AND KRISTEN JENNINGS BLACK

8 **Interpersonal Mistreatment at Work**   183
KRISTEN JENNINGS BLACK AND CHRISTOPHER J. L. CUNNINGHAM

CONTENTS

9 **Work and Nonwork Role Dynamics**   206
CHRISTOPHER J. L. CUNNINGHAM AND KRISTEN JENNINGS BLACK

10 **Physical and Environmental Demands and Resources**   232
KRISTEN JENNINGS BLACK AND CHRISTOPHER J. L. CUNNINGHAM

11 **Safety at Work**   256
KRISTEN JENNINGS BLACK AND CHRISTOPHER J. L. CUNNINGHAM

12 **Broadening OHP Impact Beyond the Workplace**   280
CHRISTOPHER J. L. CUNNINGHAM AND KRISTEN JENNINGS BLACK

Index   299

SERIES FOREWORD

Occupational health psychology (OHP) catapulted onto the IO landscape in the mid-1990s, with the establishment of the *Journal of Occupational Health Psychology* (JOHP), and the first APA/NIOSH international conference on Work, Stress, and Health. Since that time, the *JOHP* has had a significantly accelerating impact factor, and was recently ranked 12th in journal prestige by SIOP members (Highhouse et al., 2020). The same article showed that OHP submissions to the annual SIOP conference rose 80% in just the last 10 years.

It has become commonplace for prospective doctoral students to express interest in studying OHP topics as a central part of their scholarship, and for IO programs to advertise for professors specializing in OHP. It only makes sense, therefore, that the latest addition to the *Essentials* series is a textbook on OHP.

I am delighted to introduce the new book by Christopher Cunningham and Kristen Jennings Black entitled *Essentials of Occupational Health Psychology*. Chris and Kristen have set the standard for an IO textbook that not only reviews current theory and research, but also provides methods for gathering data, intervening, and applying the research to enhance well-being at work.

The goal of the *Essentials of Industrial and Organizational Psychology Series* is to produce accessible guides that cover basic and advanced concepts in a straight-forward, readable style. Each book in the series covers a specific topic, providing essentials that managers, practitioners, and other well-educated people should know. *Essentials of Occupational Health Psychology* epitomizes the goals of this series, and arms educators and practitioners with the tools they need to become experts in workplace health and safety.

Scott Highhouse
Series Editor
Bowling Green State University

PREFACE

We wrote this book to provide a broad, but accessible and practice-focused introduction to the field of OHP. Although there has been tremendous OHP-related interest and activity in recent decades, there are limited educational resources available that are translational in their focus and targeted at an audience that is more focused on intervening to address, rather than researching to understand, these topics. The material in this book comes from our nearly 30 years of combined experience as occupational health psychology (OHP) researchers, educators, interventionists, and evaluation experts. Throughout these chapters, we also summarize many more decades of theorizing and researching the most essential topics in OHP. Our main objective with this book is to provide a resource that helps you develop an understanding of the most common worker health, safety, and well-being (WHSWB) challenges and opportunities. We have also done our best to facilitate your future translation and application of this material into just about any work situation you are likely to encounter.

For most people, OHP is not necessarily a career path in its own right. Instead, it is more of a mindset, lens, or perspective that is relevant to many different types of professionals who work in organizational environments (as we discuss more fully in Chapter 1). This lens becomes invaluable when evaluating organizational decisions that are being made and are likely to impact workers, or when employees are being asked to make decisions that will impact their quality of life, family, and community. The many WHSWB challenges that can arise are complex and require us to adopt systemic, multilevel, and multidisciplinary ways of thinking and intervening. This, in turn, requires us to think and work across disciplines, as these challenges are not resolvable with one set of knowledge or skills. Our emphasis in this text is on the psychology involved in occupational health and safety challenges and how psychological research and theory factor into managing these challenges. This is important, because it is impossible to change and expect to maintain change in human thought and behavior without an understanding of psychology. Wherever possible, we also try to emphasize areas where other disciplines are likely to be informative and helpful in addressing these challenges.

We expect this book will work well for a variety of audiences, including graduate and undergraduate students, but also working professionals who are interested in learning about how to create and sustain work environments and experiences that support positive WHSWB. The chapters in this text are also well suited as supplemental reading for training and development workshops for professionals in related occupational health disciplines (e.g., public health, ergonomics, industrial hygiene, occupational medicine) who could benefit from a better understanding of the psychology associated with WHSWB.

We have structured this book so it can be used as a course text to facilitate a semester-long exploration of these topics. There are 12 chapters, targeting the topics listed in the table of contents. Each chapter targets a different and essential aspect of OHP, with a consistent emphasis on putting what is known about WHSWB into practice. First, we provide essential OHP background information over three chapters in which we discuss the history and methods of OHP, as well as the complex influence and impact of individual differences among workers. We then discuss psychological and physical health as main outcomes or targets for OHP research and practice. The six subsequent chapters each tackle major areas of OHP research and practice, including work-related stress and recovery, psychological and physical demands and resources, interpersonal mistreatment, work and nonwork role dynamics, and safety. We then conclude with a broader wrap-up chapter that highlights ways that OHP can have an impact even beyond specific organizations and work environments.

To facilitate learning and improve the utility of this book in OHP-related course settings, each chapter includes a set of targeted learning objectives to help structure student reading and in-class discussion. Additional supplements included with each chapter include focused discussion questions, pertinent media resources, and professional profiles based on interviews we conducted with 14 well-known and widely respected and skilled OHP researchers and practitioners. We hope you will find these profiles inspiring and understand that OHP professionals can take many forms. The collective wisdom of this panel of OHP subject matter experts is also summarized in a series of tables presented in Chapter 1.

As you will learn through reading this text, there is not one right answer, solution, or intervention that can fully address all WHSWB issues. However, the existing evidence base is full of helpful guidance, despite some inconsistencies in specific findings and effects across different samples and contexts. While these details are sorted out by ongoing research, it is still possible to use what we know now to start making positive changes to protect and promote WHSWB. Waiting for perfect clarity on these matters before taking action will only ensure that WHSWB risks and problems continue to grow in prevalence and complexity.

With this in mind, we want to be clear that the material and guidance we offer in this text is not based on some perfect source of wisdom. Instead we are offering contextualized guidance derived from what we understand to be true about OHP at the time of writing this book. This information comes from a larger scientific evidence base that will continue to grow and improve in its clarity and depth as research and evaluation work continues pertaining to these topics. In addition, our goal with this book was to supplement and complement other published resources (cited throughout the following chapters), which provide comprehensive summaries of OHP history and research.

We hope you are as excited to explore this book as we have been to write it. We encourage you to be engaged with this material in the classroom, but also to look for ways to apply it in your personal and professional life. We sincerely hope that this information can help you find and sustain safe, healthy, and meaningful work throughout your career.

ACKNOWLEDGEMENTS

A book like this only can happen with sustained and coordinated collective effort. We sincerely thank the editorial team at Routledge/Taylor & Francis Group, especially Christina Chronister and Danielle Dyal, and the editor of this series, Scott Highhouse, for their encouragement and support as we took on and worked through this challenging project. We are also ever so grateful for the talent and assistance of several graduate and undergraduate research assistants at The University of Tennessee at Chattanooga, especially Megan Rogers, Allie Martin, Camille Wheatley, Megan Warrenbrand, Bethany Sikkink, Katie Werth, Braden Sanford, Shelby Farrar, Hayden Curtis, and Damian Spears. These students helped us to gather, review, and organize the massive literature base that supports this text. Finally, we both are grateful for our faith, families, and friends, which collectively enabled us to navigate major life events, illnesses, and even a global pandemic while working to complete this project.

ABOUT THE AUTHORS

Christopher J. L. Cunningham, Ph.D., is an Industrial and Organizational (I-O) and Occupational Health Psychology (OHP) educator, researcher, and practitioner. Chris is a UC Foundation Professor in the Department of Psychology at The University of Tennessee at Chattanooga (UTC), where he is also the I-O Psychology Graduate Program Director. Chris earned his Ph.D. in Industrial-Organizational Psychology with a concentration in Occupational Health Psychology at Bowling Green State University. At UTC, Chris teaches graduate-level seminars on organizational psychology, OHP, consulting skills and ethics, and organizational development and change, and undergraduate courses on professional ethics and career planning. Outside of the university, Chris is a founding partner and Chief Science Officer at Logi-Serve, where he develops, validates, supports, and implements advanced talent management technologies. Chris is also the 2020–2022 President of the Society for Occupational Health Psychology, a long-standing member of the Society for Industrial and Organizational Psychology and the American Psychological Association and an editorial board member for seven well-respected journals that regularly publish OHP-related research.

Kristen Jennings Black, Ph.D., is an assistant professor in the Department of Psychology at The University of Tennessee at Chattanooga. Kristen received her Ph.D. in Industrial-Organizational Psychology from Clemson University, with a specialization in OHP. Her current research focuses on employee health and well-being, with specific interests in in high-risk work environments, workplace stress, social support, and employee engagement. Her research has been published in a number of peer-reviewed journals, including the *Journal of Occupational Health Psychology, Work & Stress, and Stress & Health.* She teaches undergraduate courses in research methods, statistics, and health psychology. At the graduate level she teaches research methods and statistics, as well as a seminar on groups and teams in organizations. Beyond teaching and research, she is a member of several professional organizations, including the Society for Industrial and Organizational Psychology and Society for Occupational Health Psychology. She serves as a reviewer for professional conferences and several peer-reviewed journals, and is an editorial board member for *Occupational Health Science.* Kristen is also regularly engaged in consulting work in the areas of employee stress, safety, and engagement.

1

UNDERSTANDING THE PSYCHOLOGY OF OCCUPATIONAL HEALTH

Christopher J. L. Cunningham and Kristen Jennings Black

This chapter provides readers with a foundational understanding of the field of occupational health psychology (OHP) and its reason for existence. Professionals in OHP focus on protecting and improving worker health, safety, and well-being (WHSWB). We review this field of OHP and show at a high level, how OHP professionals fit within a broader network of professional disciplines with expertise relevant to the protection and promotion of WHSWB. In this chapter we explain what OHP is and how OHP is most effective as part of a multidisciplinary approach to managing WHSWB. In addition, we highlight and briefly describe a number of organizations and initiatives that promote and support OHP-related research and practice.

When you are finished reading this chapter, you should be able to:

LO 1.1: Describe the history and origins of the field of OHP.
LO 1.2: Describe the current state of OHP as a research and practice discipline.
LO 1.3: Explain why and how addressing WHSWB requires multidisciplinary efforts that include psychology.
LO 1.4: Identify various institutions and organizations that support ongoing education and dissemination of OHP knowledge.

Overview of OHP

As you are reading this chapter, we can only assume that you are interested in better understanding what OHP is all about and how the theories, methods, and findings from OHP research and practice can be used to benefit workers, organizations, and maybe society more generally. By definition, OHP is a field of study and practice in which behavioral and social science theories, principles, and evidence are used to protect, promote, and generally improve worker health, safety, and well-being (WHSWB). For most OHP professionals, these three areas are broadly defined and multifaceted. *Health* is typically conceptualized along multiple continua pertaining to physical, psychological, social, and emotional forms (see Hoffmann & Tetrick, 2003 for an excellent discussion of different ways health has been operationalized). *Safety* is often focused on physical risk situations, with the goal of preventing physical harm, injury, or even death. However, safety for OHP professionals may also include psychological forms of safety.

Finally, worker *well-being* is related to health and safety, but often conceptualized even more broadly and in a multi-dimensional way to include dimensions of positive health along with the availability of essential resources that support health (e.g., shelter, food, security) and a sense of purpose and meaning (Bennett et al., 2017; Chari et al., 2018). Together, these elements contribute to the very broad concept of general wellness (e.g., Adams et al., 1997), which is also linked to a variety of *salutogenic* or health-causing factors (Antonovsky, 1979). These ultimate objectives of WHSWB are met not only when health-related symptoms are reduced, but when workers are able to positively and adaptively function in emotional, social, and cognitive ways that support thriving and even flourishing (cf., Fredrickson, 1998; Ryff & Singer, 1998).

The focus for OHP researchers and practitioners is primarily on the ways in which experiences at work, exposures to work tasks, and associated environmental factors impact WHSWB. Because worker health transcends the boundaries of a work domain, OHP professionals also often consider cross-domain (i.e., work-to-nonwork and nonwork-to-work) phenomena and effects. This crossover and spillover of health-related influences and outcomes between work and nonwork domains means that promoting and protecting WHSWB is important for workers, organizations, and society more broadly.

Defining OHP

When attempting to define a construct, phenomenon, or, in this case, professional discipline, it is helpful to not only describe or delineate what it is, but also what it is *not*. When it comes to OHP, this is easier said than done, because the knowledge, skills, and abilities of well-trained and experienced OHP professionals is often so broad. OHP supports and is influenced by many professional disciplines that target occupational health, including occupational medicine, industrial hygiene, and public health (Macik-Frey et al., 2007). OHP professionals can amplify, facilitate, and enhance the work of professionals with more domain specific KSAOs by leveraging expertise in designing and conducting research and interventions to promote and protect WHSWB. We share the hope of many in this arena that in the coming years we will see an increase in the amount of interdisciplinary collaboration on research and practice efforts to address WHSWB challenges (cf., Sauter & Hurrell, 2017).

Most OHP professionals are trained to be scientists and practitioners, which can mean a variety of different things. In our interviews with OHP professionals (noted in the Preface), we asked how a scientist-practitioner

Table 1.1 OHP Professionals' Perspectives on Being an OHP Scientist-Practitioner

Key Themes	*Essential Elements*
Science informs best practice and provides solutions with high likelihood of success	• Understanding underlying mechanisms of a tool, program, or activity • Developing interventions based on what is known • Removing need for expensive consultants
Science and practice inform and support each other	• Theories and empirical evidence help identify what is happening and why • Practice identifies key issues and real-world problems • Practice provides real-time and real-world experience • Practice helps translate science into application
Scientist-practitioner model helps to ensure comprehensive development of graduate students and early career professionals	• Teaches research-minded to always think about the "so what" and application potential • Challenges practitioner-minded to seek understanding and explanation through data and theory

model applies to their OHP-related work. The overarching theme from these responses (more detail in Table 1.1) is that OHP research and practice are strengthened and better aligned with actual WHSWB issues because of this underlying scientist-practitioner perspective.

Most OHP professionals are not licensed or otherwise certified to provide direct healthcare or clinical services to individuals. Our ability to fully address WHSWB issues is contingent on our ability to collaborate and coordinate with other occupational health professionals. For example, epidemiologists may be very effective at monitoring and describing health-related phenomena, but may struggle to explain why workers are behaving in certain ways. Likewise, industrial hygienists might be able to identify significant exposure risks in a work setting, but be more effective at addressing these risks with OHP assistance in designing, implementing, and evaluating psychologically, socially, and behaviorally oriented interventions targeting the worker in addition to the work environment.

Figure 1.1 illustrates how OHP knowledge and methods can amplify other types of efforts to improve WHSWB. The left panel in this figure is a simple representation of how various professional disciplines and silos may attempt to improve WHSWB in a general and perhaps not fully direct way. The dotted lines indicate that these efforts may be less than fully effective. The right panel in this figure illustrates how these general efforts can be amplified or strengthened when OHP knowledge and methods are included, ensuring that critical psychological and social factors are taken into account. In other words, OHP professionals can complement and supplement work by other occupational health professionals, by improving depth of understanding, deepening and broadening systems thinking about WHSWB issues, and strengthening WHSWB intervention delivery and evaluation strategies.

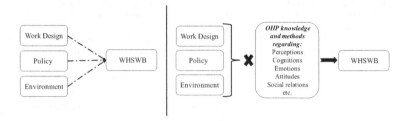

Figure 1.1 OHP as Amplifier of Efforts to Influence WHSWB

A Multi- and Inter-Disciplinary Imperative

The broad class of WHSWB issues targeted by OHP professionals is understandably complex and multifaceted. It should not be surprising, therefore, that these issues cannot be addressed from a single perspective or set of knowledge, skills, or competencies. For OHP professionals to be as impactful as possible, we need to learn to think and work in a multidisciplinary fashion (Adkins, 1999). This means understanding and collaborating with professionals in a variety of different domains that all have expertise pertinent to the challenge of managing WHSWB challenges.

This includes but is not limited to the professional disciplines summarized in Table 1.2. The definitions and societies referenced are by no means comprehensive, but high-level examples of the work done in these diverse areas to understand and address WHSWB issues. Even from just a psychological perspective, there are many different pathways and approaches to OHP-related work. Obviously, there is a strong representation of professionals with psychology training in OHP most specifically. All occupational health professionals, however, must understand and appreciate the need to work with professionals in other related occupational health and applied psychology disciplines to actually understand and address complex WHSWB challenges (Sauter & Hurrell, 2017).

OHP as a Lens or Hub

A really important and powerful way of thinking about OHP is to treat it as a lens through which we can more fully see, understand, and ultimately impact WHSWB issues. Using this lens can lead us to make and advocate for better choices and decisions in the many situations where WHSWB is likely to be affected. This applies to just about all of us, regardless of the positions we may hold in organizations. For example, any talent management professional has frequent opportunities to apply an understanding of OHP to inform decisions and take actions that keep workers healthier and safer, regardless of organizational context. We can also think of OHP as a sort of hub science, which can help to connect and amplify work that is done to protect and promote WHSWB from many different functional areas (as represented in Figure 1.2).[1]

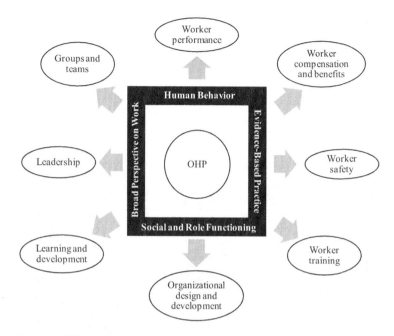

Figure 1.2 OHP as a Lens or Hub

Developing and Supporting OHP

What we identify as OHP today is a relatively new area of specialization for those who are primarily trained in some branch of applied psychology (as discussed in the next section). As noted in an informative early review by Quick (1999a), however, the field of OHP has evolved over a long history, which can be traced back at least as far as the late 1800s and the work of early occupational health focused researchers and practitioners such as Hugo Münsterberg (1913), and showing up in later work by other well-known researchers such as Kornhauser (1965; see also Zickar, 2003), Herzberg (1964), and Lazarus and Folkman (1984), among many others. The current profession of OHP really started taking shape in the late 1990s. The use of the OHP label for this area of specialization is traced back to an article by Raymond, Wood, and Patrick (1990), in which a model was outlined for OHP as a doctoral-level specialization. Around this same time, the American Psychological Association (APA) and the National Institute for Occupational Safety and Health (NIOSH) had initiated a bi-annual conference series now known as the International Conference on Work, Stress, and Health. This ongoing conference series continues to be an important forum for researchers and practitioners interested in WHSWB topics.

To be clear, there has been interest among philosophers, and behavioral and social science researchers in the psychological and social effects of and

Table 1.2 Summary of OHP-Related Professional Disciplines

Specialization *(example professional organization)*	*Focus and relevance to* *OHP research and practice?*
Industrial-Organizational (I-O) Psychology www.siop.org http://www.eawop.org	I-O Psychologists apply principles of psychology to work and organizations. They study details of workplace functioning, including recruiting, selection, training/development, retention, and succession planning. They can provide insight on how workplace dynamics and workforce planning decisions could be affected by or contribute to WHSWB.
Health Psychology https://societyforhealthpsychology.org https://ehps.net	Health Psychologists apply psychological principles to understand health and illness. They can provide a broad perspective on the interplay between biological, social, and psychological factors that affect an individual's health.
Clinical/Counseling Psychology https://www.div12.org	Clinical or Counseling psychologists are trained to diagnose and treat psychological and emotional disorders. Through one-on-one work with individuals and consultations with OHP professionals, they can recommend strategies for coping with stress and work-related accommodations for someone managing a psychological health concern.
Positive Psychology https://www.ippanetwork.org http://www.div17pospsych.com	Positive psychologists are focused on human flourishing and well-being. Through their focus on building positive and sustaining positive states, they add to practices often centered on reducing negative states. They may develop broad interventions to be administered in groups or work one-on-one with clients.
Ergonomics and Human Factors Psychology https://www.hfes.org/home https://iea.cc	Human Factors Psychologists and Ergonomists are trained to evaluate and design environments to maximize human well-being and the performance of a given system. They may be particularly helpful for OHP work involving adapting the physical environment to improve health and safety.
Industrial Hygiene https://www.aiha.org https://www.assp.org https://www.aioh.org.au	Industrial hygienists or occupational safety technicians often work directly with organizations to develop and administer safety programs that prevent or correct issues with the work environment. Because they are directly involved in workplace safety initiatives, they often have detailed knowledge on specific safety risks and intervention strategies.

Preventative/ Occupational Medicine
https://www.iomsc.net
https://acoem.org

Occupational medicine is a form of preventative medicine, which focuses on understanding risks and support strategies for the workplace. They may be involved in population-based healthcare for a particular group of workers, as well as efforts for prevention of disease and disability among at-risk workers.

Occupational Therapy
https://www.wfot.org
https://www.aota.org

Occupational therapists help individuals with injuries, diseases, or disabilities build skills for daily functioning. This includes vocational skills for returning to work or continuing to work with a chronic condition.

Physical Therapy
https://www.isprm.org
https://www.apta.org

Physical therapists help individuals to improve their mobility, relieve pain, or increase strength that results from disease of injury. They focus on helping individuals physically adjust from an injury or chronic condition.

Employment Law
https://www.eela.org
https://www.leraweb.org

Those who work in employment law (e.g., labor relations, equal opportunity representatives) help employees and employers by managing worker disputes or ensuring legal compliance within an organization. Employment law professionals can provide insight into how workplace programs in general affect the rights of workers, particularly those who are members of protected groups.

Economics
http://www.iea-world.org
https://www.aeaweb.org

Economists conduct research to understand trends and address financial and economic problems. They can help to model and predict the financial costs associated with worker health, both the costs associated with poor health and safety and potential gains from positive health and safety.

Public Health and Health Education
https://www.apha.org
https://eupha.org

Public Health researchers and educators are involved with identifying community needs, providing resources to support health, and providing education to encourage healthy lifestyles. They focus on health trends, challenges, and opportunities for larger groups (e.g., the world, a country, a community). These individuals can help to understand the broader social context around a health or well-being concern.

Note: Descriptions were informed by professional societies of the different disciplines as well as O*NET documentation of typical job tasks for each (https:// www.onetonline.org).

opportunities provided by work for at least the last couple of centuries. Going back even further, the notion of work as an important component to protecting and demonstrating human dignity is a component to the teachings of most organized religions. Much of this is discussed in the excellent and detailed historical overview of OHP provided by Schonfeld and Chang (2017). In addition to being aware of this background, it is also essential that we understand the role of many corresponding occupational health specialties that have also contributed to the development of what we now know as OHP (for an excellent summary, see Sauter & Hurrell, 2017). A detailed timeline of key events in the development and life of the OHP field is available in the summary of the history of the Society for Occupational Health Psychology (SOHP; http://sohp-online.org/field-of-ohp/history-of-ohp/).

Education and Training in OHP

Although the focus of this book is on OHP primarily as a form of applied psychological research and practice, most WHSWB issues are complex and multi-faceted. An implication of this, as we have discussed, is that an interdisciplinary and multidisciplinary approach is often needed to best address and manage the factors that influence WHSWB outcomes, including environmental factors, organizational policies and procedures, and broader societal norms and forces (Adkins, 1999). This being the case, there is not just one way to pursue and obtain an education that will set you up for a career in which OHP knowledge will be pertinent. As you will see perhaps most clearly in the Professional Profiles we have included in each chapter of this book, there are many ways to develop yourself into an OHP researcher, educator, and/or practitioner. Even if you are not trained in psychology, the material in this book and the research and practice done by OHP professionals is important to understand, as it helps to illustrate *how* and *why* people do what they do, not just *what* it is that people are doing (which is commonly the case in other occupational health professions).

Generally, OHP professionals tend to have at least a master's level degree and often some form of doctoral level training in an area of psychology (e.g., industrial-organizational, health, clinical, social) or another relevant specialty area, including public health, human factors, epidemiology, sociology, industrial hygiene, etc. There is no universal standard for education in OHP, though there have been attempts to provide best-practice guidance when it comes to curricular elements and applied training (Houdmont et al., 2008; Schneider et al., 1999; Westlander, 1994).

Formal, structured opportunities for education and training in OHP did not exist prior to the middle 1990s, when APA and NIOSH jointly sponsored a set of funding opportunities for post-doctoral training. Seven individuals were able to expand their graduate training by diving more deeply into OHP-focused topics. Later in the 1990s, APA and NIOSH again teamed up to support the development of graduate-level training programs for OHP.

The first training programs were largely supported by special training fund-ing and grant-supported research, and as such, were located in larger institu-tions that had a solid research footprint. It was through these early training mechanisms that the first recognized graduate level training programs were established. Details on these early days of OHP education are well-presented by Adkins (1999) and Quick (1999a, 1999b). It is important to note that many of the original doctoral programs to offer OHP-related training con-tinue to do so, and many additional doctoral and master's level opportunities have emerged since then, throughout the world.

Common curricular elements to OHP-related training and education programs include behavioral and social science research methods, statistical analysis techniques, applied psychological theories, and translation of empir-ical research to practice. If you are interested in finding graduate education opportunities in OHP, a good place to start are dedicated professional so-ciety websites, through SOHP (http://sohp-online.org/resources/graduate-training-in-ohp/) or the European Academy of Occupational Health Psy-chology (EA-OHP; http://www.eaohp.org/education-and-training.html).

Training in OHP often begins with formal education, but learning does not stop there. As the number of OHP researchers and practitioners continues to grow, there is more and more opportunity to benefit from lessons learned by established professionals, peers, and newcomers with fresh perspectives. An im-portant mechanism for international sharing of knowledge and experience is through the annual series of international conferences focused on WHSWB is-sues. The list of possible conferences is vast and one of the best ways to see what is happening and available is to access the list of such events managed by the International Coordinating Group for OHP (http://www.icg-ohp.org/), which is composed of representatives from a variety of related professional OHP-related societies worldwide (many of which are noted in the next subsection). Maintain-ing global connections and perspectives in our education, training, research, and interventions are essential to OHP's ability to impact universal WHSWB issues.

Rounding out this section, Table 1.3 presents a rank-ordered list of recom-mendations to students interested in pursuing an OHP-related career. These recommendations come from our panel of OHP professionals noted earlier; think of this information as collective advice you might obtain from working with multiple top-notch OHP mentors. Although we both share much of this advice with our students now, we both wish we had heard all of this advice when we were first getting excited about OHP as a possible career direction.

Supporting Institutions and Organizations

Most OHP research and practice efforts are supported by the educational and training resources already discussed, and by a number of funding sources (typically national/state/provincial level governments and other non-profit/ philanthropic groups that target these issues). The availability of funding to support this type of work is limited and competition for this funding is

Table 1.3 Recommendations to Students Interested in Pursuing an OHP Career

Main Recommendations	Specific Strategies
Get Experience	• Applied projects • Research efforts • Multidisciplinary engagements • Take every opportunity available • Go out and do it! • Use your available resources
Network	• Find mentorship • Seek out people with similar values • Stay connected to academia • Talk to employees • Talk to professionals in the field or related industries
Understand/Explore OHP	• Attend OHP conferences • Join relevant societies • Know OHP is interdisciplinary • Read/stay current on research
Succeed Academically	• Do well in school • Get relevant education • Work hard
Know Yourself	• Know your values, motivation, interests • Maintain your integrity
Develop Skills	• Communication • Writing • Research Methodology
Work Well with Others	• Respect your coworkers • Collaborate • Work well with others
Follow Your Interests	• Focus on what interest you • Have a desire to learn • Have a passion for the field • Have a passion to help others
Other	• Fix one problem at a time • Always ask questions • Have fun • Focus on making a difference

increasingly strong. This is especially true in the case of application-oriented work (e.g., intervention studies, pilot programs), which sometimes are funded directly by organizations directly.

Being successful as an OHP professional, however, requires more than good training and funding. As any OHP professional will tell you, it is also critically important to be engaged with other professionals in this domain and to maintain an active and rich professional network that can provide support and help to generate collaborative research and practice opportunities. In this section, we outline five general forms of organizations that provide both formal and informal support that can contribute to your success as an OHP professional.

Professional Societies and Associations

Professionals with a background in psychology or at least an appreciation for the value and importance of behavioral and social science principles when working to understand and address WHSWB challenges have a professional home within SOHP. This organization is relatively young but has sustained and grown its membership steadily since its founding in 2004 and 2005. The mission of SOHP is to help improve worker health, safety, and well-being, while also facilitating their productivity in work environments. This organization works towards these goals by promoting research associated with WHSWB issues, encouraging and facilitating the translation of this work into practical advice and guidance, and supporting and improving OHP-related education and training opportunities. Membership options are available at a variety of levels, including students and international affiliates. Full details about SOHP and its membership benefits are available through SOHP's website (http://www.sohp-online.org). In a more global sense, there are other professional organizations similar in some ways to SOHP, but with more of a country- or region-specific focus. This includes organizations such as the EA-OHP (http://www.eaohp.org/).

As a professional organization representing all psychologists, the APA is the largest, with more than 118,000 members, representing all manner of psychology specializations and subdisciplines. When it comes to issues of WHSWB, this organization also has a lot to offer. From the early days of OHP, several key leaders at APA helped to support and encourage the development of this profession. Two of the most visible ways in which this support is still evident today is in the ongoing International Conference on Work, Stress, and Health and the highly respected *Journal of Occupational Health Psychology*. In recent years, the APA Office of Applied Psychology has helped to emphasize the importance of applying psychology to improve workers' experiences at work as well as organizational functioning. The majority of OHP professionals who are members of APA are also members of APA's Division 14, the Society for Industrial and Organizational Psychology. This organization also supports many OHP-related initiatives through its publications and annual conference. More details about these two professional organizations are available on their websites (https://www.apa.org and https://www.siop.org).

Government-Supported Organizations

Government funding and support for OHP-related work varies quite a bit in form and function across different countries and regions of the worlds. Some governmental resources to support WHSWB exist at a state or provincial level, or perhaps at a regional level. More common, however, are country-level governmental resources such as those found in many Western European (e.g., https://osha.europa.eu/en) and Nordic countries, as well as Canada (e.g., https://www.canada.ca/en/employment-social-development/programs/health-safety.html), Japan (https://www.jisha.or.jp/english/), and Australia (e.g., https://www.worksafe.qld.gov.au/). These countries and a few others have established strong policy and procedural frameworks, guidelines, and oversight through agencies or government offices to protect and promote WHSWB. In contrast, many more recently industrialized and still developing economies have essentially no worker protections in place. Many of these countries also lack an established and sufficient set of OHP-related professionals who can address WHSWB issues. In between these two ends of the spectrum are several of the largest countries of the world, including the USA, China, and India.

Within the USA, NIOSH, a division of the American Centers for Disease Control (CDC), is the main governmentally funded research and practice organization associated with OHP. As a resource to OHP professionals, NIOSH is incredibly valuable – it is not membership-based, and its employees are continually focused on applied research targeting critical WHSWB priorities that apply to just about every occupation. A tremendous amount of valuable information is available to researchers and practitioners through the NIOSH website (https://www.cdc.gov/niosh/index.htm). In particular, you can find reports published by NIOSH on some unique case studies and interventions that are not published in research journals yet are extremely valuable for learning about risks factors or techniques for specific industries.

The work that NIOSH does is guided by its National Occupational Research Agenda NORA (for details see https://www.cdc.gov/nora/). This agenda is periodically updated with input from councils composed of stakeholders representing a wide variety of professions involved in efforts to understand and improve WHSWB. An early version of this concept of a national agenda for research in this area was shared with the broader psychology community in an article by Sauter, Murphy, and Hurrell (1990). These authors, the first of whom is featured in a professional profile in this chapter, outlined the need for a "comprehensive national strategy to protect and promote the psychological health of workers" (p. 1146). The current NORA guides the work of NIOSH as well as researchers and practitioners all over the world.

To further increase its impact on addressing WHSWB issues, in recent years NIOSH has put strong emphasis behind development of a Total Worker Health® (TWH) initiative. This effort is designed to increase the scope and impact of efforts up to this point to improve and sustain WHSWB. More details about this initiative are available at NIOSH's dedicated website to this topic (https://www.cdc.gov/niosh/twh/default.html) and in a clearly presented review by Schill and Chosewood (2013).

Within the USA, government support for addressing WHSWB issues is fairly limited. Actual laws, policies, and other forms of worker protections are sparse and not strongly supported with strong oversight powers. The main forms of governmental support for OHP-related work are through various grant and contract mechanisms designed to fund applied research efforts. Several government-related organizations support OHP research because of WHSWB concerns related to a specific worker population. For instance, the Department of Defense (https://www.defense.gov/) offers a variety of funding opportunities to support the health, safety, and well-being of military personnel. Related to this, the Department of Veterans Affairs supports efforts to help reintegrate veterans to civilian life, often including a return to civilian employment. This is indeed a great opportunity for collaboration among OHP practitioners.

There are also quite a few funding and collaboration opportunities with specialized organizations that seek to understand unique WHSWB issues pertaining to their workforces. For instance, the National Aeronautics and Space Administration (NASA) supports work that addresses WHSWB issues among the unique population of astronauts, with emphasis recently on those who will engage in long-duration space exploration missions. These workers are likely to experience a number of task-related, social, and personal challenges that can benefit from the expertise of OHP researchers and practitioners. Similarly, the Association for Healthcare Research and Quality (AHRQ) often supports OHP research, as their goal to deliver quality and safe health care hugely depends on the health and safety of their workers. Most of these formal funding opportunities, particularly those affiliated with government agencies, can be found through the American federal government's main grants resource website (https://www.grants.gov/).

Finally, the American Occupational Safety and Health Administration (https://www.osha.gov/), housed within the U.S. Department of Labor, is perhaps most known among business owners and practitioners as the federal agency charged with monitoring and enforcing safety at the federal and state level. In addition to such enforcement, OSHA provides resources for education and training materials that are targeted toward specific risks (e.g., slips, trips, and falls) and specific industries (e.g., mining, construction). OHSA also provides support for work that targets worker safety, funding efforts to train and educate workers on safe practices (e.g., Susan Harwood Training grants) and working with organizations to create policies and procedures that support worker health. OSHA particularly encourages work targeting vulnerable work populations, such as those engaged in hazardous occupations or those with low-income jobs. Additional funding opportunities for work that addresses WHSWB issues in risky occupations can often be found through the United States Department of Labor's grants website (https://dol.gov/general/grants/howto).

Non-Governmental Organizations

Other country-spanning and global influence pertaining to WHSWB comes from non-governmental agencies that provide best-practice guidance and other forms of support to organizations and communities working to protect WHSWB. Examples of such organizations include the World Health Organization (https://www.who.int/health-topics/occupational-health), International Labor Organization (https://www.ilo.org/global/), and various international standards organizations such as the International Standards Organization (ISO; https://www.iso.org/home.html), and the American National Standards Institute (ANSI; https://www.ansi.org/). In addition to providing regulations and general guidance related to a number of WHSWB matters, these organizations serve as a major resource for understanding a variety of broad occupational and societal health topics. They also serve as rich sources of work-related health data (e.g., https://ilostat.ilo.org/).

Local and global assistance with WHSWB challenges is also provided by for-profit consulting organizations and industry-specific groups, which are often contracted by organizations to address these issues (typically as a response to some sort of WHSWB incident). These organizations tend to specialize in particular segments of the WHSWB puzzle (e.g., safety, incivility, environmental controls), but generally tackle these issues with a combination of applied psychology, sociology, and organizational change management principles and techniques.

There are ongoing efforts to coordinate and collaborate across cultures through the preceding societies. One such effort, mentioned earlier, is the International Coordinating Group for OHP (http://www.icg-ohp.org/), which works to connect professional organizations of OHP professionals into a global network. Doing so helps to improve research collaboration and dissemination, and provides a forum for generating best-practice recommendations to address global WHSWB issues. Coordination through this type of network may also help with advocacy efforts to raise awareness and increase funding support for OHP-related initiatives.

Knowledge Dissemination Outlets

Research and practice-oriented information pertaining to WHSWB matters is shared by OHP and other occupational health and safety professionals through a number of well-respected and peer-reviewed journals and edited book series. Preparing for this book, we compiled more than 900 OHP-related intervention research reports and applied research manuscripts and found that these manuscripts are published across an extremely diverse set of journals. To illustrate the multidisciplinary reality of OHP research and practice, just the roughly 300 intervention-related articles we identified came from more than 150 different peer-reviewed journals, with only about 20% of these publications coming from three of the most highly regarded journals connected to the OHP domain (i.e., *Journal of Occupational Health Psychology*, *Journal of Applied Psychology*, and *International Journal of Stress Management*). The rest of these intervention studies were published in a variety of other occupational health focused outlets (e.g., *Work and Stress*, *Journal of Occupational and Environmental Medicine*, *Occupational Health Science*) and discipline-specific journals (e.g., *Journal of Clinical Anesthesia*, *Journal of Dance Medicine & Science*, *The Prison Journal*).

In addition to these types of journal outlets, there are technical reports sometimes made publicly available from governmental and government-related agencies (e.g., NIOSH, RAND, the Defense Technical Information Center [DTIC]) that provide helpful and often application-oriented information. Finally, and as mentioned earlier, a great deal of continuing education, professional networking, and knowledge sharing happens at multiple international conferences targeting WHSWB issues.

Present Challenges for OHP

Considering the relative newness of OHP as a professional domain or specialization, the amount of education, research, and practice to-date in this area is impressive. That said, this field and the broader population of professionals working to address WHSWB face several major challenges. Some of these are seemingly perpetual and have affected workers and organizations since even before the era of industrialization. Other challenges are newer, brought on by technological developments and changes to the way many of us work.

Most WHSWB challenges are connected to multiple areas of occupational health-related research and practice. For this reason, a first ongoing challenge for OHP professionals is figuring out how to increase collaboration and co-ordination within this discipline and with professionals from other related disciplines such as industrial hygiene, occupational medicine, and human factors. As just one example, the increasing presence of robotics and other forms of automation in many work environments challenges workers' sense of meaning and purpose, while also presenting new opportunities for safer operations, and new risks for exposures and injuries. Effectively working to understand and protect WHSWB as the nature of work changes will require OHP professionals to be more actively engaged in multidisciplinary collaborations.

A second ongoing challenge for OHP professionals is figuring out how to demonstrate, qualify, and quantify the impact and effects of OHP-related work in terms are meaningful to business owners and others who respond most strongly to economic indicators of effect. Positive steps have already been made toward addressing this challenge (Goetzel et al., 2004; Naydeck et al., 2008; Ozminkowski et al., 2004), but there is not a consistent and widely used approach for demonstrating economic impact that works for all types of stakeholders. There are also entire areas of OHP research and practice for which connections to financial and economic impact have never been made (though see Novotney, 2017 for examples of positive steps in this direction).

A third challenge involves engaging OHP professionals, workers, and organizations to more fully explore how to best leverage OHP-related research and theories to address the most persistent WHSWB challenges. There are some who argue that the best mechanism for this is through policy changes, often at the level of a national government or international guiding coalition. There are others who feel strongly that the best approach to addressing

OHP-related issues is directly through the work organizations, with changes being made to the practice of work driven by market and economic forces rather than policy mandates. A blended approach is likely to be most effective in most situations, but there are dramatic differences in relevant factors to consider by country, culture, and industry. Stated more simply, this major class of challenge pertains to how we can best and most effectively translate OHP knowledge into practice and actual change.

Finally, there is an ongoing challenge regarding the identity of OHP researchers and practitioners. In the USA especially, a large proportion of folks who identify with OHP are primarily trained in I-O psychology and many other occupational health and safety professionals have no training in psychology. Because addressing WHSWB issues requires a multidisciplinary approach, the challenge here is to effectively facilitate interdisciplinary and multidisciplinary collaborations that begin with education (e.g., blended curricula and other mixed training opportunities) and extend into our research and practice efforts (e.g., developing and managing multidisciplinary research and intervention teams). This blending is likely to lead to more comprehensive research and ultimately interventions, though it may also lead to further challenges in defining what OHP really is (and is not). Our experience tells us, however, that most OHP and other occupational health and safety professionals are less concerned with professional identity than with using their knowledge, skills, abilities, and general competencies to positively impact WHSWB. This is one reason why this field attracts so many professionals with a real concern for optimizing work experiences for people and society more generally.

Concluding Thoughts and Reality Check

Congratulations for working your way through this first chapter. The material we presented here provides an orientation to the breadth of OHP research and practice. We also presented a general overview of interests, challenges, and opportunities associated with WHSWB and discussed the wide variety of ways in which professionals from various disciplines can get involved with OHP. It is no small feat to protect and promote safety and a sense of meaning for workers, while simultaneously ensuring organizations function well and remain competitive. As noted throughout this chapter, there are many professional disciplines that address elements of this challenge. Professionals with OHP training can play a major role in making this all happen.

Throughout the remainder of this book, we will explore many ways in which OHP contributes to WHSWB and is important to workers, organizations, and society more generally. In most of the remaining chapters, we will directly address this issue of importance as it pertains to the topic of each chapter. For now, it is sufficient to note that although focusing on the health, safety, and well-being of workers seems like the "right" thing to do for most people, too often this argument is ignored unless presented with concrete illustrations of how/why such a focus is also financially good for the organization.

For OHP professionals to succeed in getting others to appreciate the importance and value of our research and practice efforts, we have to do a better job of not just understanding specific and isolated WHSWB issues. We have to challenge ourselves to go further and understand these issues within a broader context, which includes causal factors, conditioning variables, and outcomes at the worker level, but also includes contextual details within specific work environments. This might include the costs associated with promoting and protecting worker health using available tools and strategies. This might also include understanding more fully how long it will take for OHP-informed intervention efforts to have an impact that is noticeable on the organization's bottom line. In other words, our challenge is to demonstrate the importance of OHP research and practice with data and evidence that matter to workers, organizations, and society more generally. We take up these matters more in Chapter 2.

Media Resources

- News article examining the toxicity of work environments: http://nyti.ms/1KVRnAT
- Blog post exploring costs of difficult workers: https://blog.careerminds.com/the-true-cost-of-toxic-workers
- Trade publication article discussing the importance of "good" occupational health: https://www.personneltoday.com/hr/reflecting-on-the-role-and-value-of-good-occupational-health/

Discussion Questions

1) What is OHP and how is it related to personal, occupational, and organizational health?

2) How/why does it make sense for scientists and practitioners in any field to concern themselves with psychological and physical health in and out of work settings?

3) Pick one of the disciplines in Table 1.1 and explore the website for one of the professional societies. What are two or three unique perspectives this field brings to understanding WHSWB? How might the work of professions in this field be supported or facilitated by an OHP professional?

4) What is the role of government organizations in protecting and promoting WHSWB? How can OHP professionals help to improve governmental support for WHSWB in all occupations?

Professional Profile: Steve Sauter, Ph.D.

Country/region: United States of America (USA)

Current position title: Senior Consultant with AECOM N&C Technical Services, LLC and NIOSH

Background: I have been working to improve the health, safety, and/or well-being of workers for 43 years. I received my Ph.D. in Industrial Psychology from the University of Wisconsin-Madison and completed a Postdoctoral Fellowship at the University of Wisconsin Medical School in the Department of Preventive Medicine and Center for Environmental Toxicology. I am a member of the American Psychological Association, the Society for Occupational Health Psychology and a Fellow of the European Academy of Occupational Health Psychology.

I was fortunate to find a graduate mentor at the University of Wisconsin's Department of Psychology with a focus in occupational safety and health. I subsequently secured a Post-Doc in a medical school department with specialization in occupational and environmental health (Department of Preventive Medicine, University of Wisconsin Medical School) where I gained practical experience and added to my credentials in occupational health. I was then recruited by NIOSH to head their program in work, stress, and health.

Collectively, these experiences have prepared me well for independent practice as a consultant in occupational safety and health field. I now work primarily as a consultant to the Office of the Director at NIOSH, providing support on workplace psychosocial factors and health to the NIOSH Total Worker Health® Program and to the NIOSH Healthy Work Design Program. The primary focus of this work has been to provide guidance to the Total Worker Health® Program and to the RAND Corporation in a joint NIOSH-RAND project to develop a new survey instrument to assess worker well-being.

How my work impacts WHSWB: Historically, occupational safety and health and sister disciplines (e.g., workplace health promotion, wellness) have occupied different organizational silos. My work with the NIOSH Office of Total Worker Health has helped to foster the integration of these programs in organizations, leading to improved safety, health, and well-being outcomes. These efforts and positive outcomes have been documented in a 2019 volume co-edited by me and published by the American Psychological Association (https://www.apa.org/pubs/books/4316192).

I will comment briefly on three activities over the course of my career to help illustrate the breadth of applications in OHP. First, early in my career I worked with occupational health professionals in the University of Wisconsin's Department of Preventive Medicine to uncover significant neuropsychological deficits among fumigant-exposed grain workers, resulting in both compensation to affected workers and workplace reforms to reduce these exposures. In subsequent work at the Department of Preventive Medicine and at NIOSH I was able to show that the organization of work and workplace psychosocial conditions contributed significantly to the development of musculoskeletal disorders (MSDs) in office work, and I worked with government authorities and industry to help optimize these conditions to reduce the risk of MSDs in office workplaces. Most recently, I have been working with NIOSH and the RAND Corporation to develop a new, comprehensive measure of worker well-being that will be used by organizations and governmental agencies to gauge the overall health of workers and to target interventions to safeguard worker health.

My motivation: My career in what has become known as OHP has been fueled by: (1) recognition of the burden of occupational illness and injury, and job stress in particular, on workers, their families, organizations, and society; (2) constantly evolving needs and opportunities for application of theory and practice in psychology within the occupational safety and health field, and by receptivity and recognition of this need among occupational safety and health researchers and practitioners; and (3) the satisfaction and rewards of working together with occupational health and safety professionals to mitigate work-related hazards and protect the health and safety of working people.

Professional Profile: Arla Day, Ph.D.

Country/region: Canada
Current position title: Professor, Occupational Health Psychology at Saint Mary's University; Director, CN Centre for Occupational Health & Safety; Director, EMPOWER Partnership
Background: I have been working to improve the health, safety, and well-being of workers for over 25 years. I have a B.A. in Arts from the University of Manitoba. I went on to complete both my M.A.Sc and Ph.D. in I/O Psychology from the University of Waterloo. After my Ph.D., I accepted a faculty position at Saint Mary's University. In addition to teaching and conducting research, I also started consulting with organizations. My current work involves a mix of conducting research (e.g., developing and validating Occupational Health and Safety intervention programs), teaching and mentoring undergraduate and graduate students, and consulting with workers and organizations (e.g., giving workshops and coaching as part of Occupational Health and Safety interventions).

How my work impacts WHSWB: I conduct research to develop and validate programs that aim to create psychologically healthy workplaces by (1) supporting individual workers; (2) strengthening work groups; and (3) developing leaders. These programs involve both educational and skill development components to improve the physical and psychological health of workers. I examine the impact of individual and workplace factors on workers' health and wellbeing, and I help develop students so that they can create and run validated OHP programs (e.g., train-the-trainer sessions).

One particularly meaningful project for me is my work with the EMPOWER Partnership. This partnership involves bringing together researchers, organizations, and other stakeholders to help identify how to best support all workers, including those with chronic physical and mental health issues and/or who have chronic caregiving demands. These initiatives are designed to help individuals develop skills to manage chronic demands, help group members develop skills to manage interpersonal relationships and convey respectful behaviors at work; and help leaders develop skills to lead healthy workplaces and support the efforts of individual workers and groups. I have several precursors of EMPOWER, including ABLE (Achieving Balance in Life & Employment), which demonstrated effectiveness in improving the health and wellbeing of participants, and LEAD (Leadership Effectiveness through Accountability and Development), which used phone and web-based coaching to help develop leaders. The positive reaction to these programs from participants was overwhelming. ABLE participants reported that it helped them deal with work and life challenges and led to noticeable positive behavior changes. They argued that all employees should be able to have the opportunity to participate in this program. LEAD participants found the program helpful and would recommend it to others. The LEAD study also highlighted the importance of supporting the leaders when they are asked to engage in behavior change to help others.

My motivation: As cliché as it may sound, I got into OHP to help people. I like to inspire people in workplaces to create environments that are physically and psychologically healthy. Work is a great medium through which to help people. We often think of all of the bad aspects of work, but work also can have many positive outcomes: It can provide meaning to our lives, it can be a key aspect of our identity, it provides a valuable social context to interact with others, and it provides the potential to demonstrate our skill and mastery. It is our job as OHP researchers and practitioners to help leaders to understand how to leverage all of these positive aspects of work, while helping to minimize the negative things that work may bring. We can use this knowledge to help lead change in organizations (and even in governmental policy and the general community), ensuring that workers have opportunities for skill development, are treated fairly, and are treated with respect.

Note

1 Thanks to Jake Zerner, Modupe Omotajo, and Jeff Martin for contributing to the development of this broad concept map of areas in which OHP can have an impact. These perspectives were also strongly endorsed by the panel of OHP professionals we interviewed for this book.

Chapter References

Adkins, J. A. (1999). Promoting organizational health: The evolving practice of occupational health psychology. *Professional Psychology-Research and Practice, 30*(2), 129–137. https://doi.org/10.1037/0735-7028.30.2.129

Antonovsky, A. (1979). *Health, stress, and coping: New perspectives on mental and physical well-being.* Jossey-Bass.

Bennett, J. B., Weaver, J., Senft, M., & Neeper, M. (2017). Creating workplace well-being. In C. L. Cooper & J. C. Quick (Eds.), *The handbook of stress and health: A guide to research and practice* (pp. 570–604). John Wiley & Sons, Ltd. https://doi.org/10.1002/9781118993811.ch35

Chari, R., Chang, C. C., Sauter, S. L., Petrun Sayers, E. L., Cerully, J. L., Schulte, P., Schill, A. L., & Uscher-Pines, L. (2018). Expanding the paradigm of occupational safety and health: A new framework for worker well-being. *Journal of Occupational and Environmental Medicine, 60*(7), 589–593. https://doi.org/10.1097/JOM.0000000000001330

Fredrickson, B. L. (1998). What good are positive emotions? *Review of General Psychology, 2*(3), 300–319. https://doi.org/10.1037/1089-2680.2.3.300

Goetzel, R. Z., Long, S. R., Ozminkowski, R. J., Hawkins, K., Wang, S., & Lynch, W. (2004). Health, absence, disability, and presenteeism cost estimates of certain physical and mental health conditions affecting U.S. employers. *Journal of Occupational and Environmental Medicine, 46*(4), 398–412. https://doi.org/10.1097/01.jom.0000121151.40413.bd

Herzberg, F. (1964). The motivation-hygiene concept and problems of manpower. *Personnel Administration, 27*(1), 3–7.

Hoffmann, D. A., & Tetrick, L. E. (2003). The etiology of the concept of health: Implications for "organizing" individual and organizational health. In D. A. Hofmann & L. E. Tetrick (Eds.), *Health and safety in organizations: A multi-level perspective* (pp. 1–28). Jossey-Bass.

Houdmont, J., Leka, S., & Bulger, C. A. (2008). The definition of curriculum areas in occupational health psychology. In *Occupational Health Psychology: European Perspectives on Research, Education and Practice.* Nottingham University Press

Kornhauser, A. W. (1965). *Mental health of the industrial worker: A Detroit study.* John Wiley.

Lazarus, R. S., & Folkman, S. (1984). *Stress, appraisal, and coping.* Springer.

Macik-Frey, M., Quick, J. C., & Nelson, D. L. (2007). Advances in occupational health: From a stressful beginning to a positive future. *Journal of Management, 33*(6), 809–840. https://doi.org/10.1177/0149206307307634

Münsterberg, H. (1913). *Psychology and industrial efficiency.* Houghton Mifflin.

Naydeck, B. L., Pearson, J. A., Ozminkowski, R. J., Day, B. T., & Goetzel, R. Z. (2008, Feb). The impact of the highmark employee wellness programs on 4-year healthcare costs. *Journal of Occupational and Environmental Medicine, 50*(2), 146–156. https://doi.org/10.1097/JOM.0b013e3181617855

Novotney, A. (2017). Research zeroes in on the costs of unhealthy workplaces. *Monitor on Psychology*, November, 68–71.

Ozminkowski, R. J., Goetzel, R. Z., Santoro, J., Saenz, B. J., Eley, C., & Gorsky, B. (2004). Estimating risk reduction required to break even in a health promotion program. *American Journal of Health Promotion, 18*(4), 316–325. https://doi.org/10.4278/0890-1171-18.4.316

Quick, J. C. (1999a). Occupational health psychology: Historical roots and future directions. *Health Psychology, 18*(1), 82–88. https://doi.org/10.1037//0278-6133.18.1.82

Quick, J. C. (1999b). Occupational health psychology: The convergence of health and clinical psychology with public health and preventive medicine in an organizational context. *Professional Psychology: Research and Practice, 30*(2), 125–128. https://doi.org/10.1037/0735-7028.30.2.123

Raymond, J. S., Wood, D. W., & Patrick, W. K. (1990). Psychology doctoral training in work and health. *American Psychologist, 45*(10), 1159–1161. https://doi.org/10.1037//0003-066x.45.10.1159

Ryff, C. D., & Singer, B. (1998). The countours of positive human health. *Psychological Inquiry, 9*(1), 1–28. https://doi.org/10.1207/s15327965pli0901_1

Sauter, S. L., & Hurrell, J. J. (2017). Occupational health contributions to the development and promise of occupational health psychology. *Journal of Occupational Health Psychology, 22*(3), 251–258. https://doi.org/10.1037/ocp0000088

Sauter, S. L., Murphy, L. R., & Hurrell, J. J., Jr. (1990). Prevention of work-related psychological disorders. A national strategy proposed by the National Institute for Occupational Safety and Health (NIOSH). *American Psychologist, 45*(10), 1146–1158. https://doi.org/10.1037//0003-066x.45.10.1146

Schill, A. L., & Chosewood, L. C. (2013). The NIOSH Total Worker Health program: An overview. *Journal of Occupational and Environmental Medicine, 55*(12 Suppl), S8–11. https://doi.org/10.1097/JOM.0000000000000037

Schneider, D. L., Camara, W. J., Tetrick, L. E., & Stenberg, C. R. (1999). Training in occupational health psychology: Initial efforts and alternative models. *Professional Psychology-Research and Practice, 30*(2), 138–142. https://doi.org/10.1037/0735-7028.30.2.138

Schonfeld, I. S., & Chang, C. H. (2017). *Occupational health psychology*. Springer Publishing Company.

Westlander, G. (1994). Training of psychologists in occupational health work: Ten years of course development-experience and future perspectives. *European Work and Organizational Psychologist, 4*(2), 189–202. https://doi.org/10.1080/13594329408410483

Zickar, M. J. (2003). Remembering Arthur Kornhauser: Industrial psychology's advocate for worker well-being. *Journal of Applied Psychology, 88*(2), 363–369. https://doi.org/10.1037/0021-9010.88.2.363

2

DESIGNING AND EVALUATING OCCUPATIONAL HEALTH PSYCHOLOGY INTERVENTIONS

Christopher J. L. Cunningham and Kristen Jennings Black

Interventions developed and delivered by occupational health psychology (OHP) professionals can address a wide variety of worker health, safety, and well-being (WHSWB) issues. It is important for OHP professionals to understand and appreciate the value of well-designed, implemented, evaluated, and sustained WHSWB interventions as the natural and necessary translational extensions of OHP research. Well-designed interventions can have substantial impacts on workers, organizations, and society more generally. All successful interventions, however, result from careful consideration of a number of important factors. In this chapter we identify and discuss five essential intervention design elements, as well as a series of critical considerations when working to evaluate and sustain WHSWB interventions.

When you are finished reading this chapter, you should be able to:

LO 2.1: Identify and define five essential elements to OHP intervention design.

LO 2.2: Explain why evaluation is an essential component to any OHP intervention.

LO 2.3: Explain how different decisions affect intervention evaluation quality in real work environments.

LO 2.4: Discuss strategies for sustaining OHP interventions.

Essential Elements to Impactful OHP Intervention Design

In this book we highlight OHP as a discipline that requires the blending of behavioral and social science research and practice. Although OHP research is increasingly prevalent and strong, greater impacts on WHSWB will only come from more translation and application of OHP knowledge, theories, methodologies, and skills. The process by which this type of practical application is applied to make change happen is often referred to as an *intervention*. When this type of change process is initiated within a work setting, with the focus of addressing, improving, or preventing a WHSWB issue, this would be classified as an *occupational health intervention*.

Interventions that effectively address WHSWB issues do not just happen; they result from careful planning and a research-inspired design process. This process begins with a period of assessment or applied research that identifies a particular and specific topic for attention (often referred to as a *needs assessment*). Strong theory and empirical evidence then help us understand what is needed and what is likely to work to address the WHSWB issue at hand. Here we note a particularly relevant maxim adhered to by the originators of action research, that there can be, "No research without action, no action without research" (see Marrow, 1967 for an excellent history of these applied research and intervention pioneers).

Our ability to design and deliver impactful OHP interventions is enhanced with concepts and techniques from the fields of organizational development and change management, such as action research, and field and systems theories (Lewin, 1939, 1947). These latter two perspectives help us understand that people and phenomena within organizational environments exist, behave, and develop due to forces that collectively hold and sustain, or contribute to change. Consider various forces that might be sustaining a work environment characterized by open displays of hostility among coworkers. Some might argue that this pattern of behavior exists because it is adaptive for the organization (i.e., contributes to its profitability and competitiveness). Others might suggest that this could be a consequence of an overly masculine leadership culture. Maybe the hostility is present in the broader community and spilling over into the workplace. Multiple forces are probably at play in this type of situation and identifying them could be the focus of an in-depth organizational assessment and/or applied research effort. If we want to intervene to reduce hostility, then we have to figure out a way to rebalance or adjust one or more of these forces. We highlight the collective influence of these forces (i.e., Lewin's concept of a force field) as a conceptual way of thinking about how WHSWB interventions generally do (or fail to do) what they are designed to do.

We also have to understand that most successful interventions develop through an iterative process. When good evaluation methods complement an intervention, it is possible to gather useful information that can guide improvements and adjustments before the intervention is repeated. When designing interventions, we must train ourselves to not try to "knock it out of the park" on our first swing. A more realistic sequence of events is: proof of concept pilot study of intervention components; evaluate these components and see how they are received and if/how they work; make some adjustments, add/remove some components, and try again; evaluate this new version and keep the process going. An example of this process is found in Britt et al.'s (2018) development of an intervention to create a military unit climate

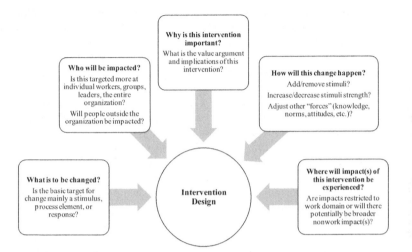

Figure 2.1 Essential Elements to OHP Intervention

that supports mental health treatment seeking, which was then tweaked and tested again to determine if an internal unit leader could be as effective as an external training (Start et al., 2020).

It is possible to earn entire graduate level degrees and certifications in intervention or program design and evaluation. We cannot provide that level of depth in this overview chapter, but in the following subsections, we discuss five essential elements to OHP intervention design summarized in Figure 2.1. There is no one-size fits all approach to intervention design that will be effective for addressing all types of WHSWB issues. Although straightforward, the five elements outlined in Figure 2.1 can be seen as building blocks that can be configured in many different ways to yield impactful interventions.

What to Change

As we mentioned in Chapter 1, most interventions to address WHSWB issues are ultimately aimed at modifying how workers respond to or are otherwise affected by work-related stimuli. Often these responses are behavioral, but they can also be psychological, physiological, or social. This breadth of potential targets for intervention requires broad expertise and thinking about how and why people do what they do in particular contexts, especially work-related contexts. This reality points to the need for multidisciplinary efforts and for specific knowledge of human psychology.

Intervention design can at least initially be approached with a very straightforward stimulus-response model originally associated with early behaviorist researchers in psychology (e.g., Pavlov, 2010; Skinner, 1963). The essence of this model is that exposure to a stimulus elicits a response. Responses met with favorable outcomes are more likely to be repeated, while responses that are met with negative or unfavorable outcomes are less likely to be repeated (and more likely to decrease in frequency and ultimately cease). With a minor extension, we can expand the utility of this model if we recognize that worker behavior results not only from exposure to a stimulus, but also through intervening processes occurring within workers' minds, bodies, and environments. This stimulus-process-response framework is not unlike the widely used input-process-output model that underlies much OHP research, evaluation, and intervention work.

We also must acknowledge that this expanded stimulus-process-response model exists within a context comprised of multiple forces operating within and between workers, their work environment, and the broader society and culture (as noted earlier in this chapter with respect to field and systems theory). This expanded and contextualized model (represented in Figure 2.2) helps to explain how OHP interventions can work by targeting stimuli, responses, and/or elements of the psychological, physical, and environmental/contextual processing that connects stimuli to responses for workers. In Figure 2.2, panel (a) summarizes the conceptual form of this model and panel (b) presents an example. In both panels, the puzzle-like background represents the various forces that play a role in sustaining or changing the relationship that is illustrated in the forefront of each panel (borrowing from Lewin, 1939). Operating from this model, the decision of what is to be changed in an OHP intervention boils down to a decision about whether the intervention will be targeting a stimulus, intervening process(es), some form of response, and/or one or more contextual factors.

Who Will Be Impacted

The next essential OHP intervention design element is who will be receiving and/or mainly impacted by the intervention. Determining this is not always easy, because intervention effects often have intended and unintended ripple effects that transcend the initial recipients. For example, training supervisors to model support should have positive intended ripple effects on subordinates, but upgrading ergonomic furniture for office workers might have unintended negative ripple effects among workers in other areas of the organization who feel they were ignored.

There are many different ways to specify who will be impacted by an OHP intervention. One of the most generalizable frameworks involves determining whether the intervention is focused on influencing a stimulus, process, response, or contextual factors that operate within individuals, groups, leaders, and/or organizations (i.e., the IGLO framework; Nielsen & Miraglia, 2017; Nielsen et al., 2017). Figure 2.3 illustrates the value of this type of multilevel intervention design thinking when preparing to address a workplace safety issue; notice how essential alignment in strategy is across these different target

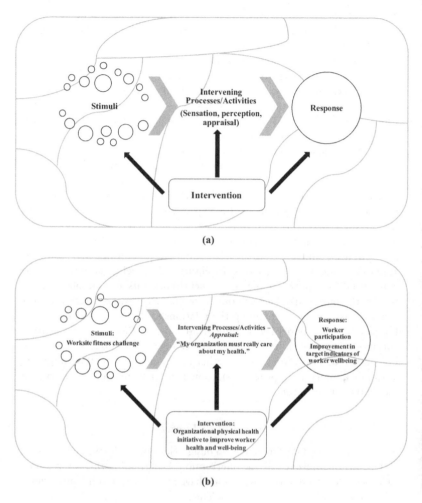

Figure 2.2 Expanded and Contextualized Stimulus-Response Model and Example

recipients (in this figure, alignment benefits are noted around the outside, while poor alignment consequences are noted in the middle).

A related complexity when establishing this element of intervention design is determining the current state of intervention recipients. More specifically, do they (i.e., the individual, the leader, the organization) recognize the specific WHSWB issue that is being targeted? Are they open to and ready for change to take place? Guidance for addressing this aspect of intervention design can be found in various models of health behavior change, including the Theory of Reasoned Action/Planned Behavior, the Integrated Behavioral Model, and the Transtheoretical Model of behavior change and its Stages of

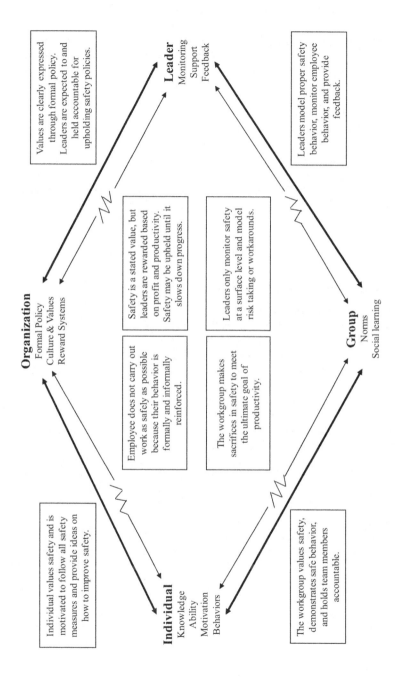

Figure 2.3 Illustration of Multilevel Intervention Alignment

Change (for helpful and accessible summaries of these perspectives, see Montaño & Kaspryzyk, 2008; Prochaska et al., 2008).

Targeting workers directly is a logical place to start with many OHP interventions. However, many WHSWB issues cannot be fully addressed without attention to corresponding group-, leader-, and organization-level factors. We explore these options for specific WHSWB issues in later chapters, but here are a couple of quick examples to illustrate. First, consider a situation in which family friendly policies and practices are being implemented as options for workers to consider; without norms within the group and support from organizational leaders, this type of intervention is unlikely to achieve its full impact. We examine these matters more in Chapter 9. Second, imagine a safety-focused intervention that trains workers to be more self-aware of safety risks while working. The effects of this intervention can be strengthened and sustained if the intervention is expanded to also teach supervisors strategies for managing worker risk-taking and positive safety behaviors. Also helpful in this scenario would be establishing safety norms at the work group and organizational level (as we explore more in Chapter 11).

Why the Intervention is Important

We cannot get far in designing effective interventions without engaging key stakeholders (e.g., organization leaders, funding sources, participants). For this reason, the third critical intervention design element is ensuring stakeholders understand why a proposed intervention is worth developing, implementing, and sustaining. Most OHP professionals are research-minded and more comfortable with theories and data, than with financial impact projections. We tend to argue that a phenomenon exists or an intervention works by using statistics to show that observed effects are not due to chance. Among non-researchers, however, this type of probability-based logic is rarely appreciated unless properly translated.

For many stakeholders, the only convincing evidence of intervention impact comes in the form of practically significant cost savings, revenue gain, and positive return on investment (ROI). Helping other stakeholders understand and appreciate the potential value of an intervention requires broad-minded thinking about where, how, and to what extent an intervention's effects will be experienced. This means making a business case or other form of impact argument for the cost, interruption, and hassle associated with an intervention.

Making the Business Case

The challenge with making the financial argument or "business case" for OHP-related interventions begins with defining what the effects of the intervention are (or are likely to be). Psychological researchers and practitioners rarely specify or determine *all* outcomes and consequences likely to result from their intervention- or research-related manipulations. This is due to our limited understanding regarding the complex interplay of psychological, environmental, and behavioral factors that influence WHSWB.

In practice, the best we can normally do is to make educated and informed estimates of whether a particular intervention's effects are likely to be small (i.e., slightly noticeable), medium (i.e., readily apparent), or large (i.e., impossible to ignore). Even then, we still must determine what these effects mean in terms of financial value to a participating worker or organization, relative to the associated costs of the intervention itself. In other words, what would a small, medium, or large reduction in health care costs be worth to a participating organization? As there are no concrete or consistent values associated with improving specific WHSWB across different contexts, our best approach is often to develop clear impact illustration models using parameters, assumptions, and values that are reasonable and transparent. We can then use interval estimation techniques or some form of sensitivity analysis to illustrate ranges of impact, rather than single point estimates that may be harder to accept or believe (e.g., Naydeck et al., 2008).

Business decisions are more easily made and supported when they are perceived as safe bets for a solid ROI, not just when there is good logic or theory behind the proposal. There are good examples of well-done cost and ROI calculations associated with WHSWB interventions beginning in the 1970s and 1980s in well-known organizations like Johnson & Johnson (e.g., Ozminkowski et al., 2002). These methods have continued to evolve in complexity and comprehensiveness (Ingels et al., 2016; Ozminkowski et al., 2004). These types of efforts rarely yield entirely consistent effect and ROI estimates, but they do tend to highlight the possibility and even likelihood for worksite health promotion and other WHSWB interventions to generate a positive (sometimes substantially positive) ROI for organizations.

Extending Beyond the Business Case

There are also opportunities with many WHSWB interventions to change the entire frame around impact estimates and value arguments. We can move beyond demonstrating financial impacts in a particular work setting, to outlining economic impact estimates for cities, regions, countries, industries, and even segments of society. Nonfinancial impacts of OHP initiatives are experienced at the level of the worker (e.g., ethically right actions to protect dignity and avoid harm) and within workers' families and communities (e.g., facilitate engagement and commitment; demonstrating social responsibility and concern for others). These become additional value and impact contexts worth considering when establishing the importance of a particular intervention. Beginning in Chapter 4, we explore major areas of OHP research and practice. In each chapter, we examine several of these different perspectives on justifying and demonstrating the impact of intervention work that targets these issues. Our hope is that you will similarly think broadly about how your own future OHP-related efforts can have impacts within and beyond an immediate work context.

How the Intervention Will Work

In addition to the what, who, and why design elements, strong OHP interventions also require a clear and workable mechanism for bringing about the desired effect(s). Before we dig deeper into this "how" element, consider a couple of relevant points. First, remember that "Everything you do is an intervention" (Schein, 1999, p. 17). This wisdom from a renowned organizational development and change management consultant is particularly true when we work to address psychologically oriented WHSWB matters. Related to this, try to remember this pragmatic sentiment from an OHP professional featured in Chapter 4, "If you assess, you must address" (S. Jex, personal communication, October 2003). The very act of gathering data from and about workers is an indication that something may need adjustment or modification. Regardless of whether we are gathering data in a work setting just to "see what is going on" or whether we are implementing a full-scale *randomized control trial* (RCT) of a well-established intervention, our involvement in a work setting signals that something may be about to change. This signal is amplified when data are gathered and reaches peak volume when a formal intervention is initiated. It is essential to remember that workers are harmed if they open up and share information, but nothing happens as a result and they feel they have not been heard or respected.

General Intervention Forms

One way of thinking about intervention mechanisms is whether they are designed to be primary, secondary, or tertiary in form (Cooper & Cartwright, 1997; Quick, 1999). *Primary interventions* are those focused on directly eliminating or reducing a particular WHSWB issue. If we consider the scenario of developing an intervention to improve safety of workers in the construction industry (a very high-risk population), a primary intervention might be to reduce risk factors that contribute to falls and other related forms of injury. This could be done by ensuring all workers had access to proper and safe equipment to use while working and by establishing safe work practices that minimize certain types of risks (e.g., messy work environment).

Secondary interventions are a bit different and are typically less directly focused on modifying a targeted WHSWB issue, event, or risk factor and more focused on addressing workers' abilities to handle it. These interventions are often the most appropriate strategy when there is an inherent risk that cannot be avoided. For example, nurses who often work in a high-stress hospital environment where stressors are uncontrollable, may benefit from secondary interventions like de-escalation training for responding to emotional patients or self-care strategies for detaching from emotional demands. Finally, *tertiary interventions* are those where the emphasis is on addressing consequences of a particular WHSWB event, either following an acute or serious exposure or after serious consequences have developed over a longer-term period of chronic exposure. In a construction setting, a tertiary intervention might be to direct workers who are injured or who have witnessed a fall and are experiencing challenging emotions and thoughts to seek out appropriate clinical or counseling forms of assistance (perhaps through a company Employee Assistant Program or other local resource).

Specific Intervention Mechanisms

Beyond loosely describing an intervention as primary, secondary, or tertiary in form, intervention design requires specific decisions about how you will influence the targeted stimulus, intervening processes, response, and/or associated contextual forces for the intervention. A variety of example intervention mechanisms can be used to address one or all components represented in Figure 2.2. OHP professionals generally have the most intervention mechanism options or "tools" when it comes to influencing intervening processes. This is because most OHP professionals are trained in psychology, the science of how and why people do what they do in particular contexts; many other professional disciplines more directly target certain types of physical and environmental stimuli (e.g., industrial hygiene) and responses (e.g., occupational medicine).

It is also possible and common for multiple mechanisms to be jointly used to optimize an intervention's likelihood of success. This type of multi-pronged intervention design is generally a good strategy for enhancing the robustness of the intervention and its likely impacts (cf., Hackman, 1986). Overlap among different intervention mechanisms is also often unavoidable, given that these mechanisms and their effects are often interlinked because of their deep connection with human psychology. For example, modifying a stimulus in the work environment can affect attitudes about work and the person's ability to manage it, which can ultimately influence behavior.

Digging into the linkages between our intervention targets and possible mechanisms more directly, when it is possible to modify or reduce exposure to harmful stimuli, then this is a prime target to consider. However, stimulus control is possible only when you have a fair amount of control over workers and the broader work environment and when enhancing controls is not going to be prohibitively expensive. It is also not always clear what the primary stimulus or set of stimuli is that is negatively affecting workers. This is one reason why stress management, for example, is so difficult to do well. Surveys regularly show that most workers report experiencing stress chronically associated with their work, but the dominant sources of that stress (i.e., the stimuli) tend to vary widely across individuals. In these types of situations, additional or different intervention mechanisms may be needed.

One option is to consider introducing a new or alternative stimulus that can counteract or minimize the effects of a noxious or problematic stimulus. Although research specifically testing this type of approach is rather sparse, we increasingly see examples of this strategy in action. As examples, consider

33

the many people who benefit from the presence of emotional support animals (Hoy-Gerlach et al., 2019) or the increasing use of job crafting as a way of facilitating worker control in how they respond to work-related demands (e.g., van Wingerden et al., 2017). Sometimes directly impacting stimuli is not possible or practical for a particular intervention. It may also be that the targeted stimulus is so strong, its effects cannot be sufficiently counteracted by the inclusion of another stimulus. In these situations, OHP professionals will typically target worker behaviors rather directly or the intervening processes that link stimuli to responses.

When the intervention target is actual behavior change, then the intervention mechanism is typically going to focus on modifying what the worker actually does in a particular context and in response to particular stimuli. This might involve use of reward strategies to gradually shape worker behavior over time, encouraging more of the behaviors you want to see and fewer of the behaviors you do not (e.g., Gonzalez-Morales et al., 2018). In some situations (e.g., safety, harassment), strong or zero-tolerance consequences may be necessary to drastically deter or eliminate problematic responses. In other situations, simple nudging through the use of clever environmental cues can be effective (e.g., Van der Meiden et al., 2019). Behavior-based interventions may not be appropriate or effective approaches to addressing all manner of WHSWB issues, but there are some areas such as safety, where these approaches can be effective at changing worker behaviors when carefully designed (e.g., Geller, 2005).

Because impacting stimuli and responses is not always possible, many OHP interventions are designed to address one or more intervening processes that link stimuli experienced by workers to some form of response. Manipulating or modifying these processes is generally accomplished using one or more of the following mechanisms: knowledge transfer, attitude/intention modification, cognitive adjustment, or social influence. *Knowledge transfer* can be an effective intervention mechanism for modifying sensation, perception, and appraisal of stimuli, as well as selection of appropriate and possible responses to such stimuli. Knowledge is a powerful mechanism for change, and workers can learn to be more alert for certain stimuli, they can learn how to more rationally appraise stimuli and their response options, they can understand real versus false risks, etc. We discuss examples of this type of intervention mechanism in most of the remaining chapters. Knowledge by itself, however, does not automatically lead to lasting behavior change for workers. This is a long-understood truth from the field of organizational development and is associated with a form of knowledge-driven change strategy known as normative-reeducation (Chin & Benne, 1969; Quinn & Sonenshein, 2008). This approach focuses on using information to influence workers and organizations to change norms or expectations about what behaviors are appropriate, rather than informing to directly change behaviors. When such norms are properly adjusted and aligned with WHSWB goals, new and improved behaviors can either develop naturally or be introduced and expected to persist.

Sometimes knowledge is not the most effective mechanism for modifying intervening processes, such as when too little is known about a particular WHSWB issue to effectively educate workers. This scenario was all too apparent when we were finishing this book and the world was struggling through several months of the COVID-19 pandemic and its effects on every aspect of work and nonwork life. In these types of situations, an alternative mechanism for influencing intervening processes may be to *modify worker attitudes and intentions*. Changing attitudes and intentions generally involves modifying how workers appraise and respond to stimuli in their work environment. There are a variety of models and frameworks in the published literature that target this type of change. As we discuss in several of the upcoming chapters, one particularly common and pragmatic example of this type of model is the theory of planned behavior/reasoned action (Ajzen, 1991; Madden et al., 1992). This model positions behavior as the result of intentions, which are affected by attitudes, which are in part shaped by our perception of norms and personal control. Leveraging this approach, an intervention could be designed to modify behavior by first changing what people perceive to be normative and acceptable. This often also involves modifying personal and shared attitudes about existing norms and establishing new behavioral intentions (Budden & Sagarin, 2007; Gollwitzer, 1999) that could help to build and sustain new norms moving forward.

Another way to influence the intervening processes linking stimuli to responses is to *readjust or change how workers cognitively process stimuli* linked to the intervention target. This might involve techniques that attempt to increase self-awareness, such as meditation or mindfulness (Bostock et al., 2019; van der Zwan et al., 2015), or some form of cognitive restructuring effort (e.g., Michel et al., 2014). Knowledge-based mechanisms can also be combined with cognitive adjustments to teach workers new strategies for identifying and responding to risk factors at work. Particularly relevant theoretical frameworks associated with this set of intervention mechanisms are goal setting theory (Locke & Latham, 2006; Ludwig & Geller, 2000) and social cognitive theory (Bandura, 1986; Wood & Bandura, 1989). These theories, and associated research, generally support setting specific and challenging (yet attainable) goals for which progress can be monitored and measured, supporting feedback along the way. Goals are also more effective and likely to be pursued more strongly when workers had a hand in setting them. In this way, workers are able weigh in on goal attainability and what is needed to achieve it (a judgment intimately linked to an individual's underlying self-efficacy; Phillips & Gully, 1997). For example, workers with higher levels of self-efficacy in a particular work setting will set more challenging goals than workers with lower levels of self-efficacy. Putting this all together, workers will be more willing and able to pursue greater change when they believe they know what to do and have the necessary resources to do it.

Sometimes changing intervening processes like sensation, perception, and appraisal requires us to *change or influence some aspect of workers' social realities*.

This fourth general intervention mechanism involves creating or manipulating the undeniably powerful force of social influence. In an intervention context, social influence can be exercised in a number of ways, including through modification of workers' own social comparators (i.e., social comparison theory; Gerber et al., 2018; Goethals, 1986) or changing social cues or information that workers turn to when trying to make sense of ambiguous stimuli (leveraging social information processing theory; Salancik & Pfeffer, 1978). Other options include building social support networks and a sense of shared responsibility, as has been done via Community Reinforcement and Family Training (CRAFT) interventions (Kirby et al., 2017) and YaHaLOM/iCover peer-based psychological first aid initiatives (Svetlitzky et al., 2019).

Finally, sometimes there is a need to *rebalance forces present in the broader context* (e.g., work environment, worker's personal life) that may prevent or enable the change effects targeted by an intervention. As shown through many of his writings and historical accounts of his work (including those cited earlier in this chapter), Lewin himself was very strongly motivated to address social conflicts and injustices. Interventions designed to address these types of forces can be seen as attempts to apply principles of field theory as discussed by Lewin (1939, 1947); see also Martin (2003) for an in-depth examination of the field theory's origins. This issue of context is also at the heart of the next essential element to OHP intervention design.

Context Matters

The fifth core element to consider when designing OHP-related interventions pertains to "where" (i.e., the contexts in which) the intervention will be delivered and expected to have its impact. This involves determining whether the intervention is limited in its delivery, form, and impact to the work domain or perhaps also extends to the nonwork domain. This also involves thinking through matters of time associated with the intervention (e.g., How long will it take to see results? How long can effects last?).

It is easy to assume that all OHP interventions are delivered in work environments for maximum work-related effect, but this is not necessarily true. For example, an intervention to provide workers with more flexibility over their working schedules is often implemented to provide workers with a stronger sense of control over their work efforts, but there is also a positive spillover effect in that more flexibility at work also provides more flexibility outside of work. This, in turn, may lead to further reciprocation, as workers feel more positive and successful about being able to manage their various (and otherwise competing) role demands. Sometimes organizations also influence worker health and well-being with interventions that are more nonwork-domain focused, such as offering workers financial management training to help reduce financial strain when increasing pay is not an option.

A challenge for OHP professionals when designing WHSWB interventions is to remember that context matters and that sometimes, limiting an intervention to one context can limit the impact of the intervention itself. For example, intervention efforts to improve workers' physical and psychological health often take the form of initiatives delivered at the worksite. This is an interesting choice of intervention mechanism, when the ultimate goal is to impact workers' health states that transcend any singular role domain. It is not always reasonable or optimal to think that large improvements in worker health can be accomplished just by encouraging people to exercise while they are at work. This is especially true when we consider that these types of interventions typically only engage around 40% of an organization's workforce (Abraham et al., 2015; Mattke et al., 2015). This is a good example of why WHSWB is often best addressed with multi-pronged, context-spanning intervention approaches.

Our point here is that determining the domain for an intervention is trickier than it may initially seem. There are good reasons to limit OHP interventions to work settings; doing so can help us control unknown and potentially influential context factors and simplify the overall intervention methodology. However, we need to be careful that this type of domain limitation decision is not made arbitrarily, or we may miss real opportunities to achieve greater impact that would be possible with more comprehensive, context-spanning efforts. Because so many WHSWB issues transcend the work environment, many OHP interventions can potentially have broader effects than may initially be considered. The work of OHP professionals is different than that of safety engineers or industrial hygienists, who might address an unreliable and risky machine or process with a change or fix that is associated with that particular issue in a specific work environment. Neither the machine/process, nor that solution, will ever leave that work setting and crossover into or be affected by nonwork factors. This is, however, the precise reality associated with most WHSWB issues OHP professionals work to address.

Evaluating and Sustaining Interventions

We hope it is apparent by now that designing WHSWB interventions is challenging. Perhaps even more challenging and important is figuring out if the intervention really worked. Evaluation should never be an afterthought or optional component to intervention design; not evaluating intervention efforts is irresponsible and borderline unethical. If we do not evaluate the effects of our interventions, we will not know whether we had any effect, made things better, or made things worse. If we do not evaluate aspects of the intervention mechanism and process, we also will not have any insight into how the intervention could be improved and how much time may really be needed for an intervention to yield its ultimate effects. Well-designed interventions lend themselves to well-designed evaluation strategies; this combination provides the basis for long-term sustainability. Evaluating the impact of our efforts to address WHSWB is important not only for the long-term impact of an intervention, but also for the long-term viability of the OHP profession (Adkins et al., 2011). Simply put, OHP cannot sustain itself as a professional discipline if we cannot prove our ability to effectively translate our science into practice.

The importance of evaluation to long-term intervention sustainability and impact is particularly evident in what is often known as the public health model. This model of intervention combined with evaluation involves: (1) identifying and defining the problem; (2) using research to understand the problem (and its causes); (3) developing, implementing, and evaluating interventions; and then (4) expanding and generalizing successful interventions for greater impact. This fourth step is one that has not been widely practiced yet among OHP professionals and it is critical one if our goal is to seriously impact WHSWB. A good example of this model in action is provided by the World Health Organization in their outline of a violence prevention approach (https://www.who.int/violenceprevention/approach/public_health/en/).

Designing an Appropriate Intervention Evaluation Strategy

Evaluation strategies for OHP interventions are about as varied as the interventions themselves. No single evaluation methodology can reasonably be used with all OHP interventions. Proper evaluation of OHP interventions requires more than a quick post-intervention check on how positive participants felt about their experiences with the intervention materials or presenter (i.e., the classic "smile sheet"; cf., Abernathy, 1999). A better and more useful general approach to intervention evaluation is to gather information regarding the intervention process itself (i.e., formative evaluation) as well as information about the effects or outcomes of the intervention (i.e., summative evaluation). There are many reasons for taking this two-pronged approach, as discussed by Nielsen and Miraglia (2017). Good evaluation strategies help to ensure effective interventions in the short- and long-term. There are four general decisions to make when designing intervention evaluation strategies: what to measure, when to measure it, how to measure it, and how to analyze and interpret the data.

What to Measure

Determining what to measure as part of an intervention evaluation is easiest when the intervention methodology and mechanisms are outlined in the form of a logic model. Typically this takes the form of a path or flow diagram that illustrates the chain of events by which the intervention is designed to have its intended effect(s). A complete logic model for an OHP intervention typically includes information about how participants are selected and assigned to conditions, and other elements pertinent to the intervention process (e.g., resources, activities), as well as targeted outcomes and impact objectives. Logic models can be developed by expanding slightly the simple intervention model from Figure 2.2 to include inputs, activities (processes), outputs, and longer-term outcomes. A helpful and detailed report that builds on and illustrates this basic concept in action is the report of the Committee on the Review of National Institute for Occupational Safety and Health

(NIOSH) Research Programs by the Institute of Medicine and National Research Council (2009; https://www.nap.edu/catalog/12639/evaluating-occupational-health-and-safety-research-programs-framework-and-next).

Another helpful example is provided by the Program Performance and Evaluation Office of the United States' Centers for Disease Control (https://www.cdc.gov/eval/logicmodels/index.htm). When developing logic models, we also recommend that you adhere as closely as possible to the Consolidated Standards of Reporting Trials (CONSORT) guidelines (http://www.consort-statement.org/) for transparency in trials-related research reporting; see McGonagle et al. (2014) for a clear example. This is especially important if you are conducting a RCT (discussed later in this section).

Measurement targets for an evaluation should align well with the targets for the intervention itself but be operationalized and defined in a way that facilitates clear and consistent measurement. Ideally, data pertaining to each evaluation target can be gathered using a multi-method (e.g., Brewer & Hunter, 2006) and multi-level approach (e.g., Glasgow & Emmons, 2007) to help minimize measurement bias and issues that can hinder accurate interpretation of your data. Evaluation targets may include reactions to intervention materials and processes themselves, but also should consider other types of indicators, metrics, and outcomes that are likely to be impacted (and maybe *not* likely impacted) by aspects of the intervention.

In a typical safety intervention, for example, we can measure workers' engagement with and reactions to the intervention, but we can also gather data regarding accident-related workers compensation claims leading up to, during, and following the intervention period. Although safer performance might seem a bit hindering at first to workers, we would hope that eventually productivity or performance metrics would not substantially decline – to know for sure, we would need to monitor these metrics. For the most comprehensive picture in this type of situation, it may also be worth monitoring other indications of worker attitudes and perceptions (e.g., engagement, perceptions of safety, perceptions of organizational/supervisor support). For most of these metrics, gathering quantitative (i.e., numerically scaled) and qualitative (i.e., open-ended response) data can help to provide a more comprehensive picture of an intervention's effects, including unexpected effects (e.g., if there are increased incident reports due to workers feeling educated and empowered to make such reports). The effects of OHP interventions do not always conform to the simplistic hypotheses we are used to seeing in published research. Designing your evaluation to gather a variety of data can only help to ensure that you can make sense of how the intervention worked (or did not work) and what can be done to improve the intervention going forward.

When to Measure

A second decision affecting intervention evaluation design is the temporal context surrounding the intervention. Although some WHSWB interventions may yield effects and a positive ROI in a matter of hours (e.g., an educational workshop), other interventions may take many months to yield measurable effects (e.g., physical health intervention). This influences our decisions about when and how frequently to gather evaluation data. This is a complex decision influenced by available resources to support short- versus long-term evaluation and by the overall level of engagement and commitment of intervention participants and their employer(s).

Unfortunately, OHP professionals rarely have perfect clarity regarding how much time is needed for an intervention's effects to emerge. We often have to adopt a logical and rational timeframe perspective when designing interventions, estimating a reasonable timeframe based on how quickly we expect people to be affected by the intervention and/or how long the intervention can feasibly be sustained (often dictated by cost and complexity). Consider these issues as they apply to a fairly common education-based intervention for safe lifting among warehouse workers. Participating workers have developed their own techniques and methods for lifting tasks that are as quick as possible and may find a new, albeit safer, technique less "natural"-feeling (at least at first). As workers learn to lift differently, there is an increased risk of injury or at least fatigue and soreness, along with slower work, which likely will dissipate once the worker acclimates to the new lifting technique. The ultimate effects of this straightforward intervention take time to develop or emerge. If we take only a short-term perspective on this intervention, we may not plan sufficiently to sustain it and see it through until the ultimate effects are reached. The point with this example, is that we can often trigger greater change and understand it more fully when we think through the time-related context for the intervention and its associated effects.

The role of time in general is a poorly understood area for most behavioral and social sciences, which partially explains why evaluation periods for WHSWB interventions vary widely from a few weeks through several years, often for no explained reason (Anger et al., 2015). Some work has started exploring these questions of time lags associated with certain phenomena (e.g., stressor-strain effects; Ford et al., 2014), but much work remains to be done before OHP professionals can confidently explain how long it will take for our interventions to have the intended effects (and concordantly design evaluation strategies that fit this timeline). For now, when questions of time arise, OHP professionals typically have to offer the psychology profession's equivalent of pleading the fifth and say, "It depends." This is not an ideal response in a business context, and we need to challenge ourselves to do better to provide time-defined boundaries around our intervention effect and cost estimates.

When we gather intervention evaluation data at a single point in time (typically post-intervention), we are conducting a point in time or cross-sectional evaluation. Often, this approach is critiqued, but there are ways to optimize it so that meaningful evaluation data can be gathered (Spector, 2019). When evaluation data are gathered over multiple time points, we are conducting a longitudinal or time series evaluation. Gathering data over time means having more data available to understand how an intervention is impacting participating workers. Multiple time points of evaluation data make it possible to consider nonlinear and growth curve-type models of change in whatever was targeted by the intervention. Evaluation data over time also can illustrate how an effect develops (process evaluation) and how that effect is sustained (ongoing outcome evaluation), as well as the extent to which this all varies by person.

Gathering evaluation data over time is not easy and comes with a heightened risk of respondent drop-off or attrition. These challenges can be mitigated somewhat by framing this type of evaluation as part of the intervention and developing incentive, communication, and other engagement strategies. Making the evaluation process manageable and relevant will also help. There are not many published examples of longitudinal evaluation of OHP-related interventions, but a good example is Frögéli et al.'s (2019) evaluation of a burnout-prevention effort among nurses over a three-year period.

How to Measure

Once the targets for an intervention evaluation and an appropriate timeframe are set, it is possible to design the specific methodology for gathering the evaluation data. These decisions involve balancing feasibility (given the domain and mechanism of the intervention) with sufficiency (to ensure necessary information is gathered). A variety of contextual factors also are relevant when establishing the focus and scope of an intervention evaluation. For example, how much environmental control do we have? Can we do this in a lab setting, or do we need to do this at the worksite? This might involve making rather simple decisions, such as whether surveys, observations, focus groups, or interviews are appropriate for a given intervention context.

Establishing the evaluation methodology can also get complex, if the decision is made to embed the intervention into a more rigorous and controlled evaluative framework. There are many options if you go this route, with perhaps the most idealized being the RCT. A RCT is essentially a methodology for simultaneously delivering and testing an intervention or change initiative. In some fields, especially medicine, RCT are often considered the only legitimate way to demonstrate cause and effect. When the opportunity and resources are present to conduct a RCT of an OHP intervention, this strong approach to evaluation needs to be considered. However, most OHP interventions occur in real-world contexts in which meeting all methodological requirements for RCT may be impractical or impossible (O'Shea et al., 2016). A few examples of this approach applied to OHP intervention evaluation are found in Hammer et al.'s (2019) evaluation of the effects of a workplace safety and health intervention, Edwardson et al.'s (2018) evaluation of an intervention to reduce sitting time at work, and Brown et al.'s (2014) evaluation of a worksite physical activity intervention.

When a full RCT is not feasible, there are RCT features that may be incorporated into alternative evaluation designs. One is that RCT provide one of the cleanest methodologies for demonstrating cause and effect because their design strictly controls for sources of variance in participant behaviors, outcomes, etc. that are supposedly not related to what is being studied. This is generally accomplished using random sampling and random assignment to ensure that multiple, roughly equivalent groups of participating workers can be established and then compared before and after (sometimes even during) the intervention itself. Sometimes we can approximate this in non-RCT designs by statistically testing for and/or controlling for differences between intervention and comparison groups. Other times, we can randomize assignment of participants at a group, instead of individual level.

Another RCT design feature that strengthens causal inferences is single- or double-blinding techniques. *Single-blind* studies are ones in which participants do not know whether they are in the intervention condition or some other placebo or control condition. In *double-blind* studies, the researcher or evaluator is also unaware of which group is in which condition. This latter feature can protect against an OHP professional influencing or biasing reactions of intervention participants when responding to an intervention evaluation. Even if this particular OHP professional is really conscientious, there is the risk associated with even an off-handed comment during the intervention delivery (e.g., "I have really put a lot of effort into this; I look forward to your feedback on the evaluation."). Just one such comment can prime participants to respond in a more positive way than they might have if responding to a dispassionate confederate or external evaluator.

Despite these strengths, RCT also present challenges to OHP-related work. First, RCT are generally conducted in strictly controlled contexts that may not mirror reality for many workers (excet in military-like settings; Cacioppo et al., 2015; Thomas et al., 2019). A consequence of such control is that the evaluation findings may not generalize to different work-related contexts. A second challenge comes if intervention access/exposure is only provided to one group of participants, while others serve as a comparison group. This type of unfairness in experience, outcome, and potential benefit simply would not be permitted in many work environments. There are techniques to help address this, such as rolling entry matching (Witman et al., 2019), though these techniques are not yet widely used and can be difficult to manage. Third, an emphasis on RCT as the best form of evaluation can reduce openness to and acceptance of intervention evaluation designs that may be as strong as possible for use in real-world contexts, but not exactly RCT in form. We have to remember that we are not always going to be able to deploy perfect evaluation designs in all intervention contexts – a simple pre-/post-intervention evaluation with a single follow-up is much better than no evaluation at all.

MEASUREMENT FORMS AND SOURCES

Whether developing new or using existing measures in our intervention evaluations, we need to ensure that we understand what exactly is being captured by each measure and whether it adequately operationalizes the "what" that we are targeting for an evaluation. These concerns support careful review and consideration of the basic psychometric properties of our selected measures (i.e., evidence for their reliability and ability to support valid inferences). When we choose proper measurement methods and forms, we increase stakeholders' faith in the quality of any conclusions ultimately based on resulting data.

To avoid perpetuating measurement-related errors and ambiguous measurement efforts from the past, we cannot simply select scales based on their names. We also need to seek out current versions and evidence for the psychometric qualities of measures as they have been used by researchers other than the ones who designed them. It is also important to confirm that participants actually interpret our evaluation items as intended. We can do this by pilot testing evaluation components with non-researchers to ensure instructions, items, and response options make sense and are interpreted as intended. This type of pre-testing can ensure that our ultimate evaluation efforts are appropriately matched to fit the context and participants for a given intervention.

Often in OHP interventions, time and access to intervention participants are limited resources. In these situations it is acceptable and sometimes necessary to use single-item measures of even key variables or constructs (Clark et al., 2011; Haddock et al., 2006). We know this guidance may contradict what you may have learned in a measurement or broader psychometric theory course, but a basic guiding principle in evaluation design is that measuring something (even imperfectly) is better than not trying to measure it at all. Further, keeping our evaluation measures clear and concise helps to keep the overall evaluation process manageable and increases the likelihood that we can get more of the data we need with lower risk for respondent fatigue and missing values.

For many reasons, it is also strongly recommended that evaluation data do not come from just one source or one method, as noted earlier when we discussed mixed-method research. Such strategies help to minimize the potential influence of method-related variance on your data and ultimate interpretations (Spector et al., 2017). In other words, gathering data from participants through self-report methods (e.g., survey, interview) is good, but additional insights can often be gleaned from reviewing existing metrics (e.g., productivity, health-related costs) and data provided by someone other than the participant (e.g., observations, supervisor ratings, significant other inputs). Increasingly, there are also measures that directly record worker behaviors or physiological indicators that may be relevant to a given OHP intervention context. Examples of this include measurement of physiological indicators of stress (Wosu et al., 2013), the use of wearable devices to track health-related data in real-time (Piwek et al., 2016), and the use of cellular phones to track behaviors (Harari et al., 2016). Creative and holistic thinking can lead to a more comprehensive picture of an intervention's effects.

Analyzing and Interpreting the Data

Evaluation data are meant to answer very concrete questions about whether and to what extent an intervention works. This is only possible if we really understand the evaluation data and have properly analyzed them. The final evaluation-related decision is determining how to most appropriately analyze and interpret the gathered evaluation data. This decision remains tentative until evaluation data are gathered, because our ultimate analysis efforts are constrained by the amount and quality of the evaluation data. There are three main components to this final intervention evaluation decision: documenting and clearly describing our data, understanding our data and working within its limitations, and preparing for complexity during analyses.

First, we have to be prepared to ultimately share our analysis approach, findings, and interpretations of these findings with others who may have no statistics training and little comfort with data. This is only possible if we carefully document and describe our data. When working with quantitative data, this means paying special attention to what each measured variable means (i.e., what a high vs. low score represents). With qualitative data, this means keeping track of the questions that are asked, the conditions in which answers were gathered, and the process by which responses were processed and analyzed (see Woo et al., 2017 for helpful guidance).

Second, our analyses and ultimate interpretations are improved when we understand our evaluation data and its limitations, and work to appropriately interpret the data. Especially when working with evaluation data gathered in actual work settings, we need to ensure that we carefully review all evaluation data to determine whether there are any confounds or complications in and about the data that need to be addressed. This includes checking for missing data, outliers, and irrelevant or impossible values likely due to participant or data collection error.

In most applied intervention settings, the first analytical challenge is describing what happened and showing that it mattered. It is often possible to do this with basic descriptive statistics and data visualizations. Once this is demonstrated, you have more time to dig into the data deeply with more advanced analyses, including tests for moderated and multilevel effects (Bliese et al., 2017; Dicke et al., 2018), as well as other more advanced techniques, such as latent profile analysis (e.g., Bennett et al., 2016).

We need to limit our analyses and interpretation to what the data support. If the evaluation only includes data gathered post-intervention (not recommended), it is not possible to reach inferences about change. Pre-intervention, baseline data is so important in any type of organizational change or intervention situation, as it is "used for both hints about what requires change and as a standard against which the effects of change that are implemented can be judged" (Schneider et al., 2011, p. 402). If the evaluation includes pre- and post-intervention data, then it may be possible to demonstrate change of some form.

Without getting too complex, we also have to understand that not all change is equivalent. Golembiewski et al. (1976) outlines alpha, beta, and gamma forms of change, illustrating the point that interventions sometimes have the effect not only of changing the targeted response or outcome, but also of modifying how participants think about, perceive, and react to associated stimuli. If ignored during our analyses, this type of perspective shift can lead to misinterpretation of observed change or effect between pre- and post-intervention evaluation data. As an example, I may rate myself as really supportive of my coworkers before a support-enhancing intervention begins, but over the course of the intervention I realize I am not as supportive as I initially thought. My post-intervention evaluation ratings indicate that I see myself as less supportive than I thought I was before the intervention. Comparing these two sets of evaluation data could lead to the conclusion that the intervention did not work, when the opposite is actually true. There are strategies for evaluating and addressing this type of response-shift bias (Howard & Dailey, 1979), with one of the simplest being the gathering of pre, post, and retrospective pre-intervention data (see Cunningham et al., 2014 for an illustration in a nursing education context).

Third, we need to prepare for and remain flexible enough to handle complexity that may arise during our analyses of intervention evaluation data. Sometimes these complexities are addressed in the early phases of cleaning and managing the data, before descriptive analyses are performed. Sometimes, though, the analytical complexities arise only when we get deeper into the analysis process. For example, if the available evaluation data exist only at the level of department or average reactions within a work unit, our analyses can really only lead to conclusions about how the intervention affected groups, not individual workers (a levels of analysis matter; Chan, 1998).

As another example, if all items used to gather evaluation data were framed in the same way (i.e., generally positive or negative in valence), then we may expect that responses to these items will also be somewhat similar, as respondents will be influenced by the underlying item valence (e.g., Biderman et al., 2011). This type of method or valence effect can be addressed to some extent with careful evaluation design choices and later with advanced statistical techniques (e.g., Morin et al., 2016). Related to this is the need to consider possible confounding variables present in the intervention context (typically demographic or contextual in nature) that might obscure how any effects of the intervention might show up in the evaluation data. One approach is to attempt to statistically control for a number of such potential confounding factors in the analyses. Instead of blindly doing this, a better approach is to consider whether any of these variables should be more substantively factored into your framework or model for understanding how the intervention actually operates (see Bernerth & Aguinis, 2016; Spector, 2020).

Finally, if we gathered evaluation data over an extended period of time, we will need to take into account how these data are likely influencing each other (i.e., auto-correlation) and how we can most accurately model the data and relationships we want to test. Analyses involving data gathered over time are also likely to be complicated by missing data due to worker attrition and incomplete responding. This is one place where more advanced statistical methods can help (when appropriate) with data imputation and making use of the data as fully as possible (e.g., Schafer & Graham, 2002).

Sustaining Interventions over Time

Most WHSWB intervention efforts do not explicitly include a detailed and strategic plan for long-term sustainability. The ultimate goal for most OHP professionals who engage in intervention work should not be a once and done achievement, but rather a program with some longevity that is scalable and capable of yielding persistent and positive impacts. Building intervention programs that can stand the test of time is not easy, but serious consideration of the following factors when designing and evaluating interventions can help.

First, the most effective interventions are delivered in the context in which they are designed to have an impact. When designing an intervention and evaluation strategy, it is important to keep this context in mind. Think about what this context can and cannot support in terms of technology, time requirements, privacy, etc. Sustainable interventions also need to easily fit within the flow and structure of workers' daily realities.

Second, it is easier to sustain WHSWB intervention effects when they result from robust, multi-pronged methods. For example, if you are trying to educate workers on a particular topic, why not also try to change motivation and appraisals that are relevant to actually implementing the learned information or skill. Additional robustness can also be added to an intervention over time by considering whether the original intervention mechanism was sufficient or whether other elements should be incorporated in future iterations.

Third, sustainable interventions are ones in which evaluation is incorporated as a design element. This is only possible when the evaluation strategy naturally functions within the structure of the intervention and is focused on gathering data that is relevant and connected to the intervention. Long-term evaluation efforts require excellent communication and coordination with participants, and a plan to sustain intervention-related personnel to monitor, analyze, interpret, and act on the data over time. This often requires establishing long-term support within the participating organization (i.e., a champion) who can help to push for sustained effects over time.

Concluding Thoughts and Reality Check

The challenges of intervention design and evaluation are complex, but absolutely essential to the work of OHP professionals. As dense as this chapter may have seemed at times, we could only scratch the surface on some of the major issues we highlighted. If you are reading this book as you work toward an undergraduate or graduate degree, you will learn more about many of the elements emphasized in this chapter through other courses focused on research methods and statistics. You can also deepen your understanding of the topics we discussed here by diving into the readings that we cited throughout this chapter.

As challenging as it may seem to tackle WHSWB issues with interventions, we want you to also be encouraged by the reality that there really are only so many ways to go about changing workers' personal and collective perceptions, thoughts, attitudes, and behaviors. Keep this in mind as you progress through the rest of this book and into your future OHP-related career. Try hard not to overcomplicate your intervention and evaluation designs. Also, remain flexible, knowing that even the most well-designed intervention and evaluation plan is likely to hit roadblocks. Perhaps most important, never lose optimism and hope that WHSWB issues at work can be at least partially improved and better managed with interventions designed to fit workers' needs in particular work contexts.

Ultimately, WHSWB challenges are unlikely to resolve themselves. If we see an issue that threatens or impairs WHSWB, our next step is to better understand the issue and then to determine if there is anything we can do to respond. If we follow the design and evaluation principles outlined in this chapter, and we base our actions on the state of OHP science as it exists, doing something is most definitely better than doing nothing. We may not have the resources and time to design and implement the most perfect intervention and evaluation ever, but in the end, helping workers and organizations even a little can mean a lot.

Media Resources

- Curated occupational health and well-being news site: https://www.personneltoday.com/occupational-health-and-wellbeing/
- Monthly research bulletins from NIOSH: https://www.cdc.gov/niosh/research-rounds/default.html

---

Discussion Questions

1) Compare and contrast primary, secondary, tertiary levels of intervention.
2) What are critical questions to ask and decisions to make that drive the choice of evaluation methods?
3) How do we develop and implement the highest quality evaluation approaches even in organizational settings that may limit our ability to collect data?
4) Although not discussed deeply in this chapter, what is the importance of adhering to basic research ethics principles and guidelines, even when working in applied settings?
5) How can we use our evaluation and research findings to guide intervention design, revision, and maintenance over time?
6) How might we develop an evaluation plan that also identifies financial impact for a physical wellness program initiative in a typical corporate environment?

Professional Profile: Karina Nielsen, Ph.D.

Country/region: United Kingdom
Current position title: Professor at the University of Sheffield
Background: I have been working to improve the health, safety, and/or wellbeing of workers for 20 years. I have a B.A. and M.S. in Psychology, both from the University of Aarhus in Denmark. I then proceeded to get a Ph.D. in Applied Psychology from the University of Nottingham. I currently belong to the Danish Psychological Association, European Academy of Occupational Health Psychology, and the European Association of Work and Organizational Psychology.

I received my Ph.D., and then I went on to be a researcher at the National Research Centre for the Working Environment in Copenhagen, then worked at the University of East Anglia and I am now employed at the University of Sheffield. My current work involves research and creating impact outside academia. I provide guidance to policy makers, professional bodies and organisations on how they may change work practices and procedures to improve employee health and wellbeing. In doing so, I focus on the processes of interventions, including the importance of employee participation, management support and integrating the intervention into existing practices. An equally important aspect is to evaluate not only whether interventions work, but how and why they work and in which circumstances so we can replicate successful elements of interventions to other settings.

How my work impacts WHSWB: I do organisational interventions, that is supporting organisations in making changes to work practices and procedures to enhance employee wellbeing. I also work with policy bodies and professional organisations providing guidance based on research. I have worked on many projects creating impact. One major project was in the Danish postal service, where we worked with management and employees to improve working conditions. As the postal service was undergoing many changes, workers experienced job insecurity and we implemented changes to minimise this job insecurity.

My motivation: I want to make workplaces better for people allowing them to thrive and function until they reach retirement age.

Chapter References

Abernathy, D. J. (1999). Thinking outside the evaluation box. *Training & Development*, *53*(2), 19–23.

Abraham, J. M., Crespin, D. J., & Rothman, A. J. (2015). Initiation and maintenance of fitness center utilization in an incentive-based employer wellness program. *Journal of Occupational and Environmental Medicine*, *57*(9), 952–957. https://doi.org/10.1097/JOM.0000000000000498

Adkins, J.A., Kelley, S.D., Bickman, L., & Weiss, H.M. (2011). Program evaluation: The bottom line in organizational health. In J.C. Quick & L.E. Tetrick (Eds.), *Handbook of occupational health psychology* (2nd ed., pp. 395–415). American Psychological Association.

Ajzen, I. (1991). The Theory of Planned Behavior. *Organizational Behavior and Human Decision Processes*, *50*, 179–211. https://doi.org/10.1016/0749-5978 (91)90020-T

Anger, W. K., Elliot, D. L., Bodner, T., Olson, R., Rohlman, D. S., Truxillo, D. M., Kuehl, K. S., Hammer, L. B., & Montgomery, D. (2015). Effectiveness of Total Worker Health interventions. *Journal of Occupational Health Psychology*, *20*(2), 226–247. https://doi.org/10.1037/a0038340

Bandura, A. (1986). *Social Foundations of Thought and Action*. Prentice Hall.

Bennett, A. A., Gabriel, A. S., Calderwood, C., Dahling, J. J., & Trougakos, J. P. (2016). Better together? Examining profiles of employee recovery experiences. *Journal of Applied Psychology*, *101*(12), 1635–1654. https://doi.org/10.1037/apl0000157

Bernerth, J. B., & Aguinis, H. (2016). A critical review and best-practice recommendations for control variable usage. *Personnel Psychology*, *69*(1), 229–283. https://doi.org/10.1111/peps.12103

Biderman, M. D., Nguyen, N. T., Cunningham, C. J. L., & Ghorbani, N. (2011). The ubiquity of common method variance: The case of the Big Five. *Journal of Research in Personality*, *45*(5), 417–429. https://doi.org/10.1016/j.jrp.2011.05.001

Bliese, P. D., Maltarich, M. A., & Hendricks, J. L. (2017). Back to basics with mixed-effects models: Nine take-away points. *Journal of Business and Psychology*, *33*(1), 1–23. https://doi.org/10.1007/s10869-017-9491-z

Bostock, S., Crosswell, A. D., Prather, A. A., & Steptoe, A. (2019). Mindfulness on-the-go: Effects of a mindfulness meditation app on work stress and well-being. *Journal of Occupational Health Psychology*, *24*(1), 127–138. https://doi.org/10.1037/ocp0000118

Brewer, J., & Hunter, A. (2006). The multimethod approach and its promise. In J. Brewer & A. Hunter (Eds.), *Foundations of multimethod research* (pp. 1–15). SAGE Publications, Inc. https://doi.org/10.4135/9781412984294

Britt, T. W., Black, K. J., Cheung, J. H., Pury, C. L. S., & Zinzow, H. M. (2018). Unit training to increase support for military personnel with mental health problems. *Work & Stress*, *32*(3), 281–296. https://doi.org/10.1080/02678373.2018.1445671

Brown, D. K., Barton, J. L., Pretty, J., & Gladwell, V. F. (2014). Walks4Work: Assessing the role of the natural environment in a workplace physical activity intervention. *Scandinavian Journal of Work, Environment & Health*, *40*(4), 390–399. https://doi.org/10.5271/sjweh.3421

Budden, J. S., & Sagarin, B. J. (2007). Implementation intentions, occupational stress, and the exercise intention-behavior relationship. *Journal of Occupational Health Psychology*, *12*(4), 391–401. https://doi.org/10.1037/1076-8998.12.4.391

Cacioppo, J. T., Adler, A. B., Lester, P. B., McGurk, D., Thomas, J. L., Chen, H. Y., & Cacioppo, S. (2015). Building social resilience in soldiers: A double dissociative randomized controlled study. *Journal of Personality and Social Psychology, 109*(1), 90–105. https://doi.org/10.1037/pspi0000022

Chan, D. (1998). Functional relations among constructs in the same content domain at different levels of analysis: A typology of composition models. *Journal of Applied Psychology, 83*(2), 234–246. https://doi.org/10.1037/0021-9010.83.2.234

Chin, R., & Benne, K. D. (1969). General strategies for effecting changes in human systems. In W. G. Bennis, K. D. Benne, & R. Chin (Eds.), *The planning of change* (pp. 32–59). Holt, Rinehart & Winston.

Clark, M. M., Warren, B. A., Hagen, P. T., Johnson, B. D., Jenkins, S. M., Werneburg, B. L., & Olsen, K. D. (2011). Stress level, health behaviors, and quality of life in employees joining a wellness center. *American Journal of Health Promotion, 26*(1), 21–25. https://doi.org/10.4278/ajhp.090821-QUAN-272

Cooper, C. L., & Cartwright, S. (1997). An intervention strategy for workplace stress. *Journal of Psychosomatic Research, 43*(1), 7–16. https://doi.org/10.1016/s0022-3999(96)00392-3

Cunningham, C. J. L., LeMay, C. C., Sarnosky, K. M., & Anderson, A. (2014). Addressing response shift bias: A cultural competence evaluation example. In *Innovations in nursing education: Building the future of nursing* (Vol. 2, pp. 39–43).

Dicke, T., Stebner, F., Linninger, C., Kunter, M., & Leutner, D. (2018). A longitudinal study of teachers' occupational well-being: Applying the job demands-resources model. *Journal of Occupational Health Psychology, 23*(2), 262–277. https://doi.org/10.1037/ocp0000070

Edwardson, C. L., Yates, T., Biddle, S. J. H., Davies, M. J., Dunstan, D. W., Esliger, D. W., Gray, L. J., Jackson, B., O'Connell, S. E., Waheed, G., & Munir, F. (2018). Effectiveness of the Stand More AT (SMArT) Work intervention: Cluster randomised controlled trial. *BMJ, 363*, k3870. https://doi.org/10.1136/bmj.k3870

Ford, M. T., Matthews, R. A., Wooldridge, J. D., Mishra, V., Kakar, U. M., & Strahan, S. R. (2014). How do occupational stressor-strain effects vary with time? A review and meta-analysis of the relevance of time lags in longitudinal studies. *Work & Stress, 28*(1), 9–30. https://doi.org/10.1080/02678373.2013.877096

Frögéli, E., Rudman, A., & Gustavsson, P. (2019). Preventing stress-related ill health among future nurses: Effects over 3 years. *International Journal of Stress Management, 26*(3), 272–286. https://doi.org/10.1037/str0000110

Geller, E. S. (2005). Behavior-based safety and occupational risk management. *Behavior Modification, 29*(3), 539–561. https://doi.org/10.1177/0145445504273287

Gerber, J. P., Wheeler, L., & Suls, J. (2018). A social comparison theory meta-analysis 60+ years on. *Psychological Bulletin, 144*(2), 177–197. https://doi.org/10.1037/bul0000127

Glasgow, R. E., & Emmons, K. M. (2007). How can we increase translation of research into practice? Types of evidence needed. *Annual Review of Public Health, 28*, 413–433. https://doi.org/10.1146/annurev.publhealth.28.021406.144145

Goethals, G. R. (1986). Social comparison theory: psychology from the lost and found. *Personality and Social Psychology Bulletin, 12*(3), 261–278.

Golembiewski, R. T., Billingsley, K., & Yeager, S. (1976). Measuring change and persistence in human affairs: Types of change generated by OD designs. *The Journal of Applied Behavioral Science, 12*(2), 133–157. https://doi.org/10.1177/002188637601200201

Gollwitzer, P. M. (1999). Implementation intentions: Strong effects of simple plans. *American Psychologist*, *54*(7), 493–503. https://doi.org/10.1037/0003-066X.54.7.493

Gonzalez-Morales, M. G., Kernan, M. C., Becker, T. E., & Eisenberger, R. (2018). Defeating abusive supervision: Training supervisors to support subordinates. *Journal of Occupational Health Psychology*, *23*(2), 151–162. https://doi.org/10.1037/ocp0000061

Hackman, J. R. (1986). The psychology of self-management in organizations. In M. S. Pallak & R. O. Perloff (Eds.), *The master lectures, Vol. 5. Psychology and work: Productivity, change, and employment* (pp. 89–136). American Psychological Association. https://doi.org/10.1037/10055-003

Haddock, C. K., Poston, W. S., Pyle, S. A., Klesges, R. C., Vander Weg, M. W., Peterson, A., & Debon, M. (2006). The validity of self-rated health as a measure of health status among young military personnel: Evidence from a cross-sectional survey. *Health and Quality of Life Outcomes*, *4*, 57. https://doi.org/10.1186/1477-7525-4-57

Hammer, L. B., Truxillo, D. M., Bodner, T., Pytlovany, A. C., & Richman, A. (2019). Exploration of the impact of organisational context on a workplace safety and health intervention. *Work & Stress*, *33*(2), 192–210. https://doi.org/10.1080/02678373.2018.1496159

Harari, G. M., Lane, N. D., Wang, R., Crosier, B. S., Campbell, A. T., & Gosling, S. D. (2016). Using smartphones to collect behavioral data in psychological science: Opportunities, practical considerations, and challenges. *Perspectives on Psychological Science*, *11*(6), 838–854. https://doi.org/10.1177/1745691616650285

Howard, G. S., & Dailey, P. R. (1979). Response-shift bias: A source of contamination of self-report measures. *Journal of Applied Psychology*, *64*(2), 144–150. https://doi.org/10.1037/0021-9010.64.2.144

Hoy-Gerlach, J., Vincent, A., & Lory Hector, B. (2019). Emotional support animals in the United States: Emergent guidelines for mental health clinicians. *Journal of Psychosocial Rehabilitation and Mental Health*, *6*(2), 199–208. https://doi.org/10.1007/s40737-019-00146-8

Ingels, J. B., Walcott, R. L., Wilson, M. G., Corso, P. S., Padilla, H. M., Zuercher, H., DeJoy, D. M., & Vandenberg, R. J. (2016). A prospective programmatic cost analysis of Fuel Your Life: A worksite translation of DPP. *Journal of Occupational and Environmental Medicine*, *58*(11), 1106–1112. https://doi.org/10.1097/JOM.0000000000000868

Kirby, K. C., Benishek, L. A., Kerwin, M. E., Dugosh, K. L., Carpenedo, C. M., Bresani, E., Haugh, J. A., Washio, Y., & Meyers, R. J. (2017). Analyzing components of Community Reinforcement and Family Training (CRAFT): Is treatment entry training sufficient? *Psychology of Addictive Behaviors*, *31*(7), 818–827. https://doi.org/10.1037/adb0000306

Lewin, K. (1939). Field theory and experiment in social psychology: Concepts and methods. *American Journal of Sociology*, *44*(6), 868–896. https://doi.org/10.1086/218177

Lewin, K. (1947). *Field theory in social science*. Harper & Row.

Locke, E. A., & Latham, G. P. (2006). New directions in goal-setting theory. *Current Directions in Psychological Science*, *15*(5), 265–268. https://doi.org/10.1111/j.1467-8721.2006.00449.x

Ludwig, T. D., & Geller, E. S. (2000). Intervening to improve the safety of delivery drivers. *Journal of Organizational Behavior Management*, *19*(4), 1–124. https://doi.org/10.1300/J075v19n04_01

Madden, T. J., Ellen, P. S., & Ajzen, I. (1992). A comparison of the Theory of Planned Behavior and the Theory of Reasoned Action. *Personality and Social Psychology Bulletin, 18*(1), 3–9. https://doi.org/10.1177/0146167292181001

Marrow, A. J. (1967). Events leading to the establishment of the National Training Laboratories. *The Journal of Applied Behavioral Science, 3*(2), 144–150. https://doi.org/10.1177/002188636700300204

Martin, J. L. (2003). What is field theory? *American Journal of Sociology, 109*(1), 1–49. https://doi.org/10.1086/375201

Mattke, S., Kapinos, K., Caloyeras, J. P., Taylor, E. A., Batorsky, B., Liu, H., Van Busum, K. R., & Newberry, S. (2015). Workplace wellness programs: Services offered, participation, and incentives. *Rand Health Quarterly, 5*(2), 7.

McGonagle, A. K., Beatty, J. E., & Joffe, R. (2014). Coaching for workers with chronic illness: Evaluating an intervention. *Journal of Occupational Health Psychology, 19*(3), 385–398. https://doi.org/10.1037/a0036601

Michel, A., Bosch, C., & Rexroth, M. (2014). Mindfulness as a cognitive-emotional segmentation strategy: An intervention promoting work-life balance. *Journal of Occupational and Organizational Psychology, 87*(4), 733–754. https://doi.org/10.1111/joop.12072

Montaño, D. E., & Kaspryzyk, D. (2008). Theory of Reasoned Action, Theory of Planned Behavior, and the Integrated Behavioral Model. In K. Glanz, B. K. Rimer, & K. Viswanath (Eds.), *Health behavior and health education: Theory, research, and practice* (4th ed., pp. 67–96). Jossey-Bass.

Morin, A. J. S., Arens, A. K., & Marsh, H. W. (2016). A bifactor exploratory structural equation modeling framework for the identification of distinct sources of construct-relevant psychometric multidimensionality. *Structural Equation Modeling: A Multidisciplinary Journal, 23*(1), 116–139. https://doi.org/10.1080/10705511.2014.961800

Naydeck, B. L., Pearson, J. A., Ozminkowski, R. J., Day, B. T., & Goetzel, R. Z. (2008). The impact of the highmark employee wellness programs on 4-year healthcare costs. *Journal of Occupational and Environmental Medicine, 50*(2), 146–156. https://doi.org/10.1097/JOM.0b013e3181617855

Nielsen, K., & Miraglia, M. (2017). What works for whom in which circumstances? On the need to move beyond the 'what works?' question in organizational intervention research. *Human Relations, 70*(1), 40–62. https://doi.org/10.1177/0018726716670226

Nielsen, K., Nielsen, M. B., Ogbonnaya, C., Känsälä, M., Saari, E., & Isaksson, K. (2017). Workplace resources to improve both employee well-being and performance: A systematic review and meta-analysis. *Work & Stress, 31*(2), 101–120. https://doi.org/10.1080/02678373.2017.1304463

O'Shea, D., O' Connell, B. H., & Gallagher, S. (2016). Randomised controlled trials in WOHP interventions: A review and guidelines for use. *Applied Psychology, 65*(2), 190–222. https://doi.org/10.1111/apps.12053

Ozminkowski, R. J., Goetzel, R. Z., Santoro, J., Saenz, B. J., Eley, C., & Gorsky, B. (2004). Estimating risk reduction required to break even in a health promotion program. *American Journal of Health Promotion, 18*(4), 316–325. https://doi.org/https://doi.org/10.4278/0890-1171-18.4.316

Ozminkowski, R. J., Ling, D., Goetzel, R. Z., Bruno, J. A., Rutter, K. R., Isaac, F., & Wang, S. (2002). Long-term impact of Johnson & Johnson's Health & Wellness Program on health care utilization and expenditures. *Journal of Occupational and Environmental Medicine, 44*(1), 21–29. https://doi.org/10.1097/00043764-200201000-00005

Pavlov, P. I. (2010). Conditioned reflexes: An investigation of the physiological activity of the cerebral cortex. *Annals of Neurosciences, 17*(3), 136–141.

Phillips, J. M., & Gully, S. M. (1997). Role of goal orientation, ability, need for achievement, and locus of control in the self-efficay and goal-setting process. *Journal of Applied Psychology, 82*(5), 792–802.

Piwek, L., Ellis, D. A., Andrews, S., & Joinson, A. (2016). The rise of consumer health wearables: Promises and barriers. *PLoS Medicine, 13*(2), e1001953. https://doi.org/10.1371/journal.pmed.1001953

Prochaska, J. O., Redding, C. A., & Evers, K. E. (2008). The Transtheoretical Model and Stages of Change. In K. Glanz, B. K. Rimer, & K. Viswanath (Eds.), *Health behavior and health education: Theory, research, and practice* (4th ed., pp. 97–147). Jossey-Bass.

Quick, J. C. (1999). Occupational health psychology: Historical roots and future directions. *Health Psychology, 18*(1), 82–88. https://doi.org/10.1037//0278-6133.18.1.82

Quinn, R. E. & Sonenshein, S. (2008). Four general strategies for changing human systems. In T. G. Cummings (Ed.), *Handbook of Organization Development* (pp. 69–78). SAGE Publications.

Salancik, G. R., & Pfeffer, J. (1978). A social information processing approach to job attitudes and task design. *Administrative Science Quarterly, 23*(2), 224–253. https://doi.org/10.2307/2392563

Schafer, J. L., & Graham, J. W. (2002). Missing data: Our view of the state of the art. *Psychological Methods, 7*(2), 142–177. https://doi.org/10.1037/1082-989X.7.2.147

Schein, E. H. (1999). *Process consultation revisited: Building the helping relationship.* Addison-Wesley.

Schneider, B., Ehrhart, M. G., & Macey, W. H. (2011). Perspectives on organizational climate and culture. In S. Zedeck (Ed.), *APA Handbook of I-O Psychology.*

Skinner, B. F. (1963). Operant behavior. *American Psychologist, 18*(8), 503–515. https://doi.org/10.1037/h0045185

Spector, P. E. (2019). Do not cross me: Optimizing the use of cross-sectional designs. *Journal of Business and Psychology.* https://doi.org/10.1007/s10869-018-09613-8

Spector, P. E. (2020). Mastering the use of control variables: The hierarchical iterative control (HIC) approach. *Journal of Business and Psychology.* https://doi.org/10.1007/s10869-020-09709-0

Spector, P. E., Rosen, C. C., Richardson, H. A., Williams, L. J., & Johnson, R. E. (2017). A new perspective on method variance: A measure-centric approach. *Journal of Management, 45*(3), 855–880. https://doi.org/10.1177/0149206316687295

Start, A. R., Amiya, R. M., Dixon, A. C., Britt, T. W., Toblin, R. L., & Adler, A. B. (2020). LINKS training and unit support for mental health: A group-randomized effectiveness trial. *Prevention Science.* https://doi.org/10.1007/s11121-020-01106-6

Svetlitzky, V., Farchi, M., Ben Yehuda, A., Start, A. R., Levi, O., & Adler, A. B. (2019). YaHaLOM training in the military: Assessing knowledge, confidence, and stigma. *Psychological Services.* https://doi.org/10.1037/ser0000360

Thomas, J. L., Bliese, P. D., Castro, C. A., Cotting, D. I., Cox, A., & Adler, A. B. (2019). Mental health training following combat: A randomized controlled trial comparing group size. *Military Behavioral Health, 7*(3), 354–362. https://doi.org/10.1080/21635781.2018.1526724

Van der Meiden, I., Kok, H., & Van der Velde, G. (2019). Nudging physical activity in offices. *Journal of Facilities Management, 17*(4), 317–330. https://doi.org/10.1108/JFM-10-2018-0063

van der Zwan, J. E., de Vente, W., Huizink, A. C., Bogels, S. M., & de Bruin, E. I. (2015). Physical activity, mindfulness meditation, or heart rate variability biofeedback for stress reduction: A randomized controlled trial. *Applied Psychophysiology and Biofeedback, 40*(4), 257–268. https://doi.org/10.1007/s10484-015-9293-x

van Wingerden, J., Bakker, A. B., & Derks, D. (2017). Fostering employee wellbeing via a job crafting intervention. *Journal of Vocational Behavior, 100*, 164–174. https://doi.org/10.1016/j.jvb.2017.03.008

Witman, A., Beadles, C., Liu, Y., Larsen, A., Kafali, N., Gandhi, S., Amico, P., & Hoerger, T. (2019). Comparison group selection in the presence of rolling entry for health services research: Rolling entry matching. *Health Services Research, 54*(2), 492–501. https://doi.org/10.1111/1475-6773.13086

Woo, S. E., O'Boyle, E. H., & Spector, P. E. (2017). Best practices in developing, conducting, and evaluating inductive research. *Human Resource Management Review, 27*(2), 255–264. https://doi.org/10.1016/j.hrmr.2016.08.004

Wood, R., & Bandura, A. (1989). Social Cognitive Theory of organizational management. *Academy of Management Review, 14*(3), 361–384. https://doi.org/10.2307/258173

Wosu, A. C., Valdimarsdottir, U., Shields, A. E., Williams, D. R., & Williams, M. A. (2013). Correlates of cortisol in human hair: Implications for epidemiologic studies on health effects of chronic stress. *Annals of Epidemiology, 23*(12), 797–811.e2. https://doi.org/10.1016/j.annepidem.2013.09.006

3

INDIVIDUAL DIFFERENCES THAT MATTER IN OHP

Christopher J. L. Cunningham and Kristen Jennings Black

In the following chapters of this text we discuss the major worker health, safety, and well-being (WHSWB) related issues commonly addressed by occupational health psychology (OHP) professionals. Most of these topics are influenced not only by work experiences, but also by individual differences within and between workers. In this chapter, we explore individual differences in terms of worker characteristics that are theoretically and empirically linked to WHSWB phenomena and effective OHP research and practice. This includes personality traits and dispositions, demographic characteristics, and emotional states that can strongly affect how individuals perceive and respond to stressors, and more generally behave in and out of work settings. We also discuss strategies for considering individual differences when designing, implementing, and evaluating WSHWB interventions.

When you are finished reading this chapter, you should be able to:

LO 3.1: Describe pertinent individual differences that may influence workers' experiences with WHSWB phenomenon (generally and in specific ways).

LO 3.2: Explain the importance of understanding individual differences relevant to WHSWB issues at work.

LO 3.3: Discuss ways in which individual differences can influence OHP intervention design and function.

LO 3.4: Identify general strengths and weaknesses associated with group-level versus individual-level OHP interventions.

Overview of Individual Differences Theory and Research Relevant to OHP

At a high level, the individual differences we see, study, and manage every day are manifestations of two main underlying sources of differentiation. Who we are (genetically, due to nature) influences our abilities and the ways in which we respond to stimuli, but so does where we are and where we have been (through nurturing and other experiences; Anreiter et al., 2018; Jayaratne et al., 2009). Over time, these factors shape us into who and how we are as working adults. This includes forming and solidifying values and beliefs, attitudes, intentions, and patterns of behaviors and expectations. Our individual differences affect sensation, perception, and response to stimuli associated with work experiences. Who we are also influences the work we pursue, the types of work-related stimuli to which we are ultimately exposed, and the ways in which we respond to these stimuli (e.g., Bolger & Zuckerman, 1995).

Although the study of individual differences has a long history in psychology, there remains a great need for more integrative and holistic treatment of individual differences in OHP research and practice (Code & Langan-Fox, 2001; Cunningham et al., 2008; Spector, 2003). Early consideration of individual differences involved theorizing about and measuring stable underlying personality traits (see Allport & Allport, 1921; Craik et al., 1993) and the importance of aligning workers' cognitive or mental abilities and traits with their work environments (i.e., establishing a proper "fit"; Edwards & Cooper, 1990; Münsterberg, 1913; Schneider, 1987). Additional worker-centric research and intervention efforts slowly emerged, including Kornhauser's (1965) efforts to understand worker well-being and Smith's (1955) work on individual differences in susceptibility to monotony in industrial settings. Individual differences have been a more prominent concern and area of focus for OHP researchers than practitioners, but this trend may shift as more refined thinking about person-focused intervention and evaluation evolves (e.g., Grice et al., 2012; Nielsen & Miraglia, 2017). Although we cannot comprehensively present this vast evidence base in one chapter, we do provide an overview of essential individual differences that can help OHP professionals understand and improve WHSWB.

Guiding Principles

As we discuss the main types of individual differences, keep the following guiding principles in mind. First, workers *are* different in ways that impact WHSWB, but not *so* different that we cannot develop solutions that work for groups of people who share certain essential characteristics. Pretending that individual differences do not exist or are irrelevant is counterproductive and not commensurate with the research and history that supports OHP. However, workers share many commonalities, particularly when it comes to work-related experiences. OHP professionals must balance desire for depth of understanding and breadth of impact against what is feasible and pragmatic. While there is no "one size fits all" WHSWB intervention, it is possible to customize interventions to target multiple objectives and multiple groups of workers with some degree of sensitivity to pertinent underlying differences. If this sounds too idealistic, consider that even the finest tailored suit or dress is not designed around millions of unique measurements, but rather a limited set of critical physical dimensions. Similar customization can be possible in research and intervention work when OHP professionals understand individual differences pertinent to specific WHSWB phenomena.

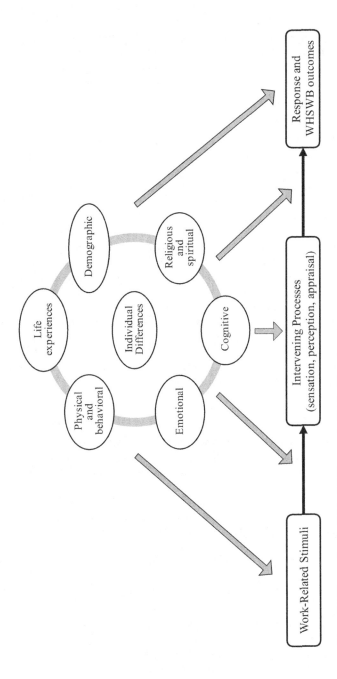

Figure 3.1 Connecting Individual Differences to Worker Experiences

Second, most individual differences are difficult to cleanly define (All-port & Allport, 1921) and even our more trait-like differences may be more malleable across time and context than we often think (Markus & Kunda, 1986; Robinson, 2009). Third, what we know about individual differences is currently limited mainly to research conducted with samples of participants from Western countries and cultures. There likely are some differences in the prevalence and impact of many of these individual differences across different cultures, as observed by Lu et al. (2011) with respect to self-efficacy among workers in China versus Western cultures. Collectively, these realities challenge OHP professionals to continue developing theories, research, and evaluation methods to more fully address ways that workers meaningfully differ.

Fourth, WHSWB and workers experience are simultaneously influenced by multiple individual differences, many forms of which we may not have even identified or labeled. In OHP research and practice, we must continually consider complex interactions among individual differences that are internal (e.g., traits/states, knowledge, experience, physical abilities) *and* external (e.g., demographic situation, work and home environment) to each worker. We conceptually illustrate these broad effects of individual differences in Figure 3.1. Despite this complexity, remember our first guiding principle and do not lose hope; it is still possible to improve WHSWB for individual workers.

In the following sections of this chapter, we provide a high-level overview of what is essential for OHP professionals to understand about workers' individual differences. We explore one overarching and five more specific forms of individual difference that are likely to be associated with WHSWB through their connections to workers' perceptions, appraisals, and reactions to work-related stimuli and experiences.

The Main Difference: Work and Life Experiences

Our life experiences provide the foundation for who we are (e.g., Cramer, 2004; Roccas & Brewer, 2002) and are part of the research and practice context for OHP professionals. As an illustration, imagine you are preparing a driving safety intervention for truck drivers. Your intervention is unlikely to be their first exposure to such information. If you design and deliver this training as if every worker is starting with no relevant prior knowledge or experience, you are less likely to engage participants and have your intended impact than if you calibrate your approach to your audience.

Although broad *life history* approaches to research and interventions (e.g., Goodson, 2001) are not common in OHP, it is important to recognize that no two workers' life histories are identical. Although it may be impossible to fully operationalize and consider such histories for every worker, we can focus attention on individual differences that are particularly salient to the WHSWB issue(s) being addressed. Examples of experiences likely to influence worker perceptions and responses to work-related stimuli include juggling multiple jobs, struggling to meet financial obligations, managing recent job or work schedule changes, and past experiences with violence, mistreatment, or trauma. In short, we need to understand that who and how workers are today is not likely the whole story.

Demographic Differences

There are many ways in which workers differ in terms of characteristics that we use to describe populations. Surprisingly few of these demographic differences are directly incorporated into dominant OHP-related theories and practice efforts. In this section we explore four particularly important classes of demographic difference.

Sex and Gender Identity

A commonly measured and discussed form of individual difference is that of sex and gender identity. The near omnipresence of this information in descriptions of research samples belies limited efforts to actually link this form of individual difference to WHSWB phenomena (but see the following for approaches to addressing this gap: Messing et al., 2003; Nelson & Burke, 2002). Further limiting OHP consideration of this individual difference are lingering inconsistencies in the use and meaning of these labels (e.g., Muehlenhard & Peterson, 2011; Unger & Crawford, 1993).

There is some evidence that underlying biological sex differences between male and female workers may influence work-related stimuli exposure, personal sensation, perception, appraisal, and response (Baum & Grunberg, 1991; Hooftman et al., 2005). As one specific example, women are substantially more likely than men to perceive weight-related discrimination at work (Roehling et al., 2007, p. 311). Also documented are sex-related differences in exposure to work-related demands (e.g., Hooftman et al., 2005) and general work-related safety (e.g., Liao et al., 2001; Wong et al., 2014). There also appear to be differences between men and women when it comes to recovery-related opportunities and experiences (e.g., Boschman et al., 2017; Saxbe et al., 2008). A wide variety of studies have also documented differences between individuals based on their masculinity versus femininity (e.g., more feminine people react more strongly and positively to social support than more masculine people do; Beehr et al., 2003). This research is not conclusive, and it has been argued that the whole concept of gendered work experiences needs much more attention (McDonough & Walters, 2001).

Some of the complexity and confusion pertaining to this common individual difference is linked to an incomplete and inconsistent understanding of how and why sex and gender identity differences exist. Such differences may actually be underlying effects of life experiences and social role pressures that have shaped workers' socially constructed identities (Bussey & Bandura, 1999; Eagly, 1987). Along these lines, some observed differences initially ascribed to sex or gender may be more closely and meaningfully linked to other associated individual differences such as marital status, age, education, and identification with a parental or family role (Michael et al., 2009; Powell & Greenhaus, 2010; Simon, 1992). We need to be more cautious and precise as OHP professionals when working with this form of individual difference. This includes not over-interpreting data that appear to indicate male versus female differences without also working to explain what such effects might mean.

Age and Life Stage

As with sex and gender, individual differences in age and period of life are neither clearly defined, nor consistently considered in OHP research and practice. With respect to chronological age, our biological and physiological systems change in complex ways that may impact WHSWB. For example, shiftworkers over the age of 40 may struggle with sleep quality after night shifts more than younger colleagues, but simultaneously report less sleepiness and sleep need (Härmä, 1996). Effects like this capture the confounding reality that experience often comes with age; extending the preceding example, older workers likely have had more time to adjust their circadian rhythms and sleep strategies to maintain an adequate level of alertness and functioning. As another example, research generally shows that youth and inexperience is associated with greater WHSWB risks (and this risk is greater for males than females; e.g., Salminen, 2004), but that older workers are more at risk for severe or fatal workplace accidents than younger workers (Peng & Chan, 2019). The only clear conclusion from findings like these is that aging is a risk factor for workers, but not an automatic or consistent one for *all* workers.

Further complicating the meaningful consideration of this form of individual difference is the much-popularized notion of "generational differences", which purportedly affect workers' perceptions, expectations, values, and behaviors within work domains (e.g., Wey Smola & Sutton, 2002). Although intriguing, theories explaining these generational differences are rather thin and fail to really explain or support clearly testable hypotheses. Many (potentially all) of these differences are more likely manifestations of underlying differences in chronological or physical age and a person's stage or period of life (for a detailed discussion of these points, see The National Academies of Sciences, Engineering, and Medicine, 2020; Rudolph & Zacher, 2020).

Different stages of life bring with them a combination of demands and resources, which most definitely impact workers' perceptions, appraisals, and responses to work-related stimuli. Ages do not cleanly and consistently correspond to individuals' actual life stages, which typically correspond to major life events such as finishing schooling, starting a first job, getting married, having children, etc. When these types of events occur, workers' priorities, challenges, and opportunities change. One particularly relevant example is the number of dependants a worker is supporting. Most workers have responsibilities outside of work to support or care for at least one other person, pet, or other dependant. Understanding workers' dependant care situation is important if we are to really protect and promote WHSWB in a way that transcends the workplace.

Race and Ethnicity

Race and ethnicity are also often emphasized as meaningful forms of individual difference. The limited theory and empirical support for this form of individual difference is even more pronounced than for the preceding demographic differences because typical sampling for OHP research and intervention evaluation does not yield subgroup sizes that permit accurate between-group comparisons. There is some empirical evidence of differences between members of certain racial and ethnic groups at a biological level (e.g., risk of cardiovascular disease, diabetes; McWilliams et al., 2009; Thomas et al., 2005) and some evidence that racial minorities may be exposed to more interpersonal stressors (e.g., selective incivility as "modern discrimination"; Cortina et al., 2011), but for most WHSWB challenges, linkages with race and ethnicity are still being explored.

Theorizing and hypothesizing about racial or ethnic differences can seem daunting, but psychological knowledge and theory can provide a starting point. For example, workers in a minority position or group may not perceive or be able to access support from others at work (e.g., James, 1997). In work settings, this can unfortunately feed into perceptions of and experiences with limited career mobility, person-environment misfit, isolation, and similar scenarios that are associated with negative work performance and worker health and well-being outcomes. Knowing how important support at work can be, OHP professionals can work to develop and test organization-based efforts to facilitate supportive relationships especially for minority workers, such as with mentoring programs and employee resource groups (Daniel, 2009; Eby et al., 2013; Gonzáles-Figueroa & Young, 2005).

Economic Situation

Although not a mainstream area of study in OHP, there is mounting evidence that workers' economic situations impact WHSWB (e.g., Ford, 2011; Sinclair & Cheung, 2016). Understanding workers' economic situations provides insight into the extent to which work commitments provide (in)sufficient financial resources and security to meet actual needs. Paradoxically, many of the most difficult occupations studied by OHP professionals also tend to pay the lowest wages. This creates a very difficult reality for workers who must handle difficult work-related demands, low perceptions of control, and limited or insufficient financial resources and reward.

There are many ways OHP professionals can conceptualize and operationalize economically oriented individual differences. This includes considering perceived income adequacy, income insecurity, and general socioeconomic status (Sinclair & Cheung, 2016). *Perceived income adequacy* is essentially whether a worker believes their income is sufficient to meet their wants and needs. A related phenomenon here is *income insecurity*, which is experienced by those who lack the financial resources to manage a major financial challenge, retirement, or serious healthcare event. A person's *socioeconomic status* is a general reflection of a person's financial security and level of access to amenities and supports that are associated with a high quality of life (e.g., access to quality education, healthcare, and food).

Economic individual differences are among the few that can be modified through organizational and societal policies, or other interventions. Such efforts might include worker training and development to make job progression, promotion, and growth possible. Other options include organizational or societal initiatives to provide a *living wage* for all workers (i.e., "wage level at which life becomes more than 'bearable' – a level at and over which capability, rather than mere subsistence, is fostered and enabled," Carr et al., 2016, p. 5). Economic individual differences can often lead to economic disparity and financial stress, which challenge workers every day over and above other work-specific demands (e.g., Edgell et al., 2012). If we are serious about protecting WHSWB, we need to address stress and strain associated with chronic economic disparities and financial hardships.

Religious and Spiritual Differences

Now we shift our focus to other OHP-related forms of individual difference that are not strictly demographic. One of the most complex and influential of these is the extent to which a person engages in the search for the sacred, either through some sort of formalized religion and/or through more personal, spiritual practices (Cunningham, 2014; Hill & Pargament, 2003; King & Koenig, 2009). Religious beliefs and practices help to form and establish underlying values and perspectives on the world, as well as common ground and/or reason for division.

This form of individual difference is linked to worker behaviors, physical and psychological health, concern for others, creativity, commitment, ethical behavior, and personality, among other critical worker behaviors (Day, 2005). Spiritual beliefs and practices have also been linked with more positive appraisals of life events, especially among older adults (Cowlishaw et al., 2012). Workers' religious and spiritual values and practices appear to support positive well-being and reduced stress and associated strain (e.g., Arnetz et al., 2013; Powell et al., 2003). This form of individual difference is also often conceptualized and studied as a special form of personal resource (discussed more in Chapter 6), which may positively influence worker engagement (e.g., Bickerton et al., 2014).

Given these many connections, it is easy to see how religion and spirituality are important individual differences for OHP attention (e.g., Miller & Thoresen, 2003). Understanding the presence, strength, and form of this type of individual difference can help to explain how and why workers do what they do at and outside of work. We are surprised that this form of individual difference does not get more attention by OHP researchers, given its strong linkage to many antecedents and outcomes emphasized in OHP research and practice (e.g., perception and appraisal of work demands and other stressors; how people treat each other in work settings).

Cognitive Differences

Workers also differ in how they perceive, process, and react to stimuli (Payne, 1991). In work settings especially, workers' *general cognitive ability* can be particularly influential, as it is one of the strongest predictors of future learning and performance potential (e.g., Schmidt & Hunter, 2004). This individual difference is rarely considered in OHP research and intervention efforts, however, because of limited theorizing about the role of cognitive ability in WHSWB phenomena and because of limited availability of generalizable and work-related assessments. Several other challenges are also associated with the use of cognitive ability assessments (e.g., frequent subgroup differences in scores for majority vs. minority group members, increased levels of respondent anxiety, time and assessment security concerns; Hausdorf et al., 2003; Reeve et al., 2009). These difficulties do not change the fact that individual differences in cognitive abilities are relevant to WHSWB, especially pertaining to safety-related awareness and behaviors (e.g., Postlethwaite et al., 2009; Vetter et al., 2018). At the very least, OHP professionals can be smarter (some irony intended) about using knowledge/skill pre-tests when designing interventions to ensure they are properly calibrated and focused.

In terms of worker perception, several trait-like cognitive differences are particularly pertinent to OHP research and practice. One such individual difference is a person's *sense of coherence*, or the extent to which that person sees the world as meaningful, manageable, and comprehensible (Jenny et al., 2017). Workers' sense of coherence appears to function as a resource itself and is strengthened by the presence of other job resources; together, these forms of resources are positively connected to desirable work-related outcomes such as worker engagement (Vogt et al., 2016). A second trait-like individual difference that influences perception and ultimately behavior is *core self-evaluations*, a composite trait incorporating locus of control, self-esteem, neuroticism, and general self-efficacy (Judge & Hurst, 2007; Kammeyer-Mueller et al., 2009). This individual difference is likely associated with many different WHSWB phenomenon, given that workers with strong core self-evaluations may be drawn to complex work situations (Judge et al., 2000). Interestingly, strong core self-evaluations are negatively associated with experienced work-family conflict (Boyar & Mosley, 2007) and positively linked to job and life satisfaction, two common indicators of worker well-being (Judge et al., 1998).

There are also cognitively oriented individual differences that may influence how workers respond or react to work-related stimuli. Early explorations of this form of individual difference focused on how workers' cognitive abilities might influence appraisal and coping efforts in response to stressors (Payne, 1991). More recent explorations along these lines have identified *mindsets* (e.g., Crum et al., 2013) and cognitive *orientations* (e.g., goal orientation; Whinghter et al., 2008) as additionally relevant cognitive differences. Also relevant here are workers' perceptions of self- and collective efficacy (e.g., Bandura, 2006; Jex & Bliese, 1999) pertaining to perceived ability to accomplish work tasks.

Emotion-Related Differences

Many WHSWB phenomena and workers' reactions to evaluations and assessments are closely related to underlying emotional states and traits. Research suggests that individual differences in emotional stability, and dispositional *positive and negative affectivity* may moderate the relationship between stressors and stress reactions (Penney et al., 2011; Penney & Spector, 2005). These emotion-related traits and states have been shown to capture meaningful variance in emotional condition that is not explained by other demographic characteristics (Crawford & Henry, 2004). More generally, positive emotional traits and states (e.g., self-efficacy, core self-evaluations, positive affectivity) are positively connected to work-related attitude, the ability to find meaning in work, and well-being, while negative emotional traits (e.g., negative affectivity, chronic anger, and hostility) are negatively related to the same outcomes (e.g., Alarcon et al., 2009; DeNeve & Cooper, 1998; Smith, 2006; Tugade & Fredrickson, 2004). Workers' underlying emotional tendencies may influence their abilities to build and sustain other emotional and cognitive resources (e.g., Fredrickson & Branigan, 2005), which can ultimately help them to flourish (i.e., thrive, not just survive, as outlined by the broaden and build theory; e.g., Fredrickson, 2000).

A separate form of emotion-related individual difference has to do with workers' ability to recognize and control their own emotional displays, and to accurately perceive and respond to emotions in others. Collectively, these elements are often referred to as *emotional intelligence* (e.g., Joseph et al., 2015). This type of individual difference can help explain how workers react to situations and other people, especially in service-related occupations where understanding and controlling emotional reactivity is necessary. We discuss the challenges of such *emotional regulation* more in Chapter 7.

Physical and Behavioral Differences

Many behavioral differences among workers are direct extensions of underlying differences in physical abilities and behavioral tendencies. For example, some people may be predisposed to *sensation seeking* (e.g., Stephenson et al., 2003) or *impulsivity* (e.g., Stanford et al., 2009). We also all have physical differences that can constrain or enable our functional capabilities and thereby influence WHSWB. For example, chronic and acute physical conditions (e.g., pain, gastrointestinal system disorders, headaches) may require accommodation and can influence workers' experiences at work. Similarly, physical attributes (e.g., height, weight, strength, cardiovascular system condition) directly affect how workers respond to work-related demands and other stimuli.

Differences in physical ability can also directly impact a person's choice of occupation and likelihood of engaging in wellness-oriented interventions or programs, based on the fit or alignment workers see between demands and their abilities to meet those demands. For example, sitting while working is extremely difficult for individuals who normally prefer to be moving and cannot otherwise maintain focus and energy. Other workers, though, may seem to not be troubled at all by having to sit while working all day (despite the associated negative physical health implications). It is important for OHP professionals to understand how individual differences and environmental characteristics of the work environment support worker functioning (topics often explored more directly by human factors professionals and discussed more in Chapter 10).

Appreciating physical and behavioral forms of individual differences is particularly important when designing WHSWB interventions. Consider an example from a personal experience leading a corporate wellness program evaluation: The organization was trying to get more of their workforce to use a new onsite fitness center and participate in organization-sponsored physical training programs. By studying workers' insurance data, they realized that workers could be classified into four or five different groups based on level of physical ability and openness to physical exercise. To accommodate these different needs and interests, this organization implemented a four-track physical health development initiative with programming to target employees who ranged in experience from never having exercised before to daily gym users or even triathletes. Workers could choose which program they wanted to join, giving them control and options over participation, and illustrating how it is possible to use knowledge of physical and behavioral individual differences to more effectively address a WHSWB issue.

Why Individual Differences at Work Matter

We hope you are starting to see how a variety of worker individual differences need to be considered when engaging in OHP research and practice. In this section, we explore three general arguments for why it is important to understand and appreciate these differences when working to protect and promote WHSWB.

Greater Understanding and Impact

There is more to individual differences than just the demographic labels we typically use to describe research samples or worker populations. When we rely on general labels and ignore deeper individual differences, we tend to form perceptions and make judgments on aggregated, higher-level concepts of types of people. As mentioned in the guiding principles at the start of this chapter, there are certainly areas where group-level means are informative, but we should not stop there or overlook opportunities to adapt and tailor our efforts to better meet the needs of workers. Trying to explain WHSWB phenomena and intervene to address these issues based on an abstract sense of an "average worker" forces us to operate in a rather abstract space, where we risk designing an intervention that might ultimately apply to no actual person (i.e., the mean may not reflect any actual person; Grice, 2011). We are more likely to impact actual workers if we conduct our research to understand, and design and deliver our interventions to positively impact actual workers.

Because Consistent is Not Automatically Fair

Appreciation for workers' individual differences supports true and meaningful efforts to enhance diversity and inclusiveness within organizations. Acknowledging, identifying, and explaining the importance of individual differences in OHP research and interventions can help organizational decision makers do likewise. Although consistency is often valued in work settings, consistency does not equal fairness or equity (cf., Walster & Walster, 1975). This is because workers differ in ways that affect perception, appraisal, and response to work-related stimuli, including organizational and managerial communication. For these reasons, consistency can still lead to bias, unfairness, and injustice at work; none of these outcomes support positive WHSWB.

Here is one example from a personal experience where consistency did not translate into fairness in an WHSWB intervention. An organization decided to implement a weight loss program modeled after popularized programs on television and in other companies as a worker- and department-level weight loss competition (i.e., who could lose the most weight over a six-month period). For the sake of consistency (and in their minds, fairness) organizational leaders implemented this program uniformly throughout the organization, despite meaningful underlying differences among workers (e.g., the workforce was split between very sedentary roles vs. very physical production line roles and there were substantial underlying differences in workers' baseline health).

This competition ran for two months before initial review of program-related data revealed that participation was not balanced across all departments. Certain departments with more sedentary work requirements had much higher proportions of obese workers with no real opportunity to be active during work shifts (due to environmental constraints). Compounding this, the whole approach of bringing attention to weight loss as a competition was quite distressing for individuals who were already struggling with body weight concerns (cf., Hunger & Major, 2015). In contrast, workers in more physically active job functions in this organization, were participating at much higher levels and generally were more physically healthy even before the intervention. This organization was also surprised by negative employee attitude and engagement data that appeared on the annual worker attitude survey that was administered three months into this initiative.

This became a teaching moment for this organization's leaders, and they learned the importance of accounting for individual differences before deploying such programs. This is a good illustration of where ignorance of individual differences can backfire and have huge consequences. This particular program was quickly cancelled, but its effects will continue to be felt for years. Imagine how difficult it will be for this organization in the future to implement any other form of health-related intervention.

More Effective Management

Acceptance and appreciation of individual differences can facilitate more effective management efforts (e.g., Nishii & Mayer, 2009; Roberson et al., 2017).

Understanding individual differences associated with WHSWB issues can also provide organizational management with helpful guidance when it comes to optimizing workers' fit to particular roles, and (re)designing work environments to foster performance, maximize attendance, and increase worker engagement. Understanding individual differences makes it possible to connect with workers and learn how to identify and meet workers where they are, as they are. Attention to individual differences among workers can also help managers communicate even difficult messages, such as news of an upcoming workforce restructuring, in a way that protects WHSWB as much as possible. In this example, workers' life stages and personal histories will influence how they respond to this information. Specifically, a worker nearing retirement is unlikely to respond the same way as one who is just beginning their career; a worker with no children or dependants is likely to respond very differently from a single, early-career professional; workers in good standing with the organization are likely to respond better than workers who may have just had a bad performance review or may require accommodations for an injury, health condition, or disability.

Personalizing every organizational communication is impractical, but instead of sending a mass email, a manager could engage with different groups of workers to share important information and respond to immediate questions and concerns. Extending the example in the previous paragraph, directly acknowledging how this type of news may be more difficult for some workers than others demonstrates support, compassion, and respect for workers' differences. Given this type of opportunity, subordinates may be more willing to share feedback and identify challenges that could ultimately improve the effectiveness of this organizational initiative. Taking steps like these is not inefficient, but rather good people management and change facilitation.

Methodological Considerations and Practical Recommendations

As noted earlier in this chapter, most OHP professionals do not solely focus on individual differences when conducting research or developing and testing interventions. Despite this, and as argued throughout this chapter, there is much to be gained if we can do a better job of considering individual differences as we tackle other WHSWB issues. Data pertaining to individual differences can take many forms. Often, we quantify and qualify these differences in a way that permits only labeling or description of person or group differences, without providing any rationale as to how and why these differences are linked to WHSWB. Other forms of individual differences, as highlighted earlier in this chapter, have more direct and meaningful connections to how and why workers do what they do. Many of these individual differences can be measured in terms of their degree of influence, intensity, strength, frequency, etc. Once such information is gathered, we then must make careful decisions about how to best analyze and interpret effects involving individual differences along with our main WHSWB variables of interest. We explore these challenges in this section.

Measuring and Monitoring Individual Differences

The default methodological approach when attempting to understand or study individual differences is to use a survey. With a little creativity, there are a variety of alternative methodologies that can also be used when measuring and monitoring the various forms of individual difference discussed earlier in this chapter. In general, this includes mixing quantitative and qualitative techniques, and using a variety of different data gathering methodologies including interviews and focus groups, surveys and assessments, archival data review, and ethnographic or observational study (Fisher & Barnes-Farrell, 2013; Schonfeld & Mazzola, 2013; Woo et al., 2017).

Individual Differences over Time

Often, individual differences are linked to OHP efforts involving trajectories of change, development, or sometimes stability over time (e.g., Raudenbush, 2001). Methodologically, this type of research involves repeated and longitudinal data collection, typically focused on information that comes from within or is directly about people. To efficiently gather this information, OHP researchers commonly use self-report data gathering techniques such as surveys, diary studies (Bolger et al., 2003; Eckenrode & Bolger, 1995), and experience sampling (Sonnentag et al., 2013). From a data collection perspective, this is pretty straightforward, but repeated measures create analysis-related challenges (as noted in Chapter 2). Much of this complexity is due to the connectedness or dependency of repeated data points gathered from the same person (source). For example, a worker's mood at 9:00 am is likely to influence their feelings at 11:30 am and even at 6:00 pm. Working with this type of dependency is complex, and certain statistical analysis techniques (e.g., repeated measures ANOVA, hierarchical linear modeling, latent growth curve analyses) make it possible to model within-person change and development. It is not necessary that all OHP professionals are experts in the use of these statistical methods (collaborating with others who can run the analyses is encouraged), but it is essential to understand that these methods exist and can be useful in specific types of research and intervention evaluation scenarios.

Person-Centered Study

There are analytical methods particularly relevant to this topic of individual differences that are increasingly used to focus attention on the person, rather than an abstract aggregated indicator of a sample or group (e.g., Wang et al., 2013). This might involve predicting or attempting to explain an outcome or effect using combinations of directly measured or observed personal characteristics or behaviors, rather than an abstract group-level mean. Although these methods differ in terms of what they require methodologically and how they function analytically, what they share is a focus on modeling effects at the level of actual people. When used in intervention evaluation, these types of techniques can provide richer process mapping that illustrates if and how an intervention is impacting specific individuals over time. Implications of this work for OHP research are becoming clearer as more studies using these techniques work their way into mainstream journals. You can explore these techniques by learning about *realist evaluation* (Nielsen & Miraglia, 2017), *latent profile/class analyses* (e.g., Bennett et al., 2016), and *observation oriented modeling* (Grice et al., 2012).

Individual Differences Can be More Than Covariates

Innumerable peer-reviewed OHP research manuscripts include a slew of demographic or personality-related individual differences as control variables or statistical covariates. Often missing are any explanation for why these variables were even measured. It is insufficient to note that such information is included to "be consistent with previous research", despite how frequently this rationale is noted (Bernerth & Aguinis, 2016).

We need to take a more conscious approach to figuring out how individual differences factor in to our WHSWB theories, research, and interventions. Helpful guidance along these lines comes from work by Paul Spector (2020; 2010), a significant contributor to OHP education, research, and practice. Including individual differences as statistical covariates in our analyses can enable us to examine other (hopefully hypothesized) effects after accounting for the influence of presumably underlying and implicitly meaningful individual differences. When we do this, though, we are implicitly indicating that these individual differences matter, just not enough to investigate in their own right. Sometimes this is appropriate, but over time, this can stunt development of our theories and our general understanding of WHSWB phenomena. Worse, in some cases, including indicators of individual difference can actually confound and even obscure the relationships that we are actually trying to identify (e.g., statistically controlling for trait negative affectivity can make it very difficult to observe an effect of momentary affect). If we cannot explain why a specific individual difference matters in a particular research or intervention context, then this information probably should not play a prominent role in our analyses or ultimate evaluations and findings.

Concluding Thoughts and Reality Check

The breadth and depth of this chapter reflects the amazing variety of ways in which workers differ from each other at an individual level. Because individual differences affect our sensation, perception, and responses to stimuli in and around our work environments, they can WHSWB and, therefore, OHP research and intervention efforts. These differences can also influence (maybe even dictate) the various resources that workers draw on to address their work demands (as we discuss more in our subsequent chapters).

There are many opportunities for OHP professionals to leverage knowledge of individual differences to improve WHSWB. This could be through improving fit or alignment between workers and their jobs, increasing the accuracy of candidate expectations before hire, and developing personalized retention and job security plans. Understanding individual differences in a richer and more comprehensive way also makes it possible to broaden our approach to measuring, studying, and generally working with the true diversity present in all organizational contexts.

Individual differences among workers are part of the broader context that connects work to other domains of our lives. Such differences also help us understand and manage WHSWB. We can be more effective as OHP professionals when we remember Bandura's (1978) concept of triadic reciprocal determinism, which connects person, environment, and behavior. More specifically, workers' behaviors at work are not just due to factors in the work environment, but also to underlying differences – who we are multi-directionally affects where we are, how we are feeling, and what we are ultimately doing.

Media Resources

- CEO reflection on being "real" and personal with your employees:
  https://www.fastcompany.com/90315587/one-ceo-makes-a-case-for-getting-more-personal-at-work
- Popular media piece on questionable value of some popular personality "tests":
  https://www.psychologytoday.com/us/blog/people-are-strange/201909/your-favorite-personality-test-is-probably-bogus
- Business article on the importance of understanding different personalities at work:
  https://www.forbes.com/sites/johnhall/2020/02/28/how-diverse-personalities-can-be-better-understood-in-your-office/?sh=1fa52ca150f6

Discussion Questions

1) What is an "individual difference" from our applied psychological perspective?
2) What are examples of individual differences that likely "matter" more (and less) when we consider WHSWB?
3) What are managerial implications of being aware of pertinent individual differences that may affect workers' experiences with health and safety issues at work?
4) Imagine you are tasked with evaluating scheduling practices at your organization and considering new approaches. What individual differences might you want to account for in trying to evaluate and develop the best possible scheduling practices?

Professional Profile: Steve Jex, Ph.D.

Country/region: USA, Florida
Current position title: Professor of Psychology, Industrial-Organizational (I-O) Psychology Ph.D. Program Director, and director of the Employee Health and Well-Being Lab at the University of Central Florida
Background: I began my career in 1989 and have worked for over 30 years to improve employee health, safety, and well-being. I received my undergraduate degree in general psychology from Central Michigan University in 1981. Following graduation, I entered the Master's program in I-O Psychology at the University of New Haven and received my degree in 1984. After completing my Master's degree I worked for over a year as a Sensory Evaluation Researcher for Cadbury Schweppes, but really had the urge to get back in to I-O Psychology so in 1985 I entered the Ph.D. program in I-O Psychology at the University of South Florida, and received my degree in 1989. I was one of the founding members of the Society for Occupational Health Psychology. I also belong to the Society for Industrial and Organizational Psychology.

I started my career as a professor immediately after receiving my Ph.D. Prior to my current position at the University of Central Florida, I held faculty positions at Central Michigan University, University of Wisconsin Oshkosh, and Bowling Green State University. During my 30-year academic career I have also done work with non-academic organizations from time to time. The most relevant to OHP were research collaborations with Walter Reed Army Institute for Research, and the National Institute for Occupational Safety and Health. Like most professors I teach, do research, and engage in service activities. As a doctoral program director, I am responsible for all aspects of the program including curriculum, program assessment, and student progress. As the director of my research lab, I set the overall research agenda and offer my expertise on a variety or research projects.

How my work impacts WHSWB: A large part of the work of my lab is focused on better understanding the dynamics of mistreatment within the workplace, so that is probably has the most impact. I am currently collaborating with two of my colleagues (Dr. Mindy Shoss and Dr. Kristin Horan) on a 5-year project focused on developing, implementing, and evaluating health and safety interventions in the hospitality and tourism industry. When I was at Bowling Green, I was involved in a 2-year project that was funded by the Ohio Bureau of Worker's Compensation. This project was focused on (1) better understanding the factors that predict injuries and accidents among nursing home employees, and (2) evaluating the impact of a stress management intervention within nursing homes. I felt this was an extremely important project because nursing home employees not only face a number of physical hazards, but also work in an environment that can be very emotionally draining.

My motivation: Since work takes up so much of people's time, it makes sense that one of the keys to overall happiness and well-being is a high-quality work life. I also think there are people out there who have very difficult jobs, and I think our field has the potential to improve their quality of life.

Chapter References

Alarcon, G., Eschleman, K. J., & Bowling, N. A. (2009). Relationships between personality variables and burnout: A meta-analysis. *Work & Stress, 23*(3), 244–263. https://doi.org/10.1080/02678370903282600

Allport, F. H., & Allport, G. W. (1921). Personality traits: Their classification and measurement. *Journal of Abnormal and Social Psychology, 16*, 6–40. https://doi.org/10.1037/h0069790

Anreiter, I., Sokolowski, H. M., & Sokolowski, M. B. (2018). Gene-environment interplay and individual differences in behavior. *Mind, Brain, and Education, 12*(4), 200–211. https://doi.org/10.1111/mbe.12158

Arnetz, B. B., Ventimiglia, M., Beech, P., DeMarinis, V., Lökk, J., & Arnetz, J. E. (2013). Spiritual values and practices in the workplace and employee stress and mental well-being. *Journal of Management, Spirituality & Religion, 10*(3), 271–281. https://doi.org/10.1080/14766086.2013.801027

Bandura, A. (1978). The self system in reciprocal determinism. *American Psychologist, 33*(4), 344–358. https://doi.org/10.1037/0003-066X.33.4.344

Bandura, A. (2006). Toward a psychology of human agency. *Perspectives on Psychological Science, 1*(2), 164–180. https://doi.org/10.1111/j.1745-6916.2006.00011.x

Baum, A., & Grunberg, N. E. (1991). Gender, stress, and health. *Health Psychology, 10*(2), 80–85. https://doi.org/10.1037/0278-6133.10.2.80

Beehr, T. A., Farmer, S. J., Glazer, S., Gudanowski, D. M., & Nair, V. N. (2003). The enigma of social support and occupational stress: Source congruence and gender role effects. *Journal of Occupational Health Psychology, 8*(3), 220–231. https://doi.org/10.1037/1076-8998.8.3.220

Bennett, A. A., Gabriel, A. S., Calderwood, C., Dahling, J. J., & Trougakos, J. P. (2016). Better together? Examining profiles of employee recovery experiences. *Journal of Applied Psychology, 101*(12), 1635–1654. https://doi.org/10.1037/apl0000157

Bernerth, J. B., & Aguinis, H. (2016). A critical review and best-practice recommendations for control variable usage. *Personnel Psychology, 69*(1), 229–283. https://doi.org/10.1111/peps.12103

Bickerton, G. R., Miner, M. H., Dowson, M., & Griffin, B. (2014). Spiritual resources and work engagement among religious workers: A three-wave longitudinal study. *Journal of Occupational and Organizational Psychology, 87*(2), 370–391. https://doi.org/10.1111/joop.12052

Bolger, N., & Zuckerman, A. (1995). A framework for studying personality in the stress process. *Journal of Personality and Social Psychology, 69*(5), 890–902. https://doi.org/10.1037/0022-3514.69.5.890

Bolger, N., Davis, A., & Rafaeli, E. (2003). Diary methods: Capturing life as it is lived. *Annual Review of Psychology, 54*, 579–616. https://doi.org/10.1146/annurev.psych.54.101601.145030

Boschman, J. S., Noor, A., Sluiter, J. K., & Hagberg, M. (2017). The mediating role of recovery opportunities on future sickness absence from a gender- and age-sensitive perspective. *PLoS One, 12*(7), e0179657. https://doi.org/10.1371/journal.pone.0179657

Boyar, S. L., & Mosley, D. C. (2007). The relationship between core self-evaluations and work and family satisfaction: The mediating role of work–family conflict and facilitation. *Journal of Vocational Behavior, 71*(2), 265–281. https://doi.org/10.1016/j.jvb.2007.06.001

Bussey, K., & Bandura, A. (1999). Social Cognitive Theory of gender development and differentiation. *Psychological Review*, *106*(4), 676–713. https://doi.org/10.1037/0033-295x.106.4.676

Carr, S. C., Parker, J., Arrowsmith, J., & Watters, P. A. (2016). The living wage: Theoretical integration and an applied research agenda. *International Labour Review*, *155*(1). https://doi.org/10.1111/j.1564-913X.2015.00029.x

Code, S., & Langan-Fox, J. (2001). Motivation, cognitions and traits: Predicting occupational health, well-being and performance. *Stress and Health*, *17*(3), 159–174. https://doi.org/10.1002/smi.897

Cortina, L. M., Kabat-Farr, D., Leskinen, E. A., Huerta, M., & Magley, V. J. (2011). Selective incivility as modern discrimination in organizations. *Journal of Management*, *39*(6), 1579–1605. https://doi.org/10.1177/0149206311418835

Cowlishaw, S., Niele, S., Teshuva, K., Browning, C., & Kendig, H. A. L. (2012). Older adults' spirituality and life satisfaction: A longitudinal test of social support and sense of coherence as mediating mechanisms. *Ageing and Society*, *33*(7), 1243–1262. https://doi.org/10.1017/s0144686x12000633

Craik, K. H., Hogan, R., & Wolfe, R. N. (Eds.). (1993). *Fifty years of personality psychology*. Springer Science+Business Media, LLC.

Cramer, P. (2004). Identity change in adulthood: The contribution of defense mechanisms and life experiences. *Journal of Research in Personality*, *38*(3), 280–316. https://doi.org/10.1016/s0092-6566(03)00070-9

Crawford, J. R., & Henry, J. D. (2004). The positive and negative affect schedule (PANAS): Construct validity, measurement properties and normative data in a large non-clinical sample. *British Journal of Clinical Psychology*, *43*(3), 245–265.

Crum, A. J., Salovey, P., & Achor, S. (2013). Rethinking stress: The role of mindsets in determining the stress response. *Journal of Personality and Social Psychology*, *104*(4), 716–733. https://doi.org/10.1037/a0031201

Cunningham, C. J. L. (2014). Religion and spirituality as factors that influence occupational stress and well-being. In *The Role of Demographics in Occupational Stress and Well Being* (pp. 135–172). https://doi.org/10.1108/s1479-355520140000012004

Cunningham, C. J. L., De La Rosa, G. M., & Jex, S. M. (2008). The dynamic influence of individual characteristics on employee well-being: A review of the theory, research, and future directions. In K. Naswall, J. Hellgren, & M. Sverke (Eds.), *The individual in the changing working life* (pp. 258–283). University Press.

Daniel, J. H. (2009). Next generation: A mentoring program for black female psychologists. *Professional Psychology: Research and Practice*, *40*(3), 299–305. https://doi.org/10.1037/a0013891

Day, N. E. (2005). Religion in the workplace: Correlates and consequences of individual behavior. *Journal of Management, Spirituality & Religion*, *2*(1), 104–135. https://doi.org/10.1080/14766080509518568

DeNeve, K. M., & Cooper, H. (1998). The happy personality: A meta-analysis of 137 personality traits and subjective well-being. *Psychological Bulletin*, *124*(2), 197–229. https://doi.org/10.1037/0033-2909.124.2.197

Eagly, A. H. (1987). *Sex differences in social behavior*. Lawrence Erlbaum Associates, Inc.

Eby, L. T., Allen, T. D., Hoffman, B. J., Baranik, L. E., Sauer, J. B., Baldwin, S., Morrison, M. A., Kinkade, K. M., Maher, C. P., Curtis, S., & Evans, S. C. (2013). An interdisciplinary meta-analysis of the potential antecedents, correlates, and consequences of protege perceptions of mentoring. *Psychological Bulletin, 139*(2), 441–476. https://doi.org/10.1037/a0029279

Eckenrode, J., & Bolger, N. (1995). Daily and within-day event measurement. In S. Cohen, R. C. Kessler, & L. U. Gordon (Eds.), *Measuring stress: A guide for health and social scientists* (pp. 80–101). Oxford University Press.

Edgell, P., Ammons, S. K., & Dahlin, E. C. (2012). Making ends meet. *Journal of Family Issues, 33*(8), 999–1026. https://doi.org/10.1177/0192513x11424261

Edwards, J. R., & Cooper, C. L. (1990). The person-environment fit approach to stress – recurring problems and some suggested solutions. *Journal of Organizational Behavior, 11*(4), 293–307. https://doi.org/10.1002/job.4030110405

Fisher, G., & Barnes-Farrell, J. L. (2013). Use of archival data in occupational health psychology research. In R. R. Sinclair, M. Wang, & L. E. Tetrick (Eds.) *Research methods in occupational health psychology: Measurement, design, and data analysis* (pp. 290–322). Routledge/Taylor & Francis.

Ford, M. T. (2011). Linking household income and work-family conflict: A moderated mediation study. *Stress and Health, 27*(2), 144–162. https://doi.org/10.1002/smi.1328

Fredrickson, B. L. (2000). Why positive emotions matter in organizations: Lessons from the broaden-and-build model. *The Psychologist-Manager Journal, 4*(2), 131–142. https://doi.org/10.1037/h0095887

Fredrickson, B. L., & Branigan, C. (2005). Positive emotions broaden the scope of attention and thought-action repertoires. *Cognition & Emotion, 19*(3), 313–332. https://doi.org/10.1080/02699930441000238

Gonzáles-Figueroa, E., & Young, A. M. (2005). Ethnic identity and mentoring among Latinas in professional roles. *Cultural Diversity and Ethnic Minority Psychology, 11*(3), 213–226. https://doi.org/10.1037/1099-9809.11.3.213

Goodson, I. (2001). The story of life history: Origins of the life history method in sociology. *Identity, 1*(2), 129–142. https://doi.org/10.1207/s1532706xid0102_02

Grice, J. W. (2011). *Observation oriented modeling: Analysis of cause in the behavioral sciences.* Elsevier Academic Press.

Grice, J. W., Barrett, P. T., Schlimgen, L. A., & Abramson, C. I. (2012). Toward a brighter future for psychology as an observation oriented science. *Behavioral Sciences, 2*(4), 1–22. https://doi.org/10.3390/bs2010001

Härmä, M. (1996). Ageing, physical fitness and shiftwork tolerance. *Applied Ergonomics, 27*(1), 25–29. https://doi.org/10.1016/0003-6870(95)00046-1

Hausdorf, P. A., LeBlanc, M. M., & Chawla, A. (2003). Cognitive ability testing and employment selection: Does test content relate to adverse impact? *Applied H.R.M. Research, 7*(2), 41–48.

Hill, P. C., & Pargament, K. I. (2003). Advances in the conceptualization and measurement of religion and spirituality. Implications for physical and mental health research. *American Psychologist, 58*(1), 64–74. https://doi.org/10.1037/0003-066x.58.1.64

Hooftman, W. E., van der Beek, A. J., Bongers, P. M., & van Mechelen, W. (2005). Gender differences in self-reported physical and psychosocial exposures in jobs with both female and male workers. *Journal of Occupational and Environmental Medicine, 47*(3), 244–252. https://doi.org/10.1097/01.jom.0000150387.14885.6b

Hunger, J. M., & Major, B. (2015). Weight stigma mediates the association between BMI and self-reported health. *Health Psychology, 34*(2), 172–175. https://doi.org/10.1037/hea0000106

James, K. (1997). Worker social identity and health-related costs for organizations: A comparative study between ethnic groups. *Journal of Occupational Health Psychology*, 2(2), 108–117.

Jayaratne, T. E., Gelman, S. A., Feldbaum, M., Sheldon, J. P., Petty, E. M., & Kardia, S. L. (2009). The perennial debate: Nature, nurture, or choice? Black and white americans' explanations for individual differences. *Review of General Psychology*, 13(1), 24–33. https://doi.org/10.1037/a0014227

Jenny, G. J., Bauer, G. F., Vinje, H. F., Vogt, K., & Torp, S. (2017). The application of salutogenesis to work. In M. B. Mittelmark, S. Sagy, M. Eriksson, G. F. Bauer, J. M. Pelikan, B. Lindstrom, & G. A. Espnes (Eds.), *The handbook of salutogenesis* (pp. 197–210). Springer. https://doi.org/10.1007/978-3-319-04600-6_20

Jex, S. M., & Bliese, P. D. (1999). Efficacy beliefs as a moderator of the impact of work-related stressors: A multilevel study. *Journal of Applied Psychology*, 84(3), 349–361. https://doi.org/10.1037/0021-9010.84.3.349

Joseph, D. L., Jin, J., Newman, D. A., & O'Boyle, E. H. (2015). Why does self-reported emotional intelligence predict job performance? A meta-analytic investigation of mixed EI. *Journal of Applied Psychology*, 100(2), 298–342. https://doi.org/10.1037/a0037681

Judge, T. A., & Hurst, C. (2007). Capitalizing on one's advantages: Role of core self-evaluations. *Journal of Applied Psychology*, 92(5), 1212–1227. https://doi.org/10.1037/0021-9010.92.5.1212

Judge, T. A., Bono, J. E., & Locke, E. A. (2000). Personality and job satisfaction: The mediating role of job characteristics. *Journal of Applied Psychology*, 85(2), 237–249. https://doi.org/10.1037/0021-9010.85.2.237

Judge, T. A., Locke, E. A., Durham, C. C., & Kluger, A. N. (1998). Dispositional effects on job and life satisfaction: The role of core evaluations. *Journal of Applied Psychology*, 83(1), 17–34. https://doi.org/10.1037/0021-9010.83.1.17

Kammeyer-Mueller, J. D., Judge, T. A., & Scott, B. A. (2009). The role of core self-evaluations in the coping process. *Journal of Applied Psychology*, 94(1), 177–195. https://doi.org/10.1037/a0013214

King, M. B., & Koenig, H. G. (2009). Conceptualising spirituality for medical research and health service provision. *BMC Health Services Research*, 9, 116. https://doi.org/10.1186/1472-6963-9-116

Kornhauser, A. W. (1965). *Mental health of the industrial worker: A Detroit study*. John Wiley.

Liao, H., Arvey, R. D., Butler, R. J., & Nutting, S. M. (2001). Correlates of work injury frequency and duration among firefighters. *Journal of Occupational Health Psychology*, 6(3), 229–242. https://doi.org/10.1037//1076-8998.6.3.229

Lu, L., Chang, Y.-Y., & Lai, S. Y.-L. (2011). What differentiates success from strain: The moderating effects of self-efficacy. *International Journal of Stress Management*, 18(4), 396–412. https://doi.org/10.1037/a0025122

Markus, H., & Kunda, Z. (1986). Stability and malleability of the self-concept. *Journal of Personality and Social Pscyhology*, 51(4), 858–866. https://doi.org/10.1037//0022-3514.51.4.858

McDonough, P., & Walters, V. (2001). Gender and health: reassessing patterns and explanations. *Social Science and Medicine*, 52(4), 547–559. https://doi.org/10.1016/s0277-9536(00)00159-3

McWilliams, J. M., Meara, E., Zaslavsky, A. M., & Ayanian, J. Z. (2009). Differences in control of cardiovascular disease and diabetes by race, ethnicity, and education:

U. S. trends from 1999 to 2006 and effects of medicare coverage. *Annals of Internal Medicine*, *150*(8), 505–515. https://doi.org/10.7326/0003-4819-150-8-200904210-00005

Messing, K., Punnett, L., Bond, M., Alexanderson, K., Pyle, J., Zahm, S., Wegman, D., Stock, S. R., & de Grosbois, S. (2003). Be the fairest of them all: Challenges and recommendations for the treatment of gender in occupational health research. *American Journal of Industrial Medicine*, *43*(6), 618–629. https://doi.org/10.1002/ajim.10225

Michael, G., Anastasios, S., Helen, K., Catherine, K., & Christine, K. (2009). Gender differences in experiencing occupational stress: The role of age, education and marital status. *Stress and Health*, *25*(5), 397–404. https://doi.org/10.1002/smi.1248

Miller, W. R., & Thoresen, C. E. (2003). Spirituality, religion, and health. An emerging research field. *American Psychologist*, *58*(1), 24–35. https://doi.org/10.1037/0003-066x.58.1.24

Muehlenhard, C. L., & Peterson, Z. D. (2011). Distinguishing between sex and gender: history, current conceptualizations, and implications. *Sex Roles*, *64*(11–12), 791–803. https://doi.org/10.1007/s11199-011-9932-5

Münsterberg, H. (1913). *Psychology and industrial efficiency*. Houghton Mifflin.

Nelson, D. L., & Burke, R. J. (2002). A framework for examining gender, work stress, and health. In D. L. Nelson & R. J. Burke (Eds.), *Gender, work stress, and health* (pp. 3–14). American Psychological Association.

Nielsen, K., & Miraglia, M. (2017). What works for whom in which circumstances? On the need to move beyond the 'what works?' question in organizational intervention research. *Human Relations*, *70*(1), 40–62. https://doi.org/10.1177/0018726716670226

Nishii, L. H., & Mayer, D. M. (2009). Do inclusive leaders help to reduce turnover in diverse groups? The moderating role of leader-member exchange in the diversity to turnover relationship. *Journal of Applied Psychology*, *94*(6), 1412–1426. https://doi.org/10.1037/a0017190

Payne, R. (1991). Individual differences in cognition and the stress process. In C. L. Cooper & R. Payne (Eds.), *Personality and stress: Individual differences in the stress process* (pp. 181–201). John Wiley.

Peng, L., & Chan, A. H. S. (2019). A meta-analysis of the relationship between ageing and occupational safety and health. *Safety Science*, *112*, 162–172. https://doi.org/10.1016/j.ssci.2018.10.030

Penney, L. M., & Spector, P. E. (2005). Job stress, incivility, and counterproductive work behavior (CWB): The moderating role of negative affectivity. *Journal of Organizational Behavior*, *26*(7), 777–796. https://doi.org/10.1002/job.336

Penney, L. M., Hunter, E. M., & Perry, S. J. (2011). Personality and counterproductive work behaviour: Using conservation of resources theory to narrow the profile of deviant employees. *Journal of Occupational and Organizational Psychology*, *84*(1), 58–77. https://doi.org/10.1111/j.2044-8325.2010.02007.x

Postlethwaite, B., Robbins, S., Rickerson, J., & McKinniss, T. (2009). The moderation of conscientiousness by cognitive ability when predicting workplace safety behavior. *Personality and Individual Differences*, *47*(7), 711–716. https://doi.org/10.1016/j.paid.2009.06.008

Powell, G. N., & Greenhaus, J. H. (2010). Sex, gender, and the work-to-family interface: Exploring negative and positive interdependencies. *Academy of Management Journal*, *53*(3), 513–534. https://doi.org/10.5465/amj.2010.51468647

Powell, L. H., Shahabi, L., & Thoresen, C. E. (2003). Religion and spirituality. Linkages to physical health. *American Psychologist*, *58*(1), 36–52. https://doi.org/10.1037/0003-066x.58.1.36

Raudenbush, S. W. (2001). Comparing personal trajectories and drawing causal inferences from longitudinal data. *Annual Review of Psychology*, *52*, 501–525. https://doi.org/10.1146/annurev.psych.52.1.501

Reeve, C. L., Heggestad, E. D., & Lievens, F. (2009). Modeling the impact of test anxiety and test familiarity on the criterion-related validity of cognitive ability tests. *Intelligence*, *37*(1), 34–41. https://doi.org/10.1016/j.intell.2008.05.003

Roberson, Q., Ryan, A. M., & Ragins, B. R. (2017). The evolution and future of diversity at work. *Journal of Applied Psychology*, *102*(3), 483–499. https://doi.org/10.1037/apl0000161

Robinson, O. C. (2009). On the social malleability of traits. *Journal of Individual Differences*, *30*(4), 201–208. https://doi.org/10.1027/1614-0001.30.4.201

Roccas, S., & Brewer, M. B. (2002). Social identity complexity. *Personality and Social Psychology Review*, *6*(2), 88–106. https://doi.org/10.1207/S15327957pspr0602_01

Roehling, M. V., Roehling, P. V., & Pichler, S. (2007). The relationship between body weight and perceived weight-related employment discrimination: The role of sex and race. *Journal of Vocational Behavior*, *71*(2), 300–318. https://doi.org/10.1016/j.jvb.2007.04.008

Rudolph, C. W., & Zacher, H. (2020). COVID-19 and careers: On the futility of generational explanations. *Journal of Vocational Behavior*, *119*, 103433. https://doi.org/10.1016/j.jvb.2020.103433

Salminen, S. (2004). Have young workers more injuries than older ones? An international literature review. *Journal of Safety Research*, *35*(5), 513–521. https://doi.org/10.1016/j.jsr.2004.08.005

Saxbe, D. E., Repetti, R. L., & Nishina, A. (2008). Marital satisfaction, recovery from work, and diurnal cortisol among men and women. *Health Psychology*, *27*(1), 15–25. https://doi.org/10.1037/0278-6133.27.1.15

Schmidt, F. L., & Hunter, J. (2004). General mental ability in the world of work: Occupational attainment and job performance. *Journal of Personality and Social Psychology*, *86*(1), 162–173. https://doi.org/10.1037/0022-3514.86.1.162

Schneider, B. (1987). E= f(P,B): The road to a Radical Approach to Person-Environment Fit. *Journal of Vocational Behavior*, *31*, 353–361. https://doi.org/10.1016/0001-8791(87)90051-0

Schonfeld, I. S., & Mazzola, J. J. (2013). Strengths and limitations of qualitative approaches to research in occupational health psychology. In R. R. Sinclair, M. Wang, & L. E. Tetrick (Eds.), *Research methods in occupational health psychology* (pp. 268–289). Routledge/Taylor & Francis.

Simon, R. W. (1992). Parental role strains, salience of parental identity and gender differences in psychological distress. *Journal of Health and Social Behavior*, *33*(1), 25–35. https://doi.org/10.2307/2136855

Sinclair, R. R., & Cheung, J. H. (2016). Money matters: Recommendations for financial stress research in occupational health psychology. *Stress and Health*, *32*(3), 181–193. https://doi.org/10.1002/smi.2688

Smith, P. C. (1955). The prediction of individual differences in susceptibility to industrial monotony. *Journal of Applied Psychology*, *39*(5), 322–329. https://doi.org/10.1037/h0043258

Smith, T. W. (2006). Personality as risk and resilience in physical health. *Current Directions in Psychological Science, 15*(5), 227–231. https://doi. org/10.1111/j.1467-8721.2006.00441.x

Sonnentag, S., Binnewies, C., & Ohly, S. (2013). Event-sampling methods in occupational health psychology. In R. R. Sinclair, M. Wang, & L. E. Tetrick (Eds.), *Research methods in occupational health psychology: Measurement, design, and data analysis* (pp. 208–228). Routledge/Taylor & Francis Group.

Spector, P. E. (2003). Individual differences in health and well-being in organizations. In D. A. Hoffmann & L. E. Tetrick (Eds.), *Health and safety in organizations: A multilevel perspective* (pp. 29–55). Jossey-Bass.

Spector, P. E. (2020). Mastering the use of control variables: The hierarchical iterative control (HIC) approach. *Journal of Business and Psychology.* https://doi. org/10.1007/s10869-020-09709-0

Spector, P. E., & Brannick, M. T. (2010). Methodological urban legends: The misuse of statistical control variables. *Organizational Research Methods, 14*(2), 287–305. https://doi.org/10.1177/1094428110369842

Stanford, M. S., Mathias, C. W., Dougherty, D. M., Lake, S. L., Anderson, N. E., & Patton, J. H. (2009). Fifty years of the Barratt Impulsiveness Scale: An update and review. *Personality and Individual Differences, 47*(5), 385–395. https://doi. org/10.1016/j.paid.2009.04.008

Stephenson, M. T., Hoyle, R. H., Palmgreen, P., & Slater, M. D. (2003). Brief measures of sensation seeking for screening and large-scale surveys. *Drug and Alcohol Dependence, 72*(3), 279–286. https://doi.org/10.1016/j.drugalcdep.2003.08.003

The National Academies of Sciences, Engineering, and Medicine. (2020). *Are generational categories meaningful distinctions for workforce management?* https://doi. org/ 10.17226/25796

Thomas, A. J., Eberly, L. E., Davey Smith, G., Neaton, J. D., & Stamler, J. (2005). Race/ethnicity, income, major risk factors, and cardiovascular disease mortality. *American Journal of Public Health, 95*(8), 1417–1423. https://doi.org/10.2105/ AJPH.2004.048165

Tugade, M. M., & Fredrickson, B. L. (2004). Resilient individuals use positive emotions to bounce back from negative emotional experiences. *Journal of Personality and Social Psychology, 86*(2), 320–333. https://doi.org/10.1037/0022-3514.86.2.320

Unger, R. K., & Crawford, M. (1993). Commentary: Sex and gender – the troubled relationship between terms and concepts. *Psychological Science, 4*(2), 122–124.

Vetter, M., Schünemann, A. L., Brieber, D., Debelak, R., Gatscha, M., Grünsteidel, F., Herle, M., Mandler, G., & Ortner, T. M. (2018). Cognitive and personality determinants of safe driving performance in professional drivers. *Transportation Research Part F: Traffic Psychology and Behaviour, 52*, 191–201. https://doi.org/10.1016/j. trf.2017.11.008

Vogt, K., Hakanen, J. J., Jenny, G. J., & Bauer, G. F. (2016). Sense of coherence and the motivational process of the job-demands-resources model. *Journal of Occupational Health Psychology, 21*(2), 194–207. https://doi.org/10.1037/a0039899

Walster, E., & Walster, G. W. (1975). Equity and social justice. *Journal of Social Issues, 31*(3), 21–43. https://doi.org/10.1111/j.1540-4560.1975.tb00001.x

Wang, M., Sinclair, R. R., Zhou, L., & Sears, L. E. (2013). Person-centered analysis: methods, applications, and implications for occupational health psychology. In R. R. Sinclair, M. Wang, & L. E. Tetrick (Eds.) *Research methods in occupational health psychology: Measurement, design, and data analysis* (pp. 349–373). Rout-ledge/Taylor & Francis.

Wey Smola, K., & Sutton, C. D. (2002). Generational differences: Revisiting gener-ational work values for the new millennium. *Journal of Organizational Behavior,* *23*(4), 363–382. https://doi.org/10.1002/job.147

Whinghter, L. J., Cunningham, C. J. L., Wang, M., & Burnfield, J. L. (2008). The moderating role of goal orientation in the workload-frustration relationship. *Journal* *of Occupational Health Psychology, 13*(3), 283–291. https://doi.org/10.1037/1076-8998.13.3.283

Wong, I. S., Smith, P. M., Mustard, C. A., & Gignac, M. A. M. (2014). For better or worse? Changing shift schedules and the risk of work injury among men and women. *Scandinavian Journal of Work, Environment & Health, 40*(6), 621–630. https://doi.org/10.5271/sjweh.3454

Woo, S. E., O'Boyle, E. H., & Spector, P. E. (2017). Best practices in developing, con-ducting, and evaluating inductive research. *Human Resource Management Review,* *27*(2), 255–264. https://doi.org/10.1016/j.hrmr.2016.08.004

4

WORKER PSYCHOLOGICAL HEALTH

Kristen Jennings Black and Christopher J. L. Cunningham

In this chapter, we explore what it means to be psychologically healthy and how work-related experiences can contribute to good, bad, and in-between forms of psychological health among workers. A worker's psychological health has many implications for their personal well-being and organizational functioning. Without a psychologically healthy workforce, there are some real limits to the potential productivity and enjoyment of work. Knowing that this connection between work and psychological health is important, we highlight intervention strategies that can increase positives states and reduce the experience of negative states. Employee Assistance Programs (EAPs), peer and climate focused interventions, and additional techniques are discussed as particularly relevant intervention strategies that can be applied to promote and protect workers' psychological health.

By the end of this chapter, you should be able to:

LO 4.1: Explain how work both affects psychological health and is affected by psychological health.

LO 4.2: Describe common negative psychological states and how they relate to work.

LO 4.3: Describe common positive psychological states and how they relate to work.

LO 4.4: Suggest and explain the value of an intervention technique for promoting psychological health, given information on a specific workplace context.

What it Means to be Psychologically Healthy

A good place to begin this chapter is by ensuring that we all understand what it means to be psychologically healthy. This concept is defined and operationalized as multiple different constructs in occupational health psychology (OHP) research and practice, including psychological health, mental health (the predominant term in clinical psychology), behavioral health, or well-being. We use "psychological health" through most of this chapter as an inclusive label that captures a wide range of related cognitive and emotional states. Our view of psychological health aligns with how the American Psychological Association (APA; 2020) defines mental health as, "a state of mind characterized by emotional well-being, good behavioral adjustment, relative freedom from anxiety and disabling symptoms, and a capacity to establish constructive relationships and cope with the ordinary demands and stresses of life" ("mental health" entry). The World Health Organization (WHO; 2018) further emphasizes that mental health means "an individual realizes his or her own abilities . . . can work productively, and is able to make a contribution to his or her community" (para. 2). An implication of these definitions is that OHP professionals can best impact worker health, safety, and well-being (WHSWB) when we view psychological health as more than the absence of symptoms or negative emotions, but also experiencing a sense of meaning and purpose, which can often come from one's work (Rosso et al., 2010).

The Complex Connection between Work and Psychological Health

The topic of psychological health is somewhat complicated to address because it seems so personal. As a result, it is easy for organizational leaders to assume that protecting workers' psychological health is not an organizational responsibility. It is true that many risk factors for poor psychological health, like genetic predispositions and nonwork stress, are not under direct organizational control. However, organizations do control the structuring and assignment of work tasks, the presence of stressors and absence of work-related resources, the safety and quality of work experiences, and the compensation and benefits packages that help support workers' general peace of mind. When we consider this, and the opportunities for work to generate a sense of purpose and meaning, it is easy to agree with Blustein's (2008) argument that "work plays a central role in the development, expression, and maintenance of psychological health" (p. 228) and that organizations have a responsibility and opportunity to protect and promote workers' psychological health.

Psychological States and Conditions

Over the past few decades, research has consistently shown connections between work-related stress and poor psychological health (Stansfield & Candy 2006), but work experiences also have the potential to generate positive psychological states (e.g., Crawford et al., 2010). These positive states are really valuable personal resources (discussed more in Chapter 6) that facilitate the development and maintenance of other resources (Hobfoll, 1989, 2001); for example, a good mood helps you focus on your work, do an exceptional job, and feel a sense of esteem that carries over into life outside of work. Incontrast, the absence of positive states and presence of negative states sets workers up to lose more resources and generally struggle when facing work demands (e.g., feeling exhausted or burned out makes it hard to do good work). These examples illustrate how work can both *contribute to* workers' psychological health and *be affected by* workers' psychological health.

Through the rest of this section, we provide an overview of primary psychological health-related constructs that OHP professionals study and work to address or change in WHSWB interventions. Our review of some of the key concepts and findings in this field starts with everyday states of affect, mood, and emotion and then moves to more specific states connected to cognitive and emotional energy and those associated with connectedness (i.e., to people and tasks) and purpose at work. We finish the section with a review of some common psychological disorders.

Affect, Mood, and Emotion

Having good or bad feelings is the most basic way that workers experience psychological states. Theories in this arena distinguish between emotion, mood, and affect (Rosenberg, 1998). An *emotion* is a fairly short-term response to a specific stimulus (e.g., joy when you get the best parking space as you pull into work). A *mood* lasts longer than an emotion, several hours or an entire day, and is not tied to a specific event. A worker's *affective state* is broader and includes more transient experiences (including emotions and moods), while *affective traits* refer to an underlying disposition toward experiencing positive or negative emotion. An important implication of all of this is that affective states, which are more stimulus-specific and short in duration, can be more easily altered than trait affect (Weiss & Cropanzano, 1996).

Affective states are typically categorized as positive (e.g., joy, excitement) or negative (e.g., worry, anger). Positive states unsurprisingly tend to be more consistently linked to favorable personal and work-related outcomes than negative states (Fredrickson, 2000; Weiss & Cropanzano, 1996). Perhaps one of the most common and relevant examples of this is the "happy-productive worker" hypothesis – it is both conceptually and empirically evident that happy workers do indeed seem to perform better on the job (Judge et al., 2001; Warr & Nielsen, 2018) and are even likely to go beyond their formal job duties (Kleine et al., 2019; Whitman et al., 2010).

Despite this evidence, even this simple hypothesis is more complex than it seems – questions remain about whether happiness *causes* good performance or *is caused* by good performance; both pathways are probably true to some extent (Judge et al., 2001). An important practical perspective to keep in mind here is that regardless of how positive states originate, they are desirable because they can be a key resource that helps workers meet work goals and offset otherwise negative effects of job demands, as outlined in the job demands and resources (JD-R) model (Demerouti et al., 2001) and conservation of resources (COR) theory (Hobfoll, 1989), and because positive emotions may broaden or improve workers' cognitive processing (cf., the Broaden and Build perspective of Fredrickson, 1998). Together, these models reinforce the idea that positive states are a resource themselves, are a mechanism for further resource gain, and can be a protective factor in the presence of stressors. In the same way, negative sates can be part of resource loss and impede the ability to gain new resources.

Energy-Related Psychological States

Workers often experience good or bad psychological health in the forms of cognitive and emotional energetic states. We can lack energy that leaves us feeling more negative emotions or we can feel a sense of energy and connection to work. Often these states are considered by OHP researchers as forms of resources within the COR theory just mentioned. We believe these states deserve more attention, though, given how directly they can be addressed with intervention efforts.

Fatigue and Boredom

Fatigue and boredom are often overlooked, yet important psychological states characterized by low cognitive and emotional energy. *Fatigue* is psychologically experienced as mental tiredness, low energy, and cognitive impairment in functioning lasting a few hours or weeks at a time. Fatigue is the body's adaptive response to being overworked, a signal to slow down from continuing to perform demanding activities that could lead to exhaustion (i.e., a state of extreme fatigue where resources feel nearly or entirely "used up"; Van Dijk & Swaen, 2003). *Chronic emotional fatigue*, essentially losing the energy to care, can prevent empathy toward others and is a primary component of employee burnout (Maslach & Jackson, 1981), which we discuss in the next section.

Boredom comes from being under-stimulated, as when work-related tasks are monotonous, repetitive, or insufficiently challenging (Smith, 1955; Vodanovich & Watt, 2016). Because boredom is perceptual in nature and individuals differ in their susceptibility to boredom, it is imprecise to consider tasks themselves as inherently boring (Loukidou et al., 2009; Smith, 1955). Instead, boredom can be viewed as the result of poor alignment between workers' abilities or skills and work demands, such that the demands require much less than what one is capable of doing. This is a challenge at the heart of organizational talent management, matching workers with the right tasks, given that the same tasks may be extremely boring to one worker, but satisfying to another.

Burnout and Engagement

Worker burnout is perhaps the most researched and popularized negative psychological state targeted by OHP professionals. In popular culture, feeling "burned out" is used to describe feeling exhausted or overwhelmed by work. By definition, though, burnout is a more specific and serious multidimensional phenomenon characterized by chronic emotional and physical exhaustion, cynicism, feelings of inefficacy, and/or depersonalization and disconnection from one's work (e.g., Demerouti et al., 2001; Maslach & Jackson, 1981). Initially this was thought to be a phenomenon unique to workers who interact heavily with people (e.g., healthcare workers, counselors), but recent perspectives suggest burnout could be experienced in many different work environments (Ahola, Honkonen, Isometsa, et al., 2006; Demerouti et al., 2001). Some key elements to understand are that burnout is a serious *chronic* state, meaning it does not just develop from one tough week at work. The global prevalence and severity of burnout has garnered the attention of the WHO (2019), which has listed burnout in the international classification of diseases (ICD-11) as a serious occupational phenomenon and is "about to embark on the development of evidence-based guidelines on mental well-being in the workplace" (para. 7), adding to their existing efforts for improving mental health in the workplace.

As our understanding of burnout increased, researchers began to also recognize an opposing phenomenon now known as *engagement*. As highlighted by Schaufeli et al. (2002), engagement is generally characterized by *vigor* (a state of high energy), *dedication* (a sense of pride or significance in relation to work), and *absorption* (being immersed in work). Though related, burnout and engagement are not simply opposing ends of the same continuum (Schaufeli et al., 2008). In fact, workers can experience engagement and burnout simultaneously (Timms et al., 2012).

There are many individual differences and work and nonwork environmental factors that may contribute to burnout and engagement. In work settings, there is evidence that both engagement and burnout are strongly connected to the balancing of work-related demands and resources. More specifically, and in-line with the JD-R model (discussed more in Chapter 6), burnout develops from a combination of high demands and insufficient resources to meet those demands; engagement is more likely to occur when there are sufficient or abundant resources to meet work demands (Alarcon, 2011; Christian et al., 2011; Demerouti et al., 2001; Lee & Ashforth, 1996).

Connectedness and Purpose at Work

Decades of social science research have documented that humans have an inherent desire to feel connected to one another and to have a sense of purpose or meaning in life. The absence of social connection and purpose is distressing, while the presence of meaningful connections is fulfilling. This is why a third essential dimension to psychological health involves connections and meaning at work.

Loneliness and Isolation

In recent years, the "loneliness epidemic" has gained the attention of all kinds of health researchers and media outlets (e.g., National Public Radio;

Chatterjee, 2018). *Loneliness* is not just an undesirable feeling, but a serious risk to our health (Holt-Lunstad, 2017). In a work context, loneliness depends on whether affiliation needs are being met in a worker's organization (Ozcelik & Barsade, 2018). Related to this, *isolation* is characterized by insufficient interaction and/or friendship with coworkers, or a general perception of insufficient work-based support from a supervisor or the broader organization (Marshall et al., 2007). Implicit in this definition is that loneliness is perceptual, so one person could feel lonely in the exact same setting that another does not. Unfortunately, loneliness is likely to increase in prevalence and severity as the nature of work changes and remote work arrangements become more common.

Meaning, Purpose, and Thriving at Work

Having a sense of meaning and purpose in life is an innate human desire, which can seriously impact our health and even our lifespan (Alimujiang et al., 2019). For many people, work is a major source of personal meaning and purpose (i.e., a life aim that creates goals for living; Rosso et al., 2010). *Meaningful work* or *meaningfulness* is concerned with the amount of significance that an individual attaches to work and can be conceptualized in a number of ways, such as perceiving the work tasks as personally meaningful, finding meaning in life through one's work, and generally having a motivation to do "good" through work (Steger et al., 2012). The *job characteristics model* (Hackman & Oldham, 1976) proposes that meaningfulness is one of three key psychological states generated by certain task characteristics (e.g., skill variety, autonomy, task significance), which together result in an internal motivation to work. Task significance, in which work "matters" for social good, is a particularly strong correlate with meaningfulness (Allan, 2017).

Positive states like thriving or flourishing often arise from purposeful work, further creating an enhanced capacity to build and sustain more resources (Fredrickson, 1998, 2000). *Flourishing* has been described as an optimal state of both experiencing positive emotions and effectively performing in one's life roles (Keyes, 2002). *Thriving* is a more specific experience, described as a feeling of personal progress, personal growth through learning, and a sense of vitality (Spreitzer et al., 2005). Spreitzer and colleagues' (2005) model of thriving at work starts with the premise that thriving is an outcome of active and purposeful behavior at work. You probably have personal examples of this, when your work has allowed you to explore new ideas, feel connected to others, or simply when you truly desire to be doing the work you are doing. Thriving at work is also linked to individual differences (e.g., proactive personality, positive affect) as well as work-related resources that promote thriving (e.g., supportive leaders and coworkers, trust and connection in workgroups; Carmeli & Spreitzer, 2009; Kleine et al., 2019).

Psychological Health Disorders Affected by Work

Research spanning 33 years and 63 different countries provided estimates that around one in five individuals experiences a serious psychological health problem or disorder in any given year (Steel et al., 2014). Although the severity of psychological disorders and their prevalence within certain populations is quite varied, the important point is that psychological health disorders are not uncommon in the general population. It should be no surprise, therefore, that these disorders are also present in every organization..

The underlying causes of psychological disorders are complex, originating from a variety of biological, psychological, and social factors, many of which are present in work environments and associated with work experiences. Table 4.1 summarizes symptoms of several common psychological health disorders and provides examples of how these conditions can affect or be affected by work experiences. Our goal with this table is to improve your general awareness of the signs and symptoms of these disorders, which you may encounter in work settings. Please note that if any long-lasting signs or symptoms like these are observed in yourself or coworkers, it is important to seek help from a mental health professional (as outlined in Figure 4.1[1]).

Not all forms of work present the same risks or challenges to workers' psychological health. At the high end of this risk spectrum, military personnel and first responders are especially likely to experience traumatic events while working, creating a direct and substantial risk for the development of psychological health disorders. Proactive interventions for high-risk contexts to promote psychological resilience are exemplified in programs like Comprehensive Soldier Fitness training (Lester et al., 2011) and peer interventions for responding to acute stress (Adler et al., 2020; Svetlitzky et al., 2019). For all organizations, it is important to recognize that regardless of whether a psychological health disorder is work-related, there is potential for work-related experiences to exacerbate symptoms or be affected by symptoms, as well as for work connections to facilitate help-seeking.

Why Worker Psychological Health Matters

As noted early in this chapter, psychological health is often seen as a personal problem. Having read up to this point, however, we hope you see that organizations have a responsibility and opportunity to promote and protect workers' psychological health. In this section, we discuss three main reasons why psychological health is an important resource for organizations.

Ripple Effects of Good and Bad States

It makes sense for organizations to work to promote worker engagement, purpose, and connectedness, as these are positive states that are associated with other positive states and attitudes, including job satisfaction and job involvement

Table 4.1 Symptoms and Work Implications of Common Psychological Disorders

Psychological Disorder	Common Symptoms	Connections to Work
Major Depressive Disorder (MDD)	• Depressed mood • Little to no interest in activities • A general lack of energy • Feelings of worthlessness • In severe cases, suicidal thoughts or planned or attempted suicide	• Symptoms may interfere with work performance (Kessler et al., 2008) • Psychosocial work demands increase risk (Wang et al., 2012); workplace social support decreases risk (Netterstrom et al., 2008)
Anxiety	• Anxiety about specific triggers (e.g., social anxiety, phobias) • A general tendency to experience excessive worry about a variety of areas of life (Generalized Anxiety Disorder; GAD)	• Symptoms may interfere with work performance (Kessler et al., 2008) • Psychosocial work demands increase GAD risk (Melchior et al., 2007)
Alcohol Abuse	• Consuming large amounts of alcohol • Difficulty cutting down on use • Being distracted by a desire for alcohol at the cost of normal activities	• Work stressors exhibit small, significant correlations with alcohol and drug use (Frone, 2008). • Consuming alcohol on the job, before work, or working with a hangover occurs at non-trivial rates (Frone, 2005). • Risk may be higher where access is easier (e.g., restaurants)
Acute Stress Response (ASR) and Post Traumatic Stress Disorder (PTSD)	• ASR includes: • Feeling anxious or dazed following a traumatic event • Symptoms lasting more than a few weeks may characterize an Acute Stress Disorder (ASD) • PTSD includes: • Symptoms last for several weeks or months after a traumatic event • Re-experiencing the event (dreams, flashbacks) • Sensitivity to cues associated with the event • Changes in thoughts or behavior patterns (irritability, easily startled)	• Could occur in response to trauma on-the-job, particularly for those in high-risk jobs (Bennett et al., 2004; Porter et al., 2018) • Could occur for an employee involved in a traumatic incident outside of work and affect work performance

Note: Disorder Descriptions from Diagnostic and Statistical Manual of Mental Disorders (DSM-5) and National Center for PTSD (https://www.ptsd.va.gov/understand/what/ptsd_basics.asp)

When an underlined immediate response is required

Observation

- Statements (or personal thoughts) that indicate serious depression like, "life just isn't worth it anymore".
- Concerns about violence, direct references to hurting oneself or others.

Response

- If you are experiencing serious thoughts of harm toward yourself or someone else, do not hesitate, ask for help immediately through local healthcare providers or your national suicide prevention hotline (if available). If you are intending to kill or seriously harm yourself or someone else make contact with a local emergency care responder (through the nearest hospital or emergency response option).
- If you suspect that someone is contemplating suicide, do not hesitate. If you feel comfortable, express your concerns to the person and directly ask if they have been thinking about harming themselves and use supportive active listening techniques. Immediately relay concerns to your supervisor or HR representative and/or immediately seek help from an emergency care responder. If someone expresses or demonstrates explicit intent to kill or seriously harm themselves or someone else, follow your company's policy (if there is one) and/or immediately seek help from an emergency care responder.

When making contact with a psychological health professional is recommended

Observation

- Changes in disposition, personality, or mood have lasted for more than a few weeks.
- Changes in work performance, energy levels, behavior in social interactions, or work attendance that have lasted for more than a few weeks.
- An inability to cope with daily stressors or tasks like normal.
- Difficulty in nonwork life and/or not engaging with a social support network.
- Consuming excess alcohol or using other drugs to cope with a stress-related problem.

Response

- If this is you, consider reaching out for support from a professional that can work with you.
- If you notice these symptoms in someone else, talk to your colleague. Ask them how they are doing, using active listening techniques.
- Assist them in finding an appropriate resource, such as a recommendation for a counselor, a representative of an Employee Assistance Program (EAP), or a supervisor or HR representative that can provide a list of resources.

References for learning more about warning signs: https://www.nami.org/Learn-More/Know-the-Warning-Signs
https://www.mhanational.org/recognizing-warning-signs
https://suicidepreventionlifeline.org/

Figure 4.1 Signs That Professional Help is Needed to Manage a Psychological Disorder (e.g., Christian et al., 2011). Positive states tend to facilitate more positive states and can also mitigate the occurrence and impact of negative states (Allan, 2017; Fredrickson, 1998). Organizations can also directly and actively work to minimize negative states from developing. This guidance is supported by many studies highlighting serious health consequences of negative work-related psychological states. For example, burnout has been related to symptoms of anxiety, depression, sleep disturbances, and alcohol dependence (Ahola, Honkonen, Pirkola, et al., 2006; Peterson et al., 2008). Poor psychological health can also contribute to physical health concerns, such as musculoskeletal pain (Peterson et al., 2008) and other psychosomatic complaints, discussed in Chapter 5.

Supporting Psychological Health is Good for Business

Meta-analyses show that negative psychological states and conditions (e.g., negative affect, burnout, depression, anxiety) can negatively impact job performance (Ford et al., 2011; Kaplan et al., 2009), work attitudes, and turnover intentions (Alarcon, 2011; Lee & Ashforth, 1996). In contrast, positive psychological states, such as positive affect, thriving, and engagement have been associated with better individual task and extra-role performance (Christian et al., 2011; Kaplan et al., 2009; Kleine et al., 2019) and business outcomes, like profit, customer satisfaction, and productivity (Harter et al., 2002).

Considering a financial argument in more detail, Goetzel and colleagues (2004) estimated that a single American worker experiencing a psychological health concern (i.e., depression, mental illness, or sadness) could cost a company $348 per year (based on an average salary). This cost was associated with absences, employer-provided costs for treatment, and lost productivity due to symptoms. Other estimates suggest that the costs of presenteeism (i.e., being at work while experiencing psychological health symptoms) are three times that of costs associated with absence or turnover (Hampson et al., 2017). With these figures, we are not advocating for workers to stay home when struggling with psychological health challenges; indeed, work can often provide much-needed resources (e.g., social support, achievement) that help with managing these challenges (Karanika-Murray & Biron, 2019). Rather, these financial estimates highlight costs that could be avoided or reduced through efforts to promote and protect workers' psychological health.

Related to this, there is increasing evidence that supporting workers' psychological health can generate a positive return on investment (ROI), potentially as high as 4.2:1 (Hampson et al., 2017). Quasi-experimental studies have found that employees who utilize their organization's EAP end up reporting fewer absences, utilizing less sick leave, and demonstrating less presenteeism (Nunes et al., 2017; Richmond et al., 2017). Although employees are not equally likely to use available psychological health resources, like EAPs, when they experience personal challenges, the available empirical evidence does suggest these programs may be valuable and worth providing and promoting (Spetch et al., 2011).

Organizations Can Initiate Larger-Scale Change

There is a real opportunity to work through organizations to protect and promote worker psychological health, and thereby positively impact workers, their families, communities, and society more broadly. Psychological health problems are expensive to address and collectively present a global public health issue; depression alone is a leading cause of disability worldwide (WHO, 2020). Beyond the economic costs, more attention to psychological health is critical in building greater awareness of common psychological health concerns and reducing stigma that can be a substantial treatment barrier (Clement et al., 2015). Proactive efforts by employers to counter stigma associated with psychological health and provide practical resources and benefits to support psychological health (e.g., EAP, educational workshops) can improve workers' access to care and likelihood of actually seeking help when needed.

Methodological Considerations and Practical Recommendations

Building on what we have discussed about essential "good" and "poor" psychological states and the impact these states can have on workers and organizations, we can now focus on how work can be designed, modified, and experienced to promote psychological health.

Measuring and Monitoring Psychological Health

Efforts to support workers' psychological health often start by gathering and analyzing a variety of data. Self-report methods are commonly used when assessing psychological states, which are inherently personal and perceptual in nature. Basic measures of positive or negative states can be easily incorporated into employee surveys, including established measures for engagement and burnout (Schaufeli et al., 2002), positive and negative affect (Watson et al., 1988), boredom (Vodanovich & Watt, 2016), or meaningfulness of work (Steger et al., 2012). There are also validated assessments for psychological disorders, but we would caution against using these measures haphazardly in an employment setting to limit unintended and illegal consequences associated with resulting scores and to ensure that they are not used without a plan for responding to indicators of a clinically significant disorder.

Beyond assessing psychological states, you may want to know about the psychological health climate that workers perceive in an organization. A positive psychological safety climate, in which workers feel that the organization values psychological health, can be associated with less job strain and fewer depressive symptoms (Bailey et al., 2015). There are also measures that individuals, particularly managers, can complete to indicate personal stigmatizing beliefs about mental health problems (Martin & Giallo, 2016).

Self-reported, cross-sectional surveys are valuable tools for efficiently gathering data to describe the state of psychological health within an organization, but other methods may also be helpful and more appropriate in some situations. Longitudinal surveys can gather evidence pertaining to stability or change in psychological states, which can be important to track if there are specific organizational events happening that might contribute to or detract from worker psychological health (e.g., critical incidents, organizational restructuring, workspace upgrades). Experience sampling or daily diaries may be necessary for understanding short-lived psychological states, such as feelings of connectedness, boredom, or frustration associated with specific work tasks or situations (e.g., Daniels et al., 2012; Dimotakis et al., 2011). Finally, organizational metrics (e.g., absences, short-term disability claims, performance ratings) can be decent indicators of psychological health concerns (Goetzel et al., 2002) and are useful when working to show intervention-related cost savings or ROI. These data need to be analyzed and interpreted with care as their meaning is not always crystal clear (e.g., a recorded absence may be health-related, or due to lack of transportation or motivation).

Intervening to Support Psychological Health

We want to close out this chapter by exploring strategies for improving worker psychological health through interventions at a variety of levels. Note that these recommendations apply to efforts to protect and promote generally positive psychological health; more severe negative psychological health matters may require person-specific, clinical interventions managed by a licensed mental health care professional.

Strategies for Individuals

Many training options and resources can help workers gain the knowledge and skills needed to be psychologically healthy. One large area of research that supports worker psychological health pertains to stress and recovery management, which we discuss more in Chapter 6. Other options target psychological health at the person level more directly. Perhaps the most researched interventions to protect worker psychological health come from interventions originally developed for military personnel, but which have clear translational potential to other work environments. As one example, the "Comprehensive Soldier Fitness (CSF) program likely represents the largest deliberate psychological intervention in history..." (Lester et al., 2011, p. 77). Based on positive psychology principles, CSF involves proactively assessing workers' emotional, social, family, and spiritual fitness, to then offer individualized modules for building fitness (i.e., resources) in each domain. The "fitness" label is a smart strategy to contextualize psychological health into a language that is particularly valued in the military, but also likely to translate well into other work settings.

Group trainings to facilitate positive psychological health have also been developed for building resources in several domains. For instance, "social fitness" has been targeted through social resilience training, which can increase social awareness skills and reduce loneliness (Cacioppo et al., 2015). Battlemind training is another example, designed to help Soldiers return to civilian life, following a combat deployment (Adler et al., 2011). Central goals of this training are to increase perceptions of control and normalize adjustment difficulties a Soldier may experience. These concepts could apply to any worker navigating an important change, like a stressful new job assignment or returning from leave.

Psychological health can be targeted in any setting with positive psychology type interventions. Kaplan and colleagues (2013) compared two such online interventions targeting gratitude and social connection among university workers. The implementation itself was pretty simple: Email reminders asked workers to pause during the day to think about things they were grateful for at work or ways they had attempted to connect with others. Both the gratitude and social connectedness interventions related to fewer work absences; the gratitude intervention also related to experiencing more positive states. These findings are really promising, as they suggest that small intentional efforts in the workplace can positively affect worker psychological health and actual behaviors.

Any type of person-level intervention, however, will only be successful when supported by others in the organization. For instance, offering a gratitude intervention in a toxic work culture (e.g., asking workers to be grateful when they feel entirely uncared for by the organization) could do more harm than good. Similarly, efforts to promote psychological health are unlikely to be successful in organizations that simultaneously impose overwhelming work demands or require workers to function in physically unhealthy environments.

Strategies for Groups

Workgroup interventions can also be an effective way to promote and protect worker psychological health. Workgroups tend to experience similar emotions (Barsade & Gibson, 2012), so a positive or negative emotional tone in a workgroup is likely to engender corresponding positive or negative worker emotions. Similarly, workgroup norms that support psychological health can positively impact worker psychological health. Unfortunately, unsupportive norms and perceived stigma around psychological health disorders can be a big barrier to treatment (Clement et al., 2015) and even reduce workers' desire to disclose psychological health concerns (Meinert, 2014). Training focused on improving workgroup climate around seeking psychological health treatment has been effective in military settings (e.g., Britt et al., 2018; Start et al., 2020) and likely could translate well into any work environment by encouraging workers to talk more openly about psychological health concerns and learn how to support each other. Along these lines, studies have found that interventions to improve workplace social support can reduce burnout and benefit psychological health (Heaney et al., 1995; Le Blanc et al., 2007).

Workgroups can also be a resource by monitoring and detecting psychological problems in peers, particularly if workers are trained on what to watch for. Acknowledging this, the Israeli Defense Forces developed the YaHa-LOM training (Svetlitzky et al., 2019) to help Soldiers provide peer support to team members experiencing an acute stress reaction while in a combat or otherwise dangerous situation. This intervention involved a few key steps to quickly try to shift a traumatized person's state of helplessness to one of effective functioning: connect with the person, ask simple questions, confirm what has happened and what will happen, and provide a direct action to reestablish a sense of mastery. Trained Soldiers reported greater knowledge and confidence for managing an acute stress reaction of a peer. This intervention, renamed iCOVER, has also been implemented in the U.S. Army (Adler et al., 2020) and the principles in this approach can apply just about anywhere – if you see a worker who seems overwhelmed, talk with them. Can you help them to re-establish a sense of control or mastery in that immediate moment? Remember, OHP interventions do not have to be complex.

Strategies for Leaders

Leader interventions are a third approach to promoting worker psychological health. For interested readers, the Chartered Institute of Personnel Development, in collaboration with Mind (based in the United Kingdom; 2018) put together a practical guide for managers to support psychological health and well-being. Here are a few key ways leaders can support worker psychological health: (1) proactively manage work in a way that supports psychological health, given the control leaders have (e.g., over work tasks; scheduling); (2) control personal emotional displays, since workers tend to experience states similar to that displayed by their leader (Johnson, 2008); and (3) minimize practical barriers to treatment seeking (e.g., time to attend appointments).

One strong strategy is to train leaders to recognize common symptoms of a psychological health concern and teach them how to appropriately respond. A strong example of this is the mental health awareness training (MHAT) developed for supervisors (Dimoff et al., 2016). The MHAT is based on the idea that knowledge and awareness of psychological health disorders tends to increase the likelihood of people speaking up when there is a problem and watching out for problems in others. The key mechanism here is knowledge transfer and changes in attitudes, with the ultimate goal of influencing the behavior of leaders and those they supervise. In a couple of different samples, the MHAT improved supervisor knowledge, attitudes, and self-efficacy around supporting mental health. In one sample, there was even a significant decrease in length of mental health disability claims in the nine months following the intervention, providing a powerful financial argument for the training's value. Data gathered from subordinates of trained supervisors affirmed that workers also perceived changes in their supervisor's support for mental health (Dimoff & Kelloway, 2019) and were more willing to seek out and use mental health resources compared to those with supervisors who were not trained. Similar short and internet-based trainings along these lines, have also shown promise for improving supervisor attitudes toward mental health concerns by increasing awareness and demonstrating the business case for proper treatment of symptoms (Shann et al., 2019).

Strategies for Organizations

As emphasized throughout this chapter, there is a lot that can be done organization-wide to support worker psychological health. Combining organizational interventions with person-level efforts can yield results greater than either approach separately (Awa et al., 2010). At a basic level, organizational policies and practices regarding the design and structuring of work matter. Consistent with JD-R theory, high workloads without sufficient corresponding resources can result in poor psychological health and burnout (Bowling et al., 2015) and can make it difficult for workers to foster social connections, which are increasingly recognized as essential to our psychological health (Holt-Lunstad, 2017).

Hollis (2019) and Stallard (2019) use the Mayo Clinic as an example of an organization that made changes to facilitate social connections among physicians. Specifically, they changed their pay system for physicians to more stable salary-based structure, to increase cooperation instead of competition among physicians. They also started a program that facilitated more frequent physician interaction. Strategic and flexible scheduling throughout an organization can not only increase opportunities for connection and collaboration, but it can also provide more opportunities for recovery which has a number of benefits for psychological health (discussed more in Chapter 6).

When significant psychological health concerns do arise, organizations need to be prepared to help, regardless of whether such concerns originate from work or nonwork factors. There are unfortunate stories of organizations attempting to ignore such issues, even to the point of terminating employees who disclose a psychological disorder (Meinert, 2014); this is not an acceptable practice (nor is it legal, in many countries). Instead, organizations have a responsibility and opportunity to provide resources to promote and protect worker psychological health. For example, providing access to an EAP that includes psychological health benefits is an example of an organization-level intervention that supports workers and also may benefit the organization by reducing absences and sick time, and increasing productive time at work (Nunes et al., 2017; Richmond et al., 2017). Such programs can also provide needed resources and even cost savings when critical organizational incidents occur (Attridge & VandePol, 2010).

Even if your organization does not have the means to provide an EAP or other formal program, any organization can make an effort to provide a list of mental health professionals and other resources for workers, and flexibility to attend appointments when needed. Organizations of all forms can also develop and sustain cultures in which psychological health is valued and protected, and those working to improve their psychological health are supported and encouraged. Such actions can make a huge difference in employees getting the help they may need when encountering a psychological health concern.

Evaluating Psychological Health Interventions

In addition to the guidance on intervention evaluation covered in Chapter 2, there are a few essential evaluation recommendations to especially keep in mind with psychological health interventions. Be creative and think broadly about data that can be monitored as indicators of workers' psychological health. Gathering data on changes in workers' psychological states is a good place to start, but ongoing measurement of worker perceptions of psychological safety, support for psychological health, and personal willingness to utilize psychological health resources or talk with peers or supervisors about psychological health concerns may also be valuable. Psychological health is a broad area of study and impact, so it is important to cast a wide net, as Dimoff and colleagues did when evaluating the MHAT (Dimoff & Kelloway, 2019; Dimoff et al., 2016). Gathering additional data from spouses, significant others, and/or family members may also be valuable, particularly in research efforts to more fully understand the impact of intervention efforts.

Because psychological health interventions may take a long time to yield their full effects, it is important to develop an evaluation strategy that involves data gathered over time. Long-term impact can be assessed in various ways, including workplace absences, disability claims, or even costs associated with prescription drug use, which was an outcome of interest in a recent study by Dahl and Pierce (2020). With any of these outcomes, we have to think realistically about how long these effects would take to occur and adjust our assessment timeline accordingly. It is also important not to interpret single data points in the absence of the broader context that longitudinal data can provide – for example, a mental health awareness training may initially result in more employees reporting psychological health concerns than were reported prior to the training. This outcome does not mean the training was ineffective or harmful; instead, these results may indicate improved recognition of symptoms and support for the disclosure of such concerns. Open-ended interview or focus group data may be really important in helping you understand some of these types of unexpected results (e.g., Shann et al., 2019).

Concluding Thoughts and Reality Check

In this chapter, we have explored how work affects our psychological health and how our psychological health affects our work. When psychological health among workers is not positive and strong, the whole organization can suffer. This makes managing workers' psychological health an organizational priority. Organizations also have an opportunity to foster positive psychological states that can yield value far beyond the boundaries of the work environment. A number of effective interventions to promote and protect worker psychological health have been developed and tested; it is possible to help workers experience positive psychological states at work.

Most OHP professionals rally behind the belief that all workers deserve the opportunity to experience full psychological health. This means not just living without symptoms, but having the opportunity to flourish and thrive. Managing worker psychological health is not easy, as many workers simultaneously manage psychological health challenges linked with factors from multiple life domains (e.g., migrant workers struggling to find stable work; low-wage workers unable to make ends meet). The breadth of psychological challenges managed by workers presents an opportunity for OHP and related mental health professionals to have a broad impact by using what we know about work and what we know about psychological health to make a lasting and meaningful impact.

Media Resources

- News story exploring how a construction company prioritizes psychological health: https://www.npr.org/sections/health-shots/2019/12/12/783300736/a-construction-company-embraces-frank-talk-about-mental-health-to-reduce-suicide
- Magazine article examining how loneliness affects workers: https://www.forbes.com/sites/nextavenue/2020/02/25/whos-lonely-at-work-and-why/#502925970379
- Trade publication article discussing the impacts of COVID-19 on workers' psychological health: https://hrexecutive.com/american-workers-are-worried-why-thats-a-problem-for-hr/?eml=20200715&oly_enc_id=0230A1877812A1U

Discussion Questions

1) Psychological health is often a very personal topic. To what extent should organizations be involved in monitoring or providing support for workers' psychological health?
2) What do we understand about the relationship between work and psychological health and well-being?
3) Pick a specific job and describe the psychological health risks that an employee in this particular job may experience, as well as opportunities for positive states they may experience from their work. Propose two or three intervention strategies to address one or two of the psychological health risks.
4) What role(s) do EAPs play in supporting workers' psychological health? Explain the value of this type of resource for an organization and its workers.

Professional Profile: Amy Adler, Ph.D.

Country/region: USA, Maryland
Current position title: Senior Science Consultant, Center for Military Psychiatry and Neuroscience and Acting Director, Research Transition Office at the Walter Reed Army Institute of Research
Background: I have a B.A. in economics from Brown University, and a M.A. in psychology and a Ph.D. in clinical psychology from the University of Kansas. I completed an APA-approved clinical internship at the Illinois Masonic Medical Center and a fellowship at Ravenswood Hospital Medical Center, both in Chicago; I then obtained my license in clinical psychology. After my doctorate, I moved to Europe and was able to get three different jobs with the US military community: working as a clinician in an Army hospital, teaching at the University of Maryland's overseas program, and conducting research. Eventually, I focused on research and that path has provided me with a range of amazing opportunities to work on issues impacting employees in high-risk occupations like the military. I am currently an associate editor of the *Journal of Occupational Health Psychology*, and am a fellow in the American Psychological Association's Society for Military Psychology.

My area of expertise is at the intersection of behavioral health, resilience, performance, traumatic stress, and leadership within high-risk occupations. I am an applied researcher, focusing on how we can use psychology to support service members effectively. I also co-manage the Army's Psychological Health, Resilience and Well-being research program, which incorporates a range of projects from epidemiology to small-team interventions. In addition, I lead a group responsible for developing interventions and conducting randomized trials, program evaluation and science implementation studies within the Army.

How my work impacts WHSWB: Everything I work on in some way is related to the health, well-being and performance of service members, and much of what we do has direct implications for other kinds of workers, particularly those serving in high-risk occupations like firefighters and police. One of our current projects is examining an intervention we call 'iCOVER.' iCOVER is a new rapid, peer-based intervention for individuals to use with team members exhibiting signs of acute stress in the midst of a dangerous occupational event like combat. The concept was originally developed by the Israel Defense Forces, and we worked with the program lead to adapt the training for the US context. Before iCOVER, US service members had no systematic technique to use if one of their team members experienced an acute stress reaction, potentially endangering themselves and others. We did not even know how often acute stress reactions happened or whether this kind of temporary and shifting stress reaction increased the risk of subsequently developing mental health problems like PTSD. Since beginning the project, we have conducted a randomized trial with US service members in a realistic training setting and are now working with deployed troops. This kind of research has the potential to have a meaningful and direct impact on worker health, safety, and well-being, extending beyond the military population, to others engaged in high-risk team-oriented occupations.

My motivation: It is satisfying to address real-world challenges that confront service members, knowing that our work has the potential to make a difference. The job entails a range of problem-solving tasks, and involves working with policy makers, military leaders, and expert scientists across a range of disciplines. The work is fast-paced and every day is different. I have been lucky to be part of dynamic teams, work closely with counterparts from other nations, and have lots of adventure along the way.

Note

1 We acknowledge and thank Dr. Ashley Howell for her review of this figure and insight based on her Clinical Psychology training and experience.

Chapter References

Adler, A. B., Bliese, P. D., McGurk, D., & Hoge, C. W. (2011). Battlemind debriefing and battlemind training as early interventions with Soldiers returning from Iraq: Randomization by platoon. *Sport, Exercise, and Performance Psychology, 1*(S), 66–83. https://doi.org/10.1037/2157-3905.1.S.66.supp

Adler, A. B., Start, A. R., Milham, L., Allard, Y. S., Riddle, D., Townsend, L., & Svetlitzky, V. (2020). Rapid response to acute stress reaction: Pilot test of iCOVER training for military units. *Psychological Trauma: Theory, Research, Practice, and Policy, 12*(4), 431–435. https://doi.org/10.1037/tra0000487

Ahola, K., Honkonen, T., Isometsa, E., Kalimo, R., Nykyri, E., Koskinen, S., Aromaa, A., & Lonnqvist, J. (2006). Burnout in the general population–Results from the Finnish Health 2000 Study. *Social Psychiatry and Psychiatric Epidemiology, 41*(1), 11–17. https://doi.org/10.1007/s00127-005-0011-5

Ahola, K., Honkonen, T., Pirkola, S., Isometsa, E., Kalimo, R., Nykyri, E., Aromaa, A., & Lonnqvist, J. (2006). Alcohol dependence in relation to burnout among the Finnish working population. *Addiction, 101*(10), 1438–1443. https://doi.org/10.1111/j.1360-0443.2006.01539.x

Alarcon, G. M. (2011). A meta-analysis of burnout with job demands, resources, and attitudes. *Journal of Vocational Behavior, 79*(2), 549–562. https://doi.org/10.1016/j.jvb.2011.03.007

Alimujiang, A., Wiensch, A., Boss, J., Fleischer, N. L., Mondul, A. M., McLean, K., Mukherjee, B., & Pearce, C. L. (2019). Association between life purpose and mortality among US adults older than 50 years. *JAMA Network Open, 2*(5), e194270. https://doi.org/10.1001/jamanetworkopen.2019.4270

Allan, B. (2017). Task significance and meaningful work: A longitudinal study. *Journal of Vocational Behavior, 102*, 174–182. https://doi.org/10.1016/j.jvb.2017.07.011

APA. (2020). *APA Dictonary of Psychology: Mental Health.* https://dictionary.apa.org/mental-health

Attridge, M., & VandePol, B. (2010). The business case for workplace critical incident response: A literature review and some employer examples. *Journal of Workplace Behavioral Health, 25*(2), 132–145. https://doi.org/10.1080/15555241003761001

Awa, W. L., Plaumann, M., & Walter, U. (2010). Burnout prevention: A review of intervention programs. *Patient Education and Counseling, 78*(2), 184–190. https://doi.org/10.1016/j.pec.2009.04.008

Bailey, T. S., Dollard, M. F., & Richards, P. A. (2015). A national standard for psychosocial safety climate (PSC): PSC 41 as the benchmark for low risk of job strain and depressive symptoms. *Journal of Occupational Health Psychology, 20*(1), 15–26. https://doi.org/10.1037/a0038166

Barsade, S. G., & Gibson, D. E. (2012). Group affect. *Current Directions in Psychological Science, 21*(2), 119–123. https://doi.org/10.1177/0963721412438352

Bennett, P., Williams, Y., Page, N., Hood, K., & Woollard, M. (2004). Levels of mental health problems among UK emergency ambulance workers. *Emergency Medicine Journal, 21*(2), 235–236. https://doi.org/10.1136/emj.2003.005645

Blustein, D. L. (2008). The role of work in psychological health and well-being: A conceptual, historical, and public policy perspective. *American Psychologist*, 63(4), 228–240. https://doi.org/10.1037/0003-066X.63.4.228

Bowling, N. A., Alarcon, G. M., Bragg, C. B., & Hartman, M. J. (2015). A meta-analytic examination of the potential correlates and consequences of workload. *Work & Stress*, 29(2), 95–113. https://doi.org/10.1080/02678373.2015.1033037

Britt, T. W., Black, K. J., Cheung, J. H., Pury, C. L. S., & Zinzow, H. M. (2018). Unit training to increase support for military personnel with mental health problems. *Work & Stress*, 32(3), 281–296. https://doi.org/10.1080/02678373.2018.1445671

Cacioppo, J. T., Adler, A. B., Lester, P. B., McGurk, D., Thomas, J. L., Chen, H. Y., & Cacioppo, S. (2015). Building social resilience in soldiers: A double dissociative randomized controlled study. *Journal of Personality and Social Psychology*, 109(1), 90–105. https://doi.org/10.1037/pspi0000022

Carmeli, A., & Spreitzer, G. M. (2009). Trust, connectivity, and thriving: Implications for innovative behaviors at work. *Journal of Creative Behavior*, 43(3), 169–191. https://doi.org/10.1002/j.2162-6057.2009.tb01313.x

Chartered Institute of Personnel and Development. (2018). *People managers' guide to mental health.* https://www.cipd.co.uk/Images/mental-health-at-work-1_tcm18-10567.pdf

Chatterjee, R. (2018, May 1, 2018). Americans are a lonely lot, and young people bear the heaviest burden. *National Public Radio.* https://www.npr.org/sections/health-shots/2020/01/23/798676465/most-americans-are-lonely-and-our-workplace-culture-may-not-be-helping

Christian, M. S., Garza, A. S., & Slaughter, J. E. (2011). Work engagement: A quantitative review and test of its relations with task and contextual performance. *Personnel Psychology*, 64, 89–136. https://doi.org/10.1111/peps.12070

Clement, S., Schauman, O., Graham, T., Maggioni, F., Evans-Lacko, S., Bezborodovs, N., Morgan, C., Rusch, N., Brown, J. S., & Thornicroft, G. (2015). What is the impact of mental health-related stigma on help-seeking? A systematic review of quantitative and qualitative studies. *Psychological Medicine*, 45(1), 11–27. https://doi.org/10.1017/S0033291714000129

Crawford, E. R., Lepine, J. A., & Rich, B. L. (2010). Linking job demands and resources to employee engagement and burnout: A theoretical extension and meta-analytic test. *Journal of Applied Psychology*, 95(5), 834–848. https://doi.org/10.1037/a0019364

Dahl, M. S., & Pierce, L. (2020). Pay-for-performance and employee mental health: Large sample evidence using employee prescription drug usage. *Academy of Management Discoveries*, 6(1), 12–38. https://doi.org/10.5465/amd.2018.0007

Daniels, K., Wimalasiri, V., Beesley, N., & Cheyne, A. (2012). Affective well-being and within-day beliefs about job demands' influence on work performance: An experience sampling study. *Journal of Occupational and Organizational Psychology*, 85(4), 666–674. https://doi.org/10.1111/j.2044-8325.2012.02062.x

Demerouti, E., Bakker, A. B., Nachreiner, F., & Schaufeli, W. B. (2001). The job demands-resources model of burnout. *Journal of Applied Psychology*, 86(3), 499–512. https://doi.org/10.1037/0021-9010.86.3.499

Dimoff, J. K., & Kelloway, E. K. (2019). With a little help from my boss: The impact of workplace mental health training on leader behaviors and employee resource utilization. *Journal of Occupational Health Psychology*, 24(1), 4–19. https://doi.org/10.1037/ocp0000126

Dimoff, J. K., Kelloway, E. K., & Burnstein, M. D. (2016). Mental Health Awareness Training (MHAT): The development and evaluation of an intervention for workplace leaders. *International Journal of Stress Management, 23*(2), 167–189. https://doi.org/10.1037/a0039479

Dimotakis, N., Scott, B. A., & Koopman, J. (2011). An experience sampling investigation of workplace interactions, affective states, and employee well-being. *Journal of Organizational Behavior, 32*(4), 572–588. https://doi.org/10.1002/job.722

Ford, M. T., Cerasoli, C. P., Higgins, J. A., & Decesare, A. L. (2011). Relationships between psychological, physical, and behavioural health and work performance: A review and meta-analysis. *Work & Stress, 25*(3), 185–204. https://doi.org/10.1080/02678373.2011.609035

Fredrickson, B. L. (1998). What good are positive emotions? *Review of General Psychology, 2*(3), 300–319. https://doi.org/10.1037/1089-2680.2.3.300

Fredrickson, B. L. (2000). Why positive emotions matter in organizations: Lessons from the broaden-and-build model. *The Psychologist-Manager Journal, 4*(2), 131–142. https://doi.org/10.1037/h0095887

Frone, M. R. (2005). Prevalence and distribution of alcohol use and impairment in the workplace: A U.S. National Survey. *Journal of Studies on Alcohol, 67*, 147–156. https://doi.org/10.15288/jsa.2006.67.147

Frone, M. R. (2008). Are work stressors related to employee substance use? The importance of temporal context assessments of alcohol and illicit drug use. *Journal of Applied Psychology, 93*(1), 199–206. https://doi.org/10.1037/0021-9010.93.1.199

Goetzel, R. Z., Long, S. R., Ozminkowski, R. J., Hawkins, K., Wang, S., & Lynch, W. (2004). Health, absence, disability, and presenteeism cost estimates of certain physical and mental health conditions affecting U.S. employers. *Journal of Occupational and Environmental Medicine, 46*(4), 398–412. https://doi.org/10.1097/01.jom.0000121151.40413.bd

Goetzel, R. Z., Ozminkowski, R. J., Sederer, L. I., & Mark, T. L. (2002). The business case for quality mental health services: Why employers should care about the mental health and well-being of their employees. *Journal of Occupational and Environmental Medicine, 44*(4), 320–330. https://doi.org/10.1097/00043764-200204000-00012

Hackman, J. R., & Oldham, G. R. (1976). Motivation through the design of work: Test of a theory. *Organizational Behavior and Human Performance, 16*, 250–279. https://doi.org/10.1016/0030-5073(76)90016-7

Hampson, E., Sonejji, U., Jacob, A., Bogdan, M., & McGahan, H. (2017). *Mental health and employers: The case for investment.* https://www2.deloitte.com/content/dam/Deloitte/uk/Documents/public-sector/deloitte-uk-mental-health-employers-monitor-deloitte-oct-2017.pdf

Harter, J. K., Schmidt, F. L., & Hayes, T. L. (2002). Business-unit-level relationship between employee satisfaction, employee engagement, and business outcomes: A meta-analysis. *Journal of Applied Psychology, 87*(2), 268–279. https://doi.org/10.1037/0021-9010.87.2.268

Heaney, C. A., Price, R. H., & Rafferty, J. (1995). Increasing coping resources at work: A field experiment to increase social support, improve work team functioning, and enhance employee mental health. *Journal of Organizational Behavior, 16*(4), 335–352. https://doi.org/10.1002/job.4030160405

Hobfoll, S. E. (1989). Conservation of resources: A new attempt at conceptualizing stress. *American Psychologist, 44*(3), 513–524. https://doi.org/10.1037/0003-066X.44.3.513

Hobfoll, S. E. (2001). The influence of culture, community, and the nested-self in the stress process: Advancing Conservation of Resources theory. *Applied Psychology: An International Review, 50*(3), 337–370. https://doi.org/10.1111/1464-0597.00062

Hollis, A. (2019). Loneliness: An increasingly recognized health problem. *Smart Brief.* https://www.smartbrief.com/original/2019/09/loneliness-increasingly-recognized-health-problem

Holt-Lunstad, J. (2017). The potential public health relevance of social isolation and loneliness: prevalence, epidemiology, and risk factors. *Public Policy & Aging Report, 27*(4), 127–130. https://doi.org/10.1093/ppar/prx030

Johnson, S. K. (2008). I second that emotion: Effects of emotional contagion and affect at work on leader and follower outcomes. *The Leadership Quarterly, 19*(1), 1–19. https://doi.org/10.1016/j.leaqua.2007.12.001

Judge, T. A., Thoresen, C. J., Bono, J. E., & Patton, G. K. (2001). The job satisfaction-job performance relationship: a qualitative and quantitative review. *Psychological Bulletin, 127*(3), 376–407. https://doi.org/10.1037/0033-2909.127.3.376

Kaplan, S., Bradley-Geist, J. C., Ahmad, A., Anderson, A., Hargrove, A. K., & Lindsey, A. (2013). A test of two positive psychology interventions to increase employee well-being. *Journal of Business and Psychology, 29*(3), 367–380. https://doi.org/10.1007/s10869-013-9319-4

Kaplan, S., Bradley, J. C., Luchman, J. N., & Haynes, D. (2009). On the role of positive and negative affectivity in job performance: A meta-analytic investigation. *Journal of Applied Psychology, 94*(1), 162–176. https://doi.org/10.1037/a0013115

Karanika-Murray, M., & Biron, C. (2019). The health-performance framework of presenteeism: Towards understanding an adaptive behaviour. *Human Relations.* https://doi.org/10.1177/0018726719827081

Kessler, R., White, L. A., Birnbaum, H., Qiu, Y., Kidolezi, Y., Mallett, D., & Swindle, R. (2008). Comparative and interactive effects of depression relative to other health problems on work performance in the workforce of a large employer. *Journal of Occupational and Environmental Medicine, 50*(7), 809–816. https://doi.org/10.1097/JOM.0b013e318169ccba

Keyes, C. L. M. (2002). The mental health continuum: From languishing to flourishing in life. *Journal of Health and Social Behavior, 43*(2), 207–222. https://doi.org/10.2307/3090197

Kleine, A. K., Rudolph, C. W., & Zacher, H. (2019). Thriving at work: A meta-analysis. *Journal of Organizational Behavior.* https://doi.org/10.1002/job.2375

Le Blanc, P. M., Hox, J. J., Schaufeli, W. B., Taris, T. W., & Peeters, M. C. (2007). Take care! The evaluation of a team-based burnout intervention program for oncology care providers. *Journal of Applied Psychology, 92*(1), 213–227. https://doi.org/10.1037/0021-9010.92.1.213

Lee, R. T., & Ashforth, B. E. (1996). A meta-analytic examination of the correlates of the three dimensions of job burnout. *Journal of Applied Psychology, 81*(2), 123–133. https://doi.org/10.1037/0021-9010.81.2.123

Lester, P. B., McBride, S., Bliese, P. D., & Adler, A. B. (2011). Bringing science to bear: An empirical assessment of the Comprehensive Soldier Fitness program. *American Psychologist, 66*(1), 77–81. https://doi.org/10.1037/a0022083

Loukidou, L., Loan-Clarke, J., & Daniels, K. (2009). Boredom in the workplace: More than monotonous tasks. *International Journal of Management Reviews*, *11*(4), 381–405. https://doi.org/10.1111/j.1468-2370.2009.00267.x

Marshall, G. W., Michaels, C. E., & Mulki, J. P. (2007). Workplace isolation: Exploring the construct and its measurement. *Psychology and Marketing*, *24*(3), 195–223. https://doi.org/10.1002/mar.20158

Martin, A. J., & Giallo, R. (2016). Confirmatory factor analysis of a questionnaire measure of managerial stigma towards employee depression. *Stress and Health*, *32*(5), 621–628. https://doi.org/10.1002/smi.2655

Maslach, C., & Jackson, S. E. (1981). The measurement of experienced burnout. *Journal of Occupational Behaviour*, *2*(2), 99–113. https://doi.org/10.1002/job.4030020205

Meinert, D. (2014). How to accommodate employees with a mental illness. *SHRM HR Today*. https://www.shrm.org/hr-today/news/hr-magazine/pages/1014-mental-health.aspx

Melchior, M., Caspi, A., Milne, B. J., Danese, A., Poulton, R., & Moffitt, T. E. (2007). Work stress precipitates depression and anxiety in young, working women and men. *Psychological Medicine*, *37*(8), 1119–1129. https://doi.org/10.1017/S0033291707000414

Netterstrom, B., Conrad, N., Bech, P., Fink, P., Olsen, O., Rugulies, R., & Stansfeld, S. (2008). The relation between work-related psychosocial factors and the development of depression. *Epidemiologic Reviews*, *30*, 118–132. https://doi.org/10.1093/epirev/mxn004

Nunes, A. P., Richmond, M. K., Pampel, F. C., & Wood, R. C. (2017). The effect of employee assistance services on reductions in employee absenteeism. *Journal of Business and Psychology*, *33*(6), 699–709. https://doi.org/10.1007/s10869-017-9518-5

Ozcelik, H., & Barsade, S. G. (2018). No employee an island: Workplace loneliness and job performance. *Academy of Management Journal*, *61*(6), 2343–2366. https://doi.org/10.5465/amj.2015.1066

Peterson, U., Demerouti, E., Bergstrom, G., Samuelsson, M., Asberg, M., & Nygren, A. (2008). Burnout and physical and mental health among Swedish healthcare workers. *Journal of Advanced Nursing*, *62*(1), 84–95. https://doi.org/10.1111/j.1365-2648.2007.04580.x

Porter, B., Hoge, C. W., Tobin, L. E., Donoho, C. J., Castro, C. A., Luxton, D. D., & Faix, D. (2018). Measuring aggregated and specific combat exposures: Associations between combat exposure measures and posttraumatic stress disorder, depression, and alcohol-related problems. *Journal of Traumatic Stress*, *31*(2), 296–306. https://doi.org/10.1002/jts.22273

Richmond, M. K., Pampel, F. C., Wood, R. C., & Nunes, A. P. (2017). The impact of employee assistance services on workplace outcomes: Results of a prospective, quasi-experimental study. *Journal of Occupational Health Psychology*, *22*(2), 170–179. https://doi.org/10.1037/ocp0000018

Rosenberg, E. L. (1998). Levels of analysis and the organization of affect. *Review of General Psychology*, *2*(3), 247–270. https://doi.org/10.1037/1089-2680.2.3.247

Rosso, B. D., Dekas, K. H., & Wrzesniewski, A. (2010). On the meaning of work: A theoretical integration and review. *Research in Organizational Behavior*, *30*, 91–127. https://doi.org/10.1016/j.riob.2010.09.001

Schaufeli, W. B., Salanova, M., Gonzalez-Roma, V., & Bakker, A. B. (2002). The measurement of engagement and burnout: A two sample confirmatory factor analytic approach. *Journal of Happiness Studies, 3,* 71–92. https://doi. org/10.1023/A:1015630930326

Schaufeli, W. B., Taris, T. W., & van Rhenen, W. (2008). Workaholism, Burnout, and Work Engagement: Three of a Kind or Three Different Kinds of Employee Well-Being? *Applied Psychology, 57*(2), 173–203. https://doi.org/10.1111/j.1464-0597.2007.00285.x

Shann, C., Martin, A., Chester, A., & Ruddock, S. (2019). Effectiveness and application of an online leadership intervention to promote mental health and reduce depression-related stigma in organizations. *Journal of Occupational Health Psychology, 24*(1), 20–35. https://doi.org/10.1037/ocp0000110

Smith, P. C. (1955). The prediction of individual differences in susceptibility to industrial monotony. *Journal of Applied Psychology, 39*(5), 322–329.

Spetch, A., Howland, A., & Lowman, R. L. (2011). EAP utilization patterns and employee absenteeism: Results of an empirical, 3-year longitudinal study in a national Canadian retail corporation. *Consulting Psychology Journal: Practice and Research, 63*(2), 110–128. https://doi.org/10.1037/a0024690

Spreitzer, G., Sutcliffe, K., Dutton, J., Sonenshein, S., & Grant, A. M. (2005). A socially embedded model of thriving at work. *Organization Science, 16*(5), 537–549. https://doi.org/10.1287/orsc.1050.0153

Stallard, M. L. (2019). America's loneliness epidemic: A hidden systemic risk to organizations. *Smart Brief.* https://www.smartbrief.com/original/2019/05/americas-loneliness-epidemic-hidden-systemic-risk-organizations

Stansfield, S., & Candy, B. (2006). Psychosocial work environment and mental health—a meta-analytic review. *Scandinavian Journal of Work, Environment & Health, 32*(6), 443–462. https://doi.org/10.5271/sjweh.1050

Start, A. R., Amiya, R. M., Dixon, A. C., Britt, T. W., Toblin, R. L., & Adler, A. B. (2020). LINKS training and unit support for mental health: A group-randomized effectiveness trial. *Prevention Science.* https://doi.org/10.1007/s11121-020-01106-6

Steel, Z., Marnane, C., Iranpour, C., Chey, T., Jackson, J. W., Patel, V., & Silove, D. (2014). The global prevalence of common mental disorders: A systematic review and meta-analysis 1980–2013. *International Journal of Epidemiology, 43*(2), 476–493. https://doi.org/10.1093/ije/dyu038

Steger, M. F., Dik, B. J., & Duffy, R. D. (2012). Measuring meaningful work. *Journal of Career Assessment, 20*(3), 322–337. https://doi.org/10.1177/1069072711436160

Svetlitzky, V., Farchi, M., Ben Yehuda, A., Start, A. R., Levi, O., & Adler, A. B. (2019). YaHaLOM training in the military: Assessing knowledge, confidence, and stigma. *Psychological Services.* https://doi.org/10.1037/ser0000360

Timms, C., Brough, P., & Graham, D. (2012). Burnt-out but engaged: The coexistence of psychological burnout and engagement. *Journal of Educational Administration, 50*(3), 327–345. https://doi.org/10.1108/09578231211223338

Van Dijk, F. J. H., & Swaen, G. M. H. (2003). Fatigue at work. *Occupational and Environmental Medicine, 60*(Supplement 1), i1–i2. https://doi.org/10.1136/oem.60.suppl_1.i1

Vodanovich, S. J., & Watt, J. D. (2016). Self-report measures of boredom: An updated review of the literature. *The Journal of Psychology: Interdisciplinary and Applied, 150*(2), 196–228. https://doi.org/10.1080/00223980.2015.1074531

Wang, J., Smailes, E., Sareen, J., Schmitz, N., Fick, G., & Patten, S. (2012). Three job-related stress models and depression: A population-based study. *Social*

Psychiatry and Psychiatric Epidemiology, 47(2), 185–193. https://doi.org/10.1007/s00127-011-0340-5

Warr, P., & Nielsen, K. (2018). Wellbeing and work performance. In E. Diener, S. Oishi, & L. Tay (Eds.), *Handbook of well-being*. DEF Publishers.

Watson, D., Clark, L. A., & Tellegen, A. (1988). Development and validation of brief measures of positive and negative affect: the PANAS scales. *Journal of Personality and Social Psychology, 54*(6), 1063–1070. https://doi.org/10.1037/0022-3514.54.6.1063

Weiss, H. M., & Cropanzano, R. (1996). Affective Events Theory: A theoretical discussion of the structure, causes, and consequences of affective experiences at work. In B. M. Staw & L. L. Cummings (Eds.), *Research in organizational behavior: An annual series of analytical essays and critical reviews* (Vol. 18, pp. 1–74). JAI Press Inc.

Whitman, D. S., Van Rooy, D. L., & Viswesvaran, C. (2010). Satisfaction, citizenship behaviors, and performance in work units: a meta-analysis of collective construct relations. *Personnel Psychology, 63*(1), 41–81. https://doi.org/10.1111/j.1744-6570.2009.01162.x

WHO. (2018). *Mental health: Strengthening our response.* https://www.who.int/en/news-room/fact-sheets/detail/mental-health-strengthening-our-response

WHO. (2019). *Burn-out an "occupational phenomenon": International Classification of Diseases.* https://www.who.int/mental_health/evidence/burn-out/en/

WHO. (2020). *Depression.* https://www.who.int/news-room/fact-sheets/detail/depression

5

WORKER PHYSICAL HEALTH

Kristen Jennings Black and Christopher J. L. Cunningham

In this chapter we provide an overview of the essential physical health out-comes that can be impacted by work, as well as how physical health can im-pact workers' performance. Given these connections, we provide examples of how physical health can be protected and potentially improved through pre-vention and promotion efforts in the workplace. More generally, we discuss how physical health fits into the broader worker health, safety, and well-being (WHSWB) framework emphasized throughout this book. In the following sections of this chapter, we explore the theory and research pertaining to physical health among workers. We also describe various indicators of and methodologies for studying physical health, and present practical suggestions organizations can use to protect, manage, and improve worker physical health.

By the end of this chapter, you should be able to:

LO 5.1: Explain how work both affects physical health (as a cause) and how it is affected by physical health (as an outcome).

LO 5.2: Describe common physical health issues associated with cer-tain types of work.

LO 5.3: Describe the challenges of working with a chronic condition and ways in which organizations can support these workers.

LO 5.4: Discuss the complex issue of where the responsibility lies for maintaining good physical health (i.e., worker vs. employer) and example strategies that organizations can use to protect and promote physical health.

How Working Affects Physical Health

Most occupational health psychology (OHP) professionals would agree that workers are physically healthy when they are free of symptoms that interfere with daily life and when they are able to successfully do the things they want to do. This rather broad definition aligns with the World Health Organi-zation's (WHO; 2014) perspective on well-being as more than the absence of symptoms. As noted in our chapter lead-in, physical health is important in OHP not just as an outcome or consequence linked to work exposures and experiences, but also as a factor contributing to individuals' work-related abilities. This dual perspective is easy to reconcile if we think of physical health as a resource that is affected by work *and* can affect a worker's ability to complete tasks.

There is a vast literature documenting how work environments and experiences directly affect physical health. This might be through exposure to environmental hazards or strenuous and repetitive tasks that create direct wear and tear on a body's systems (Gatchel & Kishino, 2011; Redden & Larkin, 2015), or in response to work-related psychosocial stressors (e.g., incivility, perceived constraints) that trigger the body's stress response and can damage workers' physiological systems over time (Ganster & Rosen, 2013; McEwen, 1998).

In the following sections, we focus on several examples of physical health conditions linked to experiences at work. Our goal is to illustrate the variety of ways work can affect physical health and help you begin to recognize many common physical health conditions that could be good targets for preventative measures. We more fully address environmental demands and work-related safety matters that contribute to physical health risks in Chapters 10 and 11.

Exposure-Related Conditions

Some physical health concerns can result from exposure to environmental stressors. *Extreme temperatures* can affect workers' physical health and functioning in outdoor work settings, but also indoor environments that get extremely hot because of the nature of the work and/or associated machinery and processes, or which stay extremely cold out of necessity (e.g., refrigeration rooms) or because climate control is impossible (e.g., large manufacturing facilities). Some of the physical effects of extreme temperatures are acute, like dehydration, joint swelling, heat stroke, burns, hypothermia, or frostbite (Redden & Larkin, 2015). Effects of extreme temperatures on outdoor workers can also have long-term consequences, like increased risk for skin cancer due to prolonged sun exposure (Stepanski & Mayer, 1998).

Another unfortunately common physical health effect associated with work in certain occupations is *hearing loss* (National Institute for Occupational Safety and Health [NIOSH], 2019). Workers in environments that are characteristically very loud, such as aircraft maintenance (Guest et al., 2010) or mining (Sun et al., 2019), may be especially susceptible to hearing loss. Some occupations may even present multiple risk factors for hearing loss (e.g., aircraft pilots exposed to loud noises and frequent pressure changes). Not as obvious, is the reality that occupational hearing loss can also result from exposure on-the-job to chemicals classified as *ototoxicants* (NIOSH, 2018), which damage the auditory system through their effects on the inner ear via the bloodstream or neural pathways.

A final example of exposure-related conditions are those linked to poor air quality and aerosolized contaminants, which can contribute to *respiratory conditions*. The most commonly- known work-related respiratory condition is *pneumoconiosis*, often referred to among miners as "black lung". Because respiratory disease has disproportionately impacted miners, NIOSH (https:// www.cdc.gov/niosh/mining/) and the Mine Safety and Health Administration (https://www.msha.gov/) have developed many resources for addressing this and other health and safety risks for this population. However, pneumoconiosis and other respiratory conditions do not just affect miners. For instance, rare lung diseases developed by non-smoking manufacturing workers have been linked to a variety of production chemicals (Stanton & Nett, 2019). Similarly, grinding and brewing of coffee is associated with mild respiratory symptoms among coffee production workers and baristas (McClelland et al., 2019). Even in typical indoor office settings, workers can develop "sick building symptoms" from poor air quality, among other factors (MacNaughton et al., 2017). In sum, it is important to consider that the air quality in *any* work environment could contribute to respiratory symptoms.

Repetitive Strain and Musculoskeletal Disorders

Physical and psychosocial demands of work can affect workers' locomotor systems (including bones, muscles, joints). *Work-related musculoskeletal disorders* (WMSD) may result when work-related demands (especially physical ones) are chronically present, and often when workers are either ill-prepared, poorly supported, or otherwise unable to properly respond to these demands. Generally, WMSD are understood as resulting from a physical load being placed on the body, affecting workers' tissues, muscles, and bones to the point at which the worker cannot sufficiently adapt or sustain the load without injury or damage to the body (NIOSH, 2001). There is great variety in WMSD, including neck and upper back pain, conditions affecting the elbow and other joints (e.g., epicondylitis, or "tennis elbow"), wrist pain (e.g., carpal tunnel syndrome), lower back pain, and pain in the knees or feet (e.g., Buckle & Devereux, 2002; Hernandez & Peterson, 2012). A variety of work characteristics (discussed more in Chapter 10) may contribute to WMSD, including: heavy lifting, overexertion, repetitive motions, prolonged standing, and vibrations from machinery.

Work-Related Stress

Many models of workplace stress (discussed in detail in Chapter 6) highlight a connection between work-related demands and physical health. Research testing these models demonstrates that the effects of stress exposures are not purely psychological and often do take rather serious physical forms. Indeed, many of the top risk factors for mortality, including high blood pressure, high blood sugar, and obesity (Ritchie & Roser, 2020) are physical in nature and linked to chronic exposures to stressors, which often come through our work-related experiences.

Psychosomatic Symptoms

Think about how you feel during and after a stressful day at work. You might experience a headache, tightness in the upper back, or digestive issues. These are all forms of psychosomatic symptoms (i.e., noticeable physical effects of a psychological experience). In addition to the examples mentioned above, psychosomatic symptoms linked to work-related stress can take many forms (e.g., gastrointestinal problems, sleep disturbances, and fatigue; Nixon et al., 2011; Spector & Jex, 1998).

Sleep Problems

Stress experiences can also negatively impact workers' sleep quality in the short-term and potentially contribute to longer-term conditions like chronic fatigue or insomnia. As you probably would expect, sleep disorders are more common among workers with non-standard work hours (Burch et al., 2005). *Shiftwork sleep disorder* is a specific label for sleep problems linked to disrupted sleep-wake cycles due to shiftwork schedules (Drake et al., 2004). Insomnia and other sleep problems are a concern in and of themselves, as well as contributing factors to other psychological and physical health conditions among workers (Vallieres et al., 2014). Put simply, our body systems require regular and sufficient sleep to function well.

Cardiovascular Health

Although not commonly seen as a psychological phenomenon, cardiovascular health is a major physical health concern for OHP professionals. Clear evidence links work-related stress experiences to cardiovascular health (Kivimaki et al., 2006) and it is no secret that work environments and demands influence our lifestyle behaviors, which also influence cardiovascular health. Cardiovascular health is consistently and negatively impacted by high work-related demands and low control (Kuper & Marmot, 2003), an imbalance between effort and reward at work (Eddy et al., 2017), and feelings of injustice at work (Kivimaki et al., 2005). Cardiovascular health is also impacted by irregular work schedules; by some estimates, night shiftworkers have a 40% higher risk for cardiovascular disease (CVD) than day shiftworkers (Boggild & Knutsson, 1999). Similarly, more cardiovascular health issues have been observed among workers in occupations that impose unusually difficult demands, like having to be constantly "on-guard" (i.e., threat-avoidant vigilance) to avoid disaster (Landsbergis et al., 2011).

Complicating this picture a bit more is that the many work-related and psychosocial risk factors for CVD are not evenly distributed across all workers and the broader population. For instance, those with a lower socioeconomic status are often more at risk for CVD (Landsbergis et al., 2011). These disproportionate health risks for certain demographic groups are often connected to disproportionate work-related demands and limited resources. These topics are often studied under the broader label of health disparities (Williams et al., 2010).

Worker Behaviors and Physical Health

In addition to the effects of psychosocial stressors, we make daily behavioral choices that affect our physical health. Sometimes these choices are about how we will behaviorally adapt or otherwise respond to stressors (Juster et al., 2010; McEwen, 1998). More generally, these health-related behaviors or lifestyle habits may include exercise, eating habits, and use of alcohol, tobacco, or other drugs. Health-related behaviors do indeed explain some of the connection between work-related stress and indicators of metabolic health risks (French et al., 2019). Many of these behaviors are established risk factors for developing physical health issues like hypertension or CVD (WHO, 2017).

Organizations as Facilitators or Hindrances

Health-related behaviors and lifestyle habits are personal by nature, but organizations play a substantial role in creating and sustaining the environmental context in which many of these behaviors occur (e.g., opportunities for physical movement, nutritional options, and self-care flexibility). In other words, where and how we work can affect the degree to which our lifestyles support or detract from our physical health. As one example, workers on extended or non-standard work schedules are less likely to exercise and more likely to have high body-mass index (BMI) scores than workers on traditional schedules (Bushnell et al., 2010). Findings like this highlight the importance of ensuring that workers have opportunities and resources to support healthy living (e.g., workers without leisure time or schedule flexibility may be unable to make optimal health-related behavior choices).

Workplace social norms and support can also influence health-related behaviors, especially with respect to substance use (Moore et al., 2012; van den Brand et al., 2019). Similarly, it is easy to understand how social norms at work can influence nutritional choices. Consider how the following questions reveal the power of social norms around eating at work: Do workers typically bring their own food or go out to lunch? Where do workers tend to eat? Are healthy options provided at a similar cost at on-site dining facilities? Is there even break time for a meal or do convenience options have to suffice from fast food or vending machines? In short, although behavioral choices are ultimately made by individual workers, OHP professionals recognize that these behaviors are influenced by factors in the work environment.

How Worker Physical Health Affects Work

We have focused on ways in which work-related experiences and exposures can impact worker physical health. It is also important to consider how one's ability to work effectively and safely is affected by underlying physical health conditions, which may or may not be work-related. This is a relatively understudied area for OHP professionals, but we can better understand how our physical health affects work through the lens of Conservation of Resources theory (COR; Hobfoll, 1989). Specifically, physical health is a resource that influences and supports our abilities to respond to stressors and acquire additional resources (e.g., objects, conditions, energies). This theory helps explain how the effects of poor physical health can compound and accumulate. For example, a worker who suffers frequent migraine headaches is likely to experience impaired concentration on work tasks, feelings of general inefficacy at work, and potentially absences from work (Collins et al., 2005). This single physical health resource depletion can then lead to multiple ripple effects.

Worker performance and organizational health can be directly impacted when a physical health condition inhibits, limits, or prevents normal functioning, and/or requires frequent time away from work. Many chronic physical health conditions are linked to some degree of work impairment (Collins et al., 2005). As a few examples, chronic musculoskeletal pain or WMSD are linked to worker self-reports of difficulty completing work tasks (de Vries et al., 2013). Workers with Type II Diabetes may be likely to retire early, take more sick days, and be less productive than their non-diabetic peers (Breton et al., 2013). Workers experiencing irritable bowel syndrome (IBS) report taking more sick days, more difficulty concentrating at work, and perceived effects on productivity than those without this condition; chronic conditions like IBS can also affect social relationships, due to concerns that symptoms will interfere (Hungin et al., 2005). Workers managing chronic physical health conditions may experience stress over perceived stigma or concerns about being treated differently because of their conditions (McGonagle & Barnes-Farrell, 2014). Chronic physical health conditions can also deplete one's psychological and energy-related resources. For example, workers with chronic pain often report negative affect and exhaustion at the end of the work day (Fragoso & McGonagle, 2018).

Although working while managing a physical health condition is not easy, there can be associated benefits. Work can serve an adaptive role for an individual who may be managing some form of noncontagious illness or physical health condition (Karanika-Murray & Biron, 2019). For instance, remaining involved in work (to the extent possible) following an injury or development of WMSD is associated with positive psychological health and long-term ability to remain at work (Howard et al., 2009). Similarly, working can provide a sense of purpose and help fill time for people who are living with or recovering from a physical illness or injury. These points are noted because they are directly aligned with underlying OHP principles that physical wellness for workers involves more than just removing or reducing symptoms.

Why Worker Physical Health Matters

Despite a widely held public belief in the importance of physical health, OHP researchers and practitioners in this domain continuously have to fight for resources (time, money, personnel, etc.) to support studies or interventions to address worker physical health. Here we explore two main sets of reasons why protecting and promoting worker physical health matters.

Healthy Workers are Good for Business

Workers struggling with physical health issues are often less able to be present and productive at work. Thus, there is real value to doing what is possible within organizations to maximize the number of workers who are in good physical health. This is especially true in physically demanding occupations, where physical abilities affect daily job requirements. This is also true for less physically demanding work, though gaining buy-in from leaders to promote and protect the general physical health of all workers can be especially challenging.

Most organizational risk management efforts are driven by the goal of managing costs. Costs associated with worker physical health are very real, including employer-provided healthcare benefits, lost productivity due to illness-related absences, and worker's compensation claims linked to occupational illness or injury. In the USA alone, occupational injuries and illnesses (primarily physical) are estimated to annually yield direct medical costs totaling $67 million and indirect costs up to $183 billion (Leigh, 2011). Other nations report similarly high costs, such as €15 billion for costs of worker illness and injury in the United Kingdom (Health and Safety Executive, 2019). There are free tools available to see these costs in more practical terms, like OSHA's "$afety pays" calculator that estimates costs of specific conditions (https://www.osha.gov/safetypays/).

One of the biggest sources of cost associated with physical health problems is lost productivity, which also occurs when workers are physically present, but not physically well (i.e., presenteeism; Johns, 2011). Meta-analytic findings report small to moderate negative correlations between physical conditions, like hypertension, sleep problems, and somatic symptoms, and measures of performance (Ford et al., 2011). There are also risks to consider where some physical health complaints, like fatigue and sleep problems, are related to safety concerns (Kao et al., 2016; Williamson et al., 2011).

Proactively addressing worker physical health with targeted promotion efforts is also justified for business and financial reasons. Programs and policies promoting worker physical health are a strategic investment in a resource that is worth protecting. Physical health promotion programs often work by encouraging positive health-related behaviors. Encouraging physical activity in particular has a number of organizational benefits from health-costs savings to better performance, both supported by empirical findings and/or a strong theoretical rationale (Calderwood et al., 2020). There is indeed evidence of positive return on investment (ROI) in physical health promotion efforts, even for smaller organizations (e.g., Goetzel et al., 2014). Participation in such programs may also positively impact worker attitudes and reduce absenteeism (Parks & Steelman, 2008).

Personal and Societal Reasons

Working while managing a physical health condition can take a toll on workers in a variety of ways, including the experience of negative emotional and cognitive states (Fragoso & McGonagle, 2018). When workers develop and engage in physically healthy lifestyles, either motivated themselves or by their organizations, these types of issues may be mitigated. There is growing evidence that participation in physical health and wellness programs does improve or sustain worker health (Merrill et al., 2011). Investing in efforts to improve worker physical health has implications for workers' broader quality of life, and even impacts that extend beyond the organization. Investing in workers' physical health can improve their abilities to fully engage with their families, communities, and society more generally. For maximum impact, OHP efforts to protect and improve worker physical health would include programs, policies, and other initiatives in and outside of work environments. Promising explorations along these lines are underway in the area of *population health* (e.g., Kindig & Stoddart, 2003).

Methodological Considerations and Practical Recommendations

Working to understand and improve worker physical health involves gathering and analyzing a variety of different forms of data. There are numerous methodological details and practical recommendations to consider when engaging in research and practice in this area of OHP.

Measuring and Monitoring Physical Health

Evaluating worker physical health in organizational settings may be for *descriptive* purposes (i.e., completing an organizational health risk appraisal or establishing a baseline understanding of worker health), *prescriptive* purposes (i.e., determining what physical health issues need the most attention), or *predictive* purposes (e.g., trying to anticipate where the greatest costs or risks are going to develop). There are multiple measurement methods to support each of these purposes. Given the complexity of physical health, the best strategy will likely involve measuring multiple indicators of your targeted physical health domain.

Physical health is a form of individual difference that can be difficult to observe, so studying it (particularly in work settings) often requires self-reported information from workers. A simple and practical method for obtaining a quick check of overall physical health is to ask individuals to rate their physical health relative to similar others (e.g., those of similar age, occupation, or other meaningful individual difference; DeSalvo et al., 2006). More detailed data may come from self-reported physical symptoms (e.g., headaches, muscle aches, digestive problems) rated in terms of prevalence, frequency, and/or severity (e.g., Spector & Jex, 1998). You could also gather data on functional impairments experienced in daily living (e.g., Rand Health Care, n.d.), or perceptions of confidence or efficacy associated with specific forms of physical effort.

It is also possible to enlist trained healthcare professionals to measure and evaluate physical health, using indicators like physiological measurements, blood tests, and other forms of laboratory-based assessment. Other indicators of health might include more objectively verifiable measures such as weight, waist circumference, or BMI. These types of indicators have all been linked to certain disorders and may even function as direct risk factors in some situations. However, all forms of physical health measurement (even these latter options) are imperfect, due to the complexity of the human body. For example, objective indicators of physical health like heart rate and blood pressure may be recorded at "unhealthy levels" due to the influence of a sudden stimuli just before the measurement. Individuals with a high degree of muscle mass can often be flagged as extremely overweight based on BMI calculations, even if they are in excellent physical health. It is important to recognize these limitations, especially if physiological indicators are being monitored as a way of incentivizing or penalizing workers who are involved in organization-sponsored physical wellness initiatives.

Worker physical health and especially its effects can also be studied using data regularly gathered by organizations for talent management purposes (e.g., absentee rates/sick days, injury rates, disability rates, and health, pharmacy, and workers' compensation benefits claims). These metrics are also imperfect and need to be considered with a clear head and broad perspective. As an illustration, sick day counts are inherently imprecise; some workers may take a sick day when they are not actually sick and other workers may come to work even while sick to save sick days. Ultimately, a mix of objective indicators and subjective self-reported data is likely to provide the most complete information for identifying physical health needs and challenges within a particular workforce.

Intervening to Promote and Improve Physical Health

Physical health interventions undertaken by OHP professionals regularly incorporate perspective and collaborators from related disciplines including public health, industrial hygiene, ergonomics, human factors, and occupational medicine. Despite this professional diversity, there are several essential strategies for change and methodological elements needed when designing, implementing, evaluating, and sustaining physical health interventions at work. In this section, we present recommendations to support successful physical health interventions. Regardless of the level of the intervention (i.e., targeted toward individuals, groups, organizations), it is important to keep in mind that attempting to influence human behavior is difficult if change efforts are not aligned with workers' own desires (e.g., Howard, 2006). Therefore, any intervention strategy is likely most successful when it helps workers achieve their own visions of good physical health, rather than an externally derived standard.

Strategies for Individuals

Employee wellness programs are perhaps the most common form of organization-based effort to promote worker physical health. These programs tend to be proactive in approach, incentivizing healthy behaviors (or disincentivizing unhealthy ones) and can involve a variety of elements, including education, goal-setting, and long-term monitoring of physical health indicators (Parks & Steelman, 2008). Simple initiatives, like walking programs, can increase worker physical activity, and even lower blood pressure and some health care costs (Swayze & Burke, 2013; Zivin et al., 2017). Programs that involve financially incentivizing healthy choices (e.g., attending educational workshops, receiving preventative screenings, engaging in exercise) are shown to maintain or improve physical health, as evidenced by biometric screenings and other self-report measures (Merrill et al., 2011). Wellness programs targeting the management of risky health-related behaviors, like smoking or excessive alcohol consumption, tend to be more challenging to successfully implement, largely due to influences of multiple individual differences among other factors (e.g., personal motivation, age, gender, income; Richmond et al., 2000; Titus et al., 2019). Efforts to change group norms may be more effective in this domain, as we discuss later.

There are a couple of practical points to consider when implementing any type of person-level physical health program. First, participation rates in organizational wellness programs are often low and in some (but not all) cases, it is healthier workers who are more likely to participate (Robroek et al., 2009). An effective program that reaches the entire working population may require a calibrated and customized approach based on knowledge of pertinent individual differences (see Chapter 3).

Second, focusing organizational attention on worker physical health can create ethically challenging scenarios. In particular, focusing on physical health issues can lead to discriminatory outcomes (Koruda, 2016; Madison, 2016), given the complex relationships between certain forms of racial, ethnic, socioeconomic, and other demographic individual differences and physical health. Organizational initiatives to promote and protect physical health need to be sensitive to the role played by underlying physical health disparities for certain groups of workers compared to others, and ensure that any program-related incentives or penalties are based on fair criteria that do not disproportionately impact certain groups. In response to these concerns around discrimination and other critiques of wellness programs, many organizations opt to reward healthy behaviors, but avoid penalizing poor health-related behaviors, metrics, or risk profiles. The likelihood of success for any physical health promotion program targeting workers will increase if it is developed with clear objectives and well-aligned strategies, and supported by a culture that promotes worker physical health.

Strategies for Groups and Leaders

Worker physical health is also affected by group affiliations and social norms surrounding health-related behaviors. Normative alcohol, tobacco, and other drug use are particularly common in certain occupational groups, including food service workers (e.g., Moore et al., 2012). Thankfully, such social and normative pressures can also support good health, encouraging smoking cessation (van den Brand et al., 2019) and discouraging problematic drinking (e.g., Wang et al., 2010). These types of successes suggest that interventions leveraging peer support and supervisor training could be effective at stimulating health-related behavior changes.

Social forces within workers' groups and teams, as well as messages from leaders, can also influence workers' decisions about reporting and attempting to work through injuries or illnesses. These forces can be counteracted by clear messaging throughout the organization that matches a culture supporting a shared desire for healthy lifestyles, without condemning those who may be perceived as unhealthy. Certain physical health conditions may even be associated with actual or perceived social consequences, experienced as stigma (McGonagle & Barnes-Farrell, 2014). Adding stigma to an "unhealthy" status could motivate some to engage in positive lifestyle habits, but could have the opposite effect on those who wish to avoid situations that are uncomfortable because of the increased attention to that stigmatized identity.

Strategies for Organizations

Sometimes worker physical health efforts are targeted at the level of the overall organization or broad work environment. Some physical health interventions are aimed at promoting worker physical health at a high level and/or supporting the unique needs of workers with a short-term or long-term illness. There are also a number of modifications that can be made to the work itself to reduce the potential for physical harm, discussed more in Chapters 10 and 11.

Typical organization-level efforts to promote and protect worker physical health involve environmental adjustments, general education and policy, and culture-level modifications. For example, organizations can provide *practical resources* that encourage healthy lifestyle choices among workers, such as on-site dining facilities offering healthy, affordable food options. Flexible workstations are another increasingly used tool for physical health promotion. Sit-stand desks can make a workspace more adaptable and reduce back pain for some workers (Agarwal et al., 2018); treadmills, bikes, and even seated elliptical machines can also be added to desk setups to facilitate worker physical activity even while seated (e.g., Carr et al., 2016). These sorts of practical resources may require a moderate financial investment (e.g., one seated elliptical machine was approximately $600 for Carr et al.), but associated gains in concentration, productivity, and health from increased physical activity are likely to outweigh this level of financial investment over time. A broad organization-wide culture change initiative might further combine the benefits of these tangible resources and other elements together, changing the overall work experience into one that rather holistically or comprehensively supports worker physical health.

The *scheduling of work* can also be a practical point of organization-level intervention to support worker physical health. Offering scheduling flexibility, where possible, can make it easier for workers to make time to be physically active, manage doctor appointments, and choose to stay home when physically unwell. Attendance expectations, as well as the overall workload, are both significant correlates with presenteeism (Miraglia & Johns, 2016). Organizations also set the tone for physical health through *formal policies and procedures* around health risks. For example, organizations with outdoor workers can provide sun protection education and resources (e.g., policies and materials) to protect these workers (e.g., Buller et al., 2018). For work that involves a good deal of physical exertion, workplace stretching programs have shown some degree of success in reducing injuries and musculoskeletal complaints (for a review, see Hess & Hecker, 2003) and can be easily added to a typical workday or included as an educational component of onboarding.

Workers returning to work after an injury, working with an injury, or actively managing a physical health condition may require accommodations to do their work well. This is another way in which organizations can support worker physical health. Quality return-to-work support might involve an experienced case manager (Shaw et al., 2008) or well-trained supervisors who know how to modify work to support workers as-needed.

Evaluating Physical Health Interventions

Regardless of the type of physical health intervention, there are a few essential principles to think about when it comes to evaluation (discussed more in Chapter 2). As we noted earlier, assessing physical health is best done with multiple indicators, ideally gathered over time to allow you to show change associated with your intervention efforts. Think about what your desired outcome truly is (e.g., changes in physical health indicators; changes in health-related attitudes; ROI) and about the time it reasonably takes to see a change in your available indicators. Some measures we have mentioned will work well for shorter-term impact evaluations (e.g., self-efficacy perceptions) while other indicators may only reflect change over an extended period of time (e.g., weight loss, sick day trends). You may need to track data for several years to see small, yet meaningful bottom-line impacts on things like workers compensation claims or absenteeism (Anger et al., 2015).

Monitoring a variety of physical health indicators and health-related attitudes will also help you understand the process and impact of your intervention, and even to see if there were any, potentially unexpected, benefits of your intervention. Sometimes, even if an intervention does not hit its original target (e.g., reducing absences), it can lead to other positive effects (e.g., improving workers' feelings of confidence at work). Something like this happened in an ergonomic intervention among kitchen workers, where workers perceived improvements to their health and workload, even though there was no significant change in the primary outcome of interest, musculoskeletal pain (Haukka et al., 2008; Pehkonen et al., 2009). Other studies find that physical health interventions impact concentration or productivity, even though changes in more objective health data are modest (e.g., Carr et al., 2016).

Finally, because so many factors can influence worker physical health, our ability to evaluate these types of interventions is dramatically improved if we can use evaluation methodologies and designs that strengthen the quality of our inferences, such as randomized control group or quasi-experimental designs (Anger et al., 2015). Some evaluations of physical health interventions randomly assign individuals to groups (Carr et al., 2016), while others may randomly assign entire organizations or workgroups to specific conditions for study (Buller et al., 2018; Haukka et al., 2008). Regardless of the level of assignment, gathering data from a comparison group allows us to more convincingly demonstrate effects of an intervention compared to a non-intervention or simpler alternative. Data from these types of evaluations create a strong argument for investing in a particular intervention, especially if the evaluation period spans enough time to document the longevity and stability of an intervention's effects.

Concluding Thoughts and Reality Check

Many aspects of work and our work environment can affect our physical health. Being physically healthy is an important resource for effectively carrying out daily work. Influencing worker physical health is not an easy task

119

for organizations, but ignoring the importance of physical health can be tremendously costly and challenging. Organizations can rarely control workers' health and health-related behaviors with direct mandates (at least, not without triggering serious backlash). Instead, organizations can provide resources that make achieving and maintaining physical health easier. Often these resources are as simple as allowing sufficient breaks and flexible scheduling so workers can eat nutritiously and get some physical activity during their work period, or even creating an environment where it is ok to miss work when sick or make time for regular doctor visits.

Organizations have a prime opportunity to structure and design work in ways that limit damage to worker physical health, while also facilitating maintenance and improvement of physical health. Workers, organizations, and society as a whole benefit when we can create and sustain physically healthy workforces.

Media Resources

- Physical health and returns to work following COVID-19: https://hbswk.hbs.edu/item/who-guarantees-your-workplace-is-safe-for-return
- Critiques of workplace wellness programs: https://slate.com/technology/2016/09/workplace-wellness-programs-are-a-sham.html
- Yale wellness program lawsuit: https://www.wnpr.org/post/union-workers-file-civil-action-lawsuit-against-yale-over-employee-wellness-program

---

Discussion Questions

1) Describe the complex relationships between work and worker physical health.
2) Imagine you are assessing the state of worker physical health for a commercial fast food chain. How would you go about measuring physical health in this population? What potential physical health conditions could be connected to work-related factors?
3) What would be the incentive for a company to offer a wellness program? How could they encourage participation from all employees?
4) Imagine that you are a consultant to an organization that wants to know why their smoking cessation program has been unsuccessful. What are some of the factors that may be limiting the success of this program and how could you address them?

Professional Profile: Ryan Olson, Ph.D.

Country/region: USA, Oregon

Current position title: Professor at Oregon Institute of Occupational Health Sciences, Oregon Health & Science University (OHSU), OHSU-Portland State University (PSU) School of Public Health, and PSU Department of Psychology

Background: I have been working to improve the health, safety, and/or well-being of workers for 20 years. I have a B.S. in Psychology from Utah State University. I also have an M.A. in Industrial-Organizational Psychology and Ph.D. in Applied Behavior Analysis from Western Michigan University. I belong to the American Psychological Association, the American Academy of Sleep Medicine, and the Association for Behavior Analysis International. I also participate in specialized societies or divisions of those, and other parent organizations, including the Society for Occupational Health Psychology, Society for Industrial-Organizational Psychology, and the Organizational Behavior Management Network.

My first faculty position after graduate school was serving as an Assistant Professor in the Psychology Department at Santa Clara University – a wonderful place with supportive colleagues! I then moved to OHSU where I have spent the past 15 years working at what is now called the Oregon Institute of Occupational Health Sciences. My current work involves conducting occupational safety and health research with working populations in need, focusing on isolated or 'lone' workers in demanding occupations (e.g., commercial drivers, home care workers).

How my work impacts WHSWB: My work has included interventions that have reduced workplace exposures to physical hazards, improved supportive supervisory behaviors, and improved injury prevention and health promoting behaviors. I led the team that developed and evaluated the Safety and Health Involvement for Truck drivers (SHIFT) intervention. We designed the intervention by selecting evidence-based tactics that were amenable for adaptation to engage a dispersed workforce (web-based competition with individual and social comparison feedback; goal setting and self-monitoring; training on healthy body weight management, eating, exercise, and sleep; and motivational interviewing via cell phone). At the current date, SHIFT is one of two programs globally, evaluated with a randomized controlled design, to produce medically meaningful weight loss among truck drivers.

My motivation: Work environments have tremendous potential to advance and enrich human health and well-being. However, workplaces can also put people at-risk for injury, degraded health, or poor mental health and well-being. Human organizations also represent our greatest levers for generating positive or negative impacts on environmental and planetary health. I am driven by the belief that if I apply scientific methods to testing potential solutions to occupational and environmental problems, I might play a role in advancing life-sustaining and health-advancing conditions on earth.

Chapter References

Agarwal, S., Steinmaus, C., & Harris-Adamson, C. (2018). Sit-stand workstations and impact on low back discomfort: A systematic review and meta-analysis. *Ergonomics*, *61*(4), 538–552. https://doi.org/10.1080/00140139.2017.1402960

Anger, W. K., Elliot, D. L., Bodner, T., Olson, R., Rohlman, D. S., Truxillo, D. M., Kuehl, K. S., Hammer, L. B., & Montgomery, D. (2015). Effectiveness of total worker health interventions. *Journal of Occupational Health Psychology*, *20*(2), 226–247. https://doi.org/10.1037/a0038340

Boggild, H., & Knutsson, A. (1999). Shift work, risk factors and cardiovascular disease. *Scandinavian Journal of Work, Environment & Health*, *25*(2), 85–99. https://doi.org/https://www.jstor.org/stable/40966872

Breton, M. C., Guenette, L., Amiche, M. A., Kayibanda, J. F., Gregoire, J. P., & Moisan, J. (2013). Burden of diabetes on the ability to work: A systematic review. *Diabetes Care*, *36*(3), 740–749. https://doi.org/10.2337/dc12-0354

Buckle, P. W., & Devereux, J. J. (2002). The nature of work-related neck and upper limb musculoskeletal disorders. *Applied Ergonomics*, *33*(3), 207–217. https://doi.org/10.1016/s0003-6870(02)00014-5

Buller, D. B., Walkosz, B. J., Buller, M. K., Wallis, A., Andersen, P. A., Scott, M. D., Eye, R., Liu, X., & Cutter, G. (2018). Results of a randomized trial on an intervention promoting adoption of occupational sun protection policies. *American Journal of Health Promotion*, *32*(4), 1042–1053. https://doi.org/10.1177/0890117117704531

Burch, J. B., Yost, M. G., Johnson, W., & Allen, E. (2005). Melatonin, sleep, and shift work adaptation. *Journal of Occupational & Environmental Medicine*, *47*(9), 893–901. https://doi.org/10.1097/01.jom.0000177336.21147.9f

Bushnell, T., Colombi, A., Caruso, C. C., & Tak, S. (2010). Work schedules and health behavior outcomes at a large manufacturer. *Industrial Health*, *48*, 395–405. https://doi.org/https://doi.org/10.2486/indhealth.MSSW-03

Calderwood, C., ten Brummelhuis, L. L., Patel, A. S., Watkins, T., Gabriel, A. S., & Rosen, C. C. (2020). Employee physical activity: A multidisciplinary integrative review. *Journal of Management*. https://doi.org/10.1177/0149206320940413

Carr, L. J., Leonhard, C., Tucker, S., Fethke, N., Benzo, R., & Gerr, F. (2016). Total Worker Health intervention increases activity of sedentary workers. *American Journal of Preventive Medicine*, *50*(1), 9–17. https://doi.org/10.1016/j.amepre.2015.06.022

Collins, J. J., Baase, C. M., Sharda, C. E., Ozminkowski, R. J., Nicholson, S., Billotti, G. M., Turpin, R. S., Olson, M., & Berger, M. L. (2005). The assessment of chronic health conditions on work performance, absence, and total economic impact for employers. *Journal of Occupational and Environmental Medicine*, *47*(6), 547–557. https://doi.org/10.1097/01.jom.0000166864.58664.29

de Vries, H. J., Reneman, M. F., Groothoff, J. W., Geertzen, J. H., & Brouwer, S. (2013). Self-reported work ability and work performance in workers with chronic nonspecific musculoskeletal pain. *Journal of Occupational Rehabilitation*, *23*(1), 1–10. https://doi.org/10.1007/s10926-012-9373-1

DeSalvo, K. B., Fisher, W. P., Tran, K., Bloser, N., Merrill, W., & Peabody, J. (2006). Assessing measurement properties of two single-item general health measures. *Quality of Life Research*, *15*(2), 191–201. https://doi.org/10.1007/s11136-005-0887-2

Drake, C. L., Roehrs, T., Richardson, G., Walsh, J. K., & Roth, T. (2004). Shift work sleep disorder: Prevalence and consequences beyond that of symptomatic day workers. *Sleep*, *27*(8), 1453–1462. https://doi.org/https://doi.org/10.1093/sleep/27.8.1453

Eddy, P., Wertheim, E. H., Kingsley, M., & Wright, B. J. (2017). Associations between the effort-reward imbalance model of workplace stress and indices of cardiovascular health: A systematic review and meta-analysis. *Neuroscience and Biobehavioral Reviews*, *83*, 252–266. https://doi.org/10.1016/j.neubiorev.2017.10.025

Ford, M. T., Cerasoli, C. P., Higgins, J. A., & Decesare, A. L. (2011). Relationships between psychological, physical, and behavioural health and work performance: A review and meta-analysis. *Work & Stress*, *25*(3), 185–204. https://doi.org/10.1080/02678373.2011.609035

Fragoso, Z. L., & McGonagle, A. K. (2018). Chronic pain in the workplace: A diary study of pain interference at work and worker strain. *Stress and Health*, *34*(3), 416–424. https://doi.org/10.1002/smi.2801

French, K. A., Allen, T. D., & Henderson, T. G. (2019). Challenge and hindrance stressors and metabolic risk factors. *Journal of Occupational Health Psychology*, *24*(3), 307–321. https://doi.org/10.1037/ocp0000138

Ganster, D. C., & Rosen, C. C. (2013). Work stress and employee health: A multidisciplinary review. *Journal of Management*, *39*(5), 1085–1122. https://doi.org/10.1177/0149206313475815

Gatchel, R. J., & Kishino, N. (2011). Pain, musculoskeletal injuries, and return to work. In J. C. Quick & L. E. Tetrick (Eds.), *Handbook of occupational health psychology* (2nd ed., pp. 265–275). American Psychological Association.

Goetzel, R. Z., Tabrizi, M., Henke, R. M., Benevent, R., Brockbank, C. V., Stinson, K., Trotter, M., & Newman, L. S. (2014). Estimating the return on investment from a health risk management program offered to small Colorado-based employers. *Journal of Occupational and Environmental Medicine*, *56*(5), 554–560. https://doi.org/10.1097/JOM.0000000000000152

Guest, M., Boggess, M., Attia, J., D'Este, C., Brown, A., Gibson, R., Tavener, M., Gardner, I., Harrex, W., Horsley, K., & Ross, J. (2010). Hearing impairment in F-111 maintenance workers: The study of health outcomes in aircraft maintenance personnel (SHOAMP) general health and medical study. *American Journal of Industrial Medicine*, *53*(11), 1159–1169. https://doi.org/10.1002/ajim.20867

Haukka, E., Leino-Arjas, P., Viikari-Juntura, E., Takala, E. P., Malmivaara, A., Hopsu, L., Mutanen, P., Ketola, R., Virtanen, T., Pehkonen, I., Holtari-Leino, M., Nykanen, J., Stenholm, S., Nykyri, E., & Riihimaki, H. (2008). A randomised controlled trial on whether a participatory ergonomics intervention could prevent musculoskeletal disorders. *Occupational and Environmental Medicine*, *65*(12), 849–856. https://doi.org/10.1136/oem.2007.034579

Health and Safety Executive. (2019). *Health and Safety Statistics: Key figures for Great Britian (2018/19)*. https://www.hse.gov.uk/statistics/

Hernandez, A. M., & Peterson, A. L. (2012). Work-related musculoskeletal disorders and pain. In *Handbook of Occupational Health and Wellness* (pp. 63–85). https://doi.org/10.1007/978-1-4614-4839-6_4

Hess, J. A., & Hecker, S. (2003). Stretching at work for injury prevention: Issues, evidence, and recommendations. *Applied Occupational and Environmental Hygiene*, *18*(5), 331–338. https://doi.org/10.1080/10473220301367

Hobfoll, S. E. (1989). Conservation of resources: A new attempt at conceptualizing stress. *American Psychologist*, *44*(3), 513–524. https://doi.org/10.1037/0003-066X.44.3.513

Howard, A. (2006). Positive and negative emotional attractors and intentional change. *Journal of Management Development*, 25(7), 657–670. https://doi.org/10.1108/02621710610678472

Howard, K. J., Mayer, T. G., & Gatchel, R. J. (2009). Effects of presenteeism in chronic occupational musculoskeletal disorders: Stay at work is validated. *Journal of Occupational and Environmental Medicine*, 51(6), 724–731. https://doi.org/10.1097/JOM.0b013e3181a297b5

Hungin, A. P., Chang, L., Locke, G. R., Dennis, E. H., & Barghout, V. (2005). Irritable bowel syndrome in the United States: Prevalence, symptom patterns and impact. *Alimentary Pharmacology and Therapeutics*, 21(11), 1365–1375. https://doi.org/10.1111/j.1365-2036.2005.02463.x

Johns, G. (2011). Attendance dynamics at work: The antecedents and correlates of presenteeism, absenteeism, and productivity loss. *Journal of Occupational Health Psychology*, 16(4), 483–500. https://doi.org/10.1037/a0025153

Juster, R. P., McEwen, B. S., & Lupien, S. J. (2010). Allostatic load biomarkers of chronic stress and impact on health and cognition. *Neuroscience and Biobehavioral Reviews*, 35(1), 2–16. https://doi.org/10.1016/j.neubiorev.2009.10.002

Kao, K. Y., Spitzmueller, C., Cigularov, K., & Wu, H. (2016). Linking insomnia to workplace injuries: A moderated mediation model of supervisor safety priority and safety behavior. *Journal of Occupational Health Psychology*, 21(1), 91–104. https://doi.org/10.1037/a0039144

Karanika-Murray, M., & Biron, C. (2019). The health-performance framework of presenteeism: Towards understanding an adaptive behaviour. *Human Relations*. https://doi.org/10.1177/0018726719827081

Kindig, D., & Stoddart, G. (2003). What is population health? *American Journal of Public Health*, 93(3), 380–383. https://doi.org/10.2105/AJPH.93.3.380

Kivimaki, M., Ferrie, J. E., Brunner, E., Head, J., Shipley, M. J., Vahtera, J., & Marmot, M. G. (2005). Justice at work and reduced risk of coronary heart disease among employees – The Whitehall II Study. *Archives of Internal Medicine*, 165(19), 2245–2251. https://doi.org/10.1001/archinte.165.19.2245

Kivimaki, M., Virtanen, M., Elovainio, M., Kouvonen, A., Vaananen, A., & Vahtera, J. (2006). Work stress in the etiology of coronary heart disease: A meta-analysis. *Scandinavian Journal of Work, Environment & Health*, 32(6), 431–442. https://doi.org/10.5271/sjweh.1049

Koruda, E. (2016). More carrot, less stick: Workplace wellnessprograms and the discriminatory impact of financial and health-based incentives. *Boston College Journal of Law and Social Justice*, 36(1), 131–157.

Kuper, H., & Marmot, M. (2003). Job strain, job demands, decision latitude, and risk of coronary heart disease within the Whitehall II study. *Journal of Epidemiology and Community Health*, 57(2), 147–153. https://doi.org/10.1136/jech.57.2.147

Landsbergis, P. A., Schnall, P. L., Belkic, K. L., Baker, D., Schwartz, J. E., & Pickering, T. G. (2011). Workplace and cardiovascular disease: Relevance and potential role for occupational health psychology. In J. C. Quick & L. E. Tetrick (Eds.), *Handbook of occupational health psychology* (2nd ed., pp. 243–264). American Psychological Association.

Leigh, J. P. (2011). Economic burden of occupational injury and illness in the United States. *The Milbank Quarterly*, 89, 728–772. https://doi.org/10.1111/j.1468-0009.2011.00648.x

MacNaughton, P., Satish, U., Laurent, J. G. C., Flanigan, S., Vallarino, J., Coull, B., Spengler, J. D., & Allen, J. G. (2017). The impact of working in a green certified building on cognitive function and health. *Building and Environment*, *114*, 178–186. https://doi.org/10.1016/j.buildenv.2016.11.041

Madison, K. M. (2016). The risks of using workplace wellness programs to foster a culture of health. *Health Affairs*, *35*(11), 2068–2074. https://doi.org/10.1377/hlthaff.2016.0729

McClelland, T., Boylstein, R. J., Martin, S. B., & Beaty, M. (2019). *Evaluation of exposures and respiratory health at a coffee roasting and packaging facility and two off-site retail cafés*. https://www.cdc.gov/niosh/hhe/reports/pdfs/2016-0109-3343.pdf

McEwen, B. S. (1998). Stress, adaptation, and disease: Allostasis and allostatic load *Annals of the New York Academy of Sciences*, *840*(1), 33–44. https://doi.org/10.1111/j.1749-6632.1998.tb09546.x

McGonagle, A. K., & Barnes-Farrell, J. L. (2014). Chronic illness in the workplace: Stigma, identity threat and strain. *Stress and Health*, *30*(4), 310–321. https://doi.org/10.1002/smi.2518

Merrill, R. M., Aldana, S. G., Garrett, J., & Ross, C. (2011). Effectiveness of a workplace wellness program for maintaining health and promoting healthy behaviors. *Journal of Occupational and Environmental Medicine*, *53*(7), 782–787. https://doi.org/10.1097/JOM.0b013e318220c2f4

Miraglia, M., & Johns, G. (2016). Going to work ill: A meta-analysis of the correlates of presenteeism and a dual-path model. *Journal of Occupational Health Psychology*, *21*(3), 261–283. https://doi.org/10.1037/ocp0000015

Moore, R. S., Ames, G. M., Duke, M. R., & Cunradi, C. B. (2012). Food service employee alcohol use, hangovers and norms during and after work hours. *Journal of Substance Abuse*, *17*(3), 269–276. https://doi.org/10.3109/14659891.2011.580414

NIOSH. (2001). *National Occupational Research Agenda for Musculoskeletal Disorders: Research Topics for the Next Decade A Report by the NORA Musculoskeletal Disorders Team*. https://www.cdc.gov/niosh/docs/2001-117/pdfs/2001-117.pdf?id=10.26616/NIOSHPUB2001117

NIOSH. (2018). *Preventing Hearing Loss Caused by Chemical (Ototoxicity) and Noise Exposure*. https://www.cdc.gov/niosh/docs/2018-124/default.html

NIOSH. (2019). *Occupational Hearing Loss (OHL) Surveillance*. https://www.cdc.gov/niosh/topics/ohl/default.html

Nixon, A. E., Mazzola, J. J., Bauer, J., Krueger, J. R., & Spector, P. E. (2011). Can work make you sick? A meta-analysis of the relationships between job stressors and physical symptoms. *Work & Stress*, *25*(1), 1–22. https://doi.org/10.1080/02678373.2011.569175

Parks, K. M., & Steelman, L. A. (2008). Organizational wellness programs: A meta-analysis. *Journal of Occupational Health Psychology*, *13*(1), 58–68. https://doi.org/10.1037/1076-8998.13.1.58

Pehkonen, I., Takala, E. P., Ketola, R., Viikari-Juntura, E., Leino-Arjas, P., Hopsu, L., Virtanen, T., Haukka, E., Holtari-Leino, M., Nykyri, E., & Riihimaki, H. (2009). Evaluation of a participatory ergonomic intervention process in kitchen work. *Applied Ergonomics*, *40*(1), 115–123. https://doi.org/10.1016/j.apergo.2008.01.006

Rand Health Care. (n.d.). *36-Item Short Form Survey (SF-36)*. https://www.rand.org/health-care/surveys_tools/mos/36-item-short-form/scoring.html

Redden, E. S., & Larkin, G. B. (2015). Environmental conditions and physical stressors. In *APA Handbook of Human Systems Integration*. (pp. 193–209). https://doi.org/10.1037/14528-013

Richmond, R., Kehoe, L., Heather, N., & Wodak, A. (2000). Evaluation of a work-place brief intervention for excessive alcohol consumption: the workscreen project. *Preventative Medicine, 30*(1), 51–63. https://doi.org/10.1006/pmed.1999.0587

Ritchie, H., & Roser, M. (2020). *Causes of Death.* https://ourworldindata.org/causes-of-death

Robroek, S. J., van Lenthe, F. J., van Empelen, P., & Burdorf, A. (2009). Deter-minants of participation in worksite health promotion programmes: A systematic review. *International Journal of Behavioral Nutrition and Physical Activity, 6,* 26. https://doi.org/10.1186/1479-5868-6-26

Shaw, W., Hong, Q. N., Pransky, G., & Loisel, P. (2008). A literature review describ-ing the role of return-to-work coordinators in trial programs and interventions designed to prevent workplace disability. *Journal of Occupational Rehabilitation, 18*(1), 2–15. https://doi.org/10.1007/s10926-007-9115-y

Spector, P. E., & Jex, S. M. (1998). Development of four self-report measures of job stressors and strain: Interpersonal conflict at work scale, organizational con-straints scale, quantitative workload inventory, and physical symptoms inven-tory. *Journal of Occupationanl Health Psychology, 3*(4), 356–367. https://doi.org/10.1037/1076-8998.3.4.356

Stanton, M. L., & Nett, R. J. (2019). *Evaluation of exposures and respiratory health concerns in a paper converting equipment manufacturing facility* (2012-0055-3337). https://www.cdc.gov/niosh/hhe/reports/pdfs/2012-0055-3337.pdf

Stepanski, B. M., & Mayer, J. A. (1998). Solar protection behaviors among out-door workers. *Journal of Occupational & Environmental Medicine, 40*(1), 43–48. https://doi.org/10.1097/00043764-199801000-00009

Sun, K., Azman, A. S., Camargo, H. E., & Dempsey, P. G. (2019). Risk assessment of recordable occupational hearing loss in the mining industry. *International Journal of Audiology,* 1–8. https://doi.org/10.1080/14992027.2019.1622041

Swayze, J. S., & Burke, L. A. (2013). Employee wellness program outcomes: A case study. *Journal of Workplace Behavioral Health, 28*(1), 46–61. https://doi.org/10.1080/15555240.2013.755448

Titus, A. R., Kalousova, L., Meza, R., Levy, D. T., Thrasher, J. F., Elliott, M. R., Lantz, P. M., & Fleischer, N. L. (2019). Smoke-free policies and smoking cessation in the United States, 2003–2015. *International Journal of Environmental Research and Public Health, 16*(17). https://doi.org/10.3390/ijerph16173200

Vallieres, A., Azaiez, A., Moreau, V., LeBlanc, M., & Morin, C. M. (2014). Insom-nia in shift work. *Sleep Medicine, 15*(12), 1440–1448. https://doi.org/10.1016/j.sleep.2014.06.021

van den Brand, F. A., Nagtzaam, P., Nagelhout, G. E., Winkens, B., & van Schayck, C. P. (2019). The association of peer smoking behavior and social support with quit success in employees who participated in a smoking cessation intervention at the workplace. *International Journal of Environmental Research and Public Health, 16*(16). https://doi.org/10.3390/ijerph16162831

Wang, M., Liu, S., Zhan, Y., & Shi, J. (2010). Daily work-family conflict and alcohol use: testing the cross-level moderation effects of peer drinking norms and social support. *Journal of Applied Psychology, 95*(2), 377–386. https://doi.org/10.1037/a0018138

Williams, D. R., Mohammed, S. A., Leavell, J., & Collins, C. (2010). Race, socioec-onomic status, and health: complexities, ongoing challenges, and research oppor-tunities. *Annals of the New York Academy of Sciences, 1186,* 69–101. https://doi.org/10.1111/j.1749-6632.2009.05339.x

Williamson, A., Lombardi, D. A., Folkard, S., Stutts, J., Courtney, T. K., & Connor, J. L. (2011). The link between fatigue and safety. *Accident Analysis and Prevention*, *43*(2), 498–515. https://doi.org/10.1016/j.aap.2009.11.011

WHO (2014). *Basic Documents.* http://apps.who.int/gb/bd/PDF/bd48/basic-documents-48th-edition-en.pdf#page=7

WHO (2017). *Cardiovascular Diseases (CVDs).* https://www.who.int/news-room/fact-sheets/detail/cardiovascular-diseases-(cvds)

Zivin, K., Sen, A., Plegue, M. A., Maciejewski, M. L., Segar, M. L., AuYoung, M., Miller, E. M., Janney, C. A., Zulman, D. M., & Richardson, C. R. (2017). Comparative effectiveness of wellness programs: Impact of incentives on healthcare costs for obese enrollees. *American Journal of Preventive Medicine*, *52*(3), 347–352. https://doi.org/10.1016/j.amepre.2016.10.006

6

WORK-RELATED STRESS AND RECOVERY

Christopher J. L. Cunningham and Kristen Jennings Black

Stress is connected to most worker health, safety, and well-being (WHSWB) matters. In this overview chapter we review major theories, perspectives, and research findings that help to explain workers' experiences with the continual and interlinked process of work-related stress and recovery. This process is deeply connected to workers' experiences and functional capabilities in and outside of work. We focus on essential stress and recovery concepts and introduce general strategies and reasons for improving workers' abilities to handle work demands, including strategies to increase resources such as autonomy (e.g., job crafting) and support (e.g., various types of supportive supervisor training).

When you are finished reading this chapter, you should be able to:

LO 6.1: Define work-related stress and recovery phenomena in terms of demands and resources.

LO 6.2: Describe how stress and recovery are components to a shared and ongoing cycle or process.

LO 6.3: Explain why it is important to improve workers' stress and recovery management abilities.

LO 6.4: Explain general forms of intervention to address stress and recovery.

Overview of Work-Related Stress and Recovery

Work-related stress (and associated recovery) collectively represent the largest area of research within the Occupational Health Psychology (OHP) domain (Bennett et al., 2018; Houdmont et al., 2008; Sonnentag & Frese, 2012). This makes sense, given that the theories, models, frameworks, and methods used in work-related stress and recovery research also are used to explain and address most other WHSWB phenomena. Our goal in this orienting chapter is to present an overview of work-related stress and recovery that supports our more focused discussion of specific work-related stressors, demands, and resources in Chapters 7 through 10.

Essential Concepts and Terminology

Stress is studied in a variety of professional disciplines including engineering, medicine, and psychology. Across these domains, stress is generally considered as a force that operates on an object or, in the OHP domain, a worker or group of workers. Several essential concepts and terms need to be understood to facilitate clear communication about these matters. Most of these key stress and recovery terms and concepts emerged from early thinking about stress as a stimulus and response phenomenon, as outlined in the general OHP framework we introduced in Chapter 2.

Stressors, Demands, and Resources

Stressors are the stimuli that workers sense, perceive, and appraise, and which ultimately trigger some sort of response. These stimuli are often assumed to be external to the person (e.g., workload, interpersonal conflict, environmental exposures), but they can also be internal (e.g., self-doubt). These various forms of stressors can also be seen as different types of *demands*, as we discuss in the following chapters. When we are exposed to demands, we experience a state of discrepancy in which our perceived or actual reality does not match our desired homeostatic reality. We respond to address these discrepancies with some form of adaptation or action. By definition, demands are "the requirement of work or of the expenditure of a resource" (Merriam-Webster, n.d.a). This mirrors the way OHP professionals commonly conceptualize work experiences, as requiring constant juggling or balancing of work-related demands and the *resources* (i.e., a source of supply or support; an ability to meet and handle a situation; Merriam-Webster, n.d.b) needed to meet those demands. Use of these terms also illustrates why efforts to somehow vanquish the "enemy" of work-related stressors are often misguided. Work-related demands are intimately linked to worker motivation (e.g., LePine et al., 2004; Lucas et al., 2004), so a more rational objective is to work toward minimizing workers' risks of chronic demand overload, and maximizing workers' access to necessary resources.

Nearly any work environment characteristic or experience can impose a demand on a worker if exposure to it (or the absence of it) requires the worker to adaptively respond. Even the absence of normally helpful work-related resources (e.g., support) can function as demands. When focusing on work-related stress, OHP professionals tend to target a variety of defined and measurable (i.e., operationalized) demand constructs. A relatively simple classification of such demands is summarized in Table 6.1; we discuss these forms of demands and associated resources in subsequent chapters. Note that many of the example stressors or demands listed in this table can take a variety of forms (e.g., be both psychological and physical/environmental in form); we have attempted to summarize them here as they are predominantly studied and addressed by OHP professionals.

130

Table 6.1 General Forms of Work-Related Demands

Stressor/ Demand Form	Examples
Psychological	Cognitive, emotional, physiological, and behavioral states or conditions that differ from personal homeostasis or preferred level or form:
	• Perceived insecurity (Greenhalgh & Rosenblatt, 1984) • Low perceived self-efficacy (Schwarzer & Hallum, 2008) • Emotional labor (Liu et al., 2008) • Lack of control or autonomy in work (Spector, 1986) • Constraints (Pindek & Spector, 2016) • Workload (i.e., too much, too little, too difficult work; Meijman & Mulder, 1998)
Social	States of discomfort that originate from the (in)action of others or perceived relational expectations of oneself:
	• Responsibility for others (Hurrell & McLaney, 1988) • Inadequate support from colleagues (Pejtersen et al., 2010) • Negative interpersonal interactions (Spector & Jex, 1998)
Role-Related	Conflicting information or inconsistent demands concerning one's work:
	• Role ambiguity and conflict (Kahn et al., 1964; King & King, 1990)
Physical and Environmental	Factors in the work environment that impact worker functioning and more generally WHSWB:
	• Noise (McBain, 1961) • Poor air quality (Wargocki & Wyon, 2017) • Extreme temperatures (Pilcher et al., 2002) • Lighting (Lamb & Kwok, 2016) • Poor demands-abilities fit (Park et al., 2012)

Stress

Stress is the general psychological and physiological arousal experienced in response to one or more demands. If the subsequent cognitive, emotional, and behavioral response it triggers successfully addresses, reduces, or removes the demand, then this stress experience decreases in intensity. Although similar, stress and *anxiety* (discussed more in Chapter 4) are not identical phenomena. Stress is often a response to a particular demand or threat, while anxiety tends to be a more persistent reaction to a stress experience. Both conditions can include rumination (i.e., ongoing cognitive and emotional processing) and chronic stress experiences can make this worse (e.g., Michl et al., 2013). While many OHP professionals are engaged in helping workers with issues of work-related stress experiences, persistent difficulties with anxiety are more commonly addressed by clinically trained psychological and mental health professionals.

Stress affects workers physiologically and psychologically, and not always in negative ways. The physiological arousal associated with stress can energize and arouse our senses and ability to process information. Stress can also motivate us at a psychological and behavioral level, functioning a little bit like a natural stimulant. Going a bit deeper, stress experiences trigger the activation of the body's autonomic nervous system. This includes activating the sympathetic nervous system, which is responsible for arousing our body systems to prepare for action (e.g., increasing heart rate, focusing our attention, increasing blood pressure; Nixon et al., 2011; Sapolsky, 1998). This can be adaptive when facing legitimate physical threats (e.g., if you are in a fight, you want to turn your attention toward the greatest threat), but most work-related stressors do not present clear-cut threats to our survival and the associated response can be more damaging than adaptive. This is especially true for the psychological and social stressors that we need to perceive and appraise before they become an arousing stimulus (e.g., a harsh email, last-minute report request from your manager). When a stress experience is passing, the parasympathetic nervous system typically activates, allowing us to begin the process of calming and recovery (e.g., Tugade & Fredrickson, 2004). Over time, there is evidence that exposure to work-related demands can impair the body's ability to adaptively respond, as shown by reduced vagal nerve control of the heart, which has been linked to increased risk of disease and general poor health (e.g., Jarczok et al., 2013).

A couple of widely researched theories help to illustrate common scenarios in which workers are likely to experience stress. We review a few of those theories here and explore additional theories that more clearly incorporate the role of resources later. First, the *effort-reward imbalance* theory is often used to help explain work-related scenarios in which workers are likely to experience a stress reaction due to a perceived imbalance or discrepancy between the level of effort they are putting into their work and the level, form, and/or value of the reward they are getting in return (e.g., Siegrist, 2002). In a work context, perceived effort is linked to imposed work-related demands, while perceived reward (or lack thereof) can be seen as an indication of how meeting that demand will result in resource gain (or loss).

Second, it can also be helpful to conceptualize work-related stress as arising when workers perceive a mismatch or poor "fit" between who they are and what they can do, and what a particular work situation requires. This can happen when workers are not properly screened, developed, or supported so they can actually meet their work-related demands. This concept of *person-environment fit* (in a variety of forms) has a long history of study in the fields of industrial and organizational psychology and OHP (Edwards, 2008; Hoffman & Woehr, 2006; Kristof-Brown et al., 2005). Poor person-environment fit at work imposes major demands on workers because every time they are in that environment, it feels wrong and difficult, no matter what resources they may leverage. In contrast, when workers fit their environments well, there is a real opportunity for work to be resource replenishing instead of only resource depleting.

Strain

When our response to stress experiences is ineffective or depletes us of resources, we may experience *strain*. Strain can take many forms (e.g., cognitive, emotional, physical, behavioral) and can range in severity from not so bad (e.g., headache, light fatigue) to very serious (e.g., burnout, depression). As noted earlier, stress-related arousal can be adaptive, but it also is associated with a variety of other less-positive consequences. At a cognitive level, stress tends to focus us on the source of the stressors (even if we do not really know the exact source). Such stress-induced attentional narrowing may limit worker ability to think broadly (Prinet & Sarter, 2016; Wichary et al., 2016). This can be beneficial in certain types of high-vigilance work situations (e.g., air traffic controllers, emergency dispatchers), but detrimental to performance in occupations that require creativity and innovation, holistic thinking, and open-mindedness. Emotionally, workers dealing with stress are less likely to regulate and respond well emotionally (Raio et al., 2013). In addition, severe strain consequences such as burnout have been significantly linked with reduced empathy for others (Wilkinson et al., 2017).

Behaviorally, work-related stress is linked to basically all the "bad" and problematic behaviors that are generally not good for work or WHSWB (e.g., withdrawal, presenteeism, absenteeism, poor nutritional choices, chronic pain, and drug use). Work-related stress is not the only cause of these behaviors, but this linkage is important not to ignore because all of these types of behaviors tend to lead to additional demands and reduced resources, potentially exacerbating or perpetuating workers' stress experiences (e.g., Allen & Armstrong, 2006; Clark et al., 2011; Clarke, 2012). Behavioral reactions to stress are especially problematic because they tend to build on each other (e.g., regularly going out for drinks after a hard day of work can turn into a socially supported addiction).

We discussed earlier that stress triggers physiological and biological responses that can be adaptive to some extent, but harmful when over-activated (e.g., Sapolsky, 1998). Some of these nuances are further explained in the *allostatic load model* (Juster et al., 2010; McEwen, 1998), which describes how our stress responses can impose a damaging load on our body's systems. Our immediate or primary responses to stressors (e.g., elevated heart rate, cortisol release) can result in lasting secondary responses or health effects (e.g., high blood pressure) and tertiary responses or disease endpoints if stress experiences occur too frequently or for extended periods of time. Research demonstrates linkages between work-related stressors or demands and these types of stress responses (Ganster & Rosen, 2013), including health complaints like headaches, gastrointestinal problems, and sleep disturbances (Nixon et al., 2011). There is also evidence of rather immediate connections between daily stressors and emotional and physiological responses (Ilies et al., 2016), as well as strong evidence that the effects of chronic stressors on health seem to accumulate over time (Ford et al., 2014). Chronic stress experiences is also linked to changes in our brains (Lucas et al., 2004; McEwen, 2000) and even cells, predisposing our bodies to react more quickly and more strongly to future stressors (e.g., McVicar et al., 2014). Understanding even the basic physiology and biology associated with stress is essential to appreciating the close connection between work-related stress and recovery. As you are hopefully seeing by now, responding to strains reactively is potentially more difficult than trying to proactively optimize workers' demand and resource situation to minimize chronic stress reactions.

Extending the Stress Process to Include Recovery

From an OHP perspective, *recovery* is the process by which we can prevent strain by replenishing resources that we deplete when responding to demands. In this way, recovery can offset effects of work-related stress experiences and help workers to stay engaged, be proactive, and generally perform better at work (Sonnentag, 2003). Work-related stress and recovery are fascinating phenomena in and of themselves, but there is greater power and utility for OHP professionals when we recognize that both phenomena can be seen as components to a single, ongoing cycle (represented in Figure 6.1; e.g., van Hooff et al., 2011). This perspective helps us understand workers' continual challenge of managing work-related demands and resources, and can facilitate the development of helpful strategies and interventions to make this more possible. As already noted, because many effects of stress are cumulative (e.g., Godin et al., 2005) and do not dissolve or disappear quickly, workers' abilities to respond adaptively and effectively to today's demands are influenced by how (un)successful they were at responding to preceding demands. A practical implication is that understanding workers' perceived demands and resource levels can improve our ability to better direct and management worker effort, while also protecting WHSWB.

It is so important to understand recovery as much as we understand stress because workers typically have more control over themselves, their lifestyle

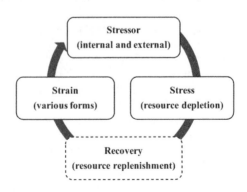

Figure 6.1 The Stress and Recovery Cycle

choices, their routines, etc. than they do over their work-related demands. Given this reality, workers are almost guaranteed to experience stress at work and ultimate strain, unless they are regularly engaging in recovery. The dominant theories supporting stress and recovery research and practice all implicitly recognize that recovery takes time (Fritz & Ellis, 2015) and needs to happen, else the effects of stress build to unsustainable levels and longer-term consequences. These theories also emphasize that resources as essential to workers' abilities to function. For example, the *effort-recovery model* (Meijman & Mulder, 1998) emphasizes a need to disconnect from or otherwise stop exerting effort in response to work-related demands to replenish spent resources when a period of resource expenditure has ended. Within this model, the main sources of demands are linked to work assignments and environmental factors, including one's social relationships. A related perspective is emphasized in the *stressor-detachment model* (Sonnentag, Kuttler, et al., 2010), which highlights how workers' reactions to stress (i.e., "strain reactions") often persist even after a worker's exposure to a demand is minimized. Because of this, these strain reactions can persist and even build on each other over time, impacting worker physical and psychological health unless there is physical and psychological detachment from work (Sonnentag & Fritz, 2015).

The goal for most interventions in this area becomes teaching workers how to recover as well as possible. This seems like a simple thing to do, but it is not because workers differ in terms of their demands and corresponding resource needs. Much remains to be learned about how workers can optimally recover depleted resources. A general principle is that not all recovery activities are equally valuable. For example, active forms of recovery (i.e., requiring some level of resource investment) may do a better job of replenishing resources than passive forms of recovery (Rook & Zijlstra, 2006; Sonnentag & Jelden, 2009). A more commonly applied perspective on recovery suggests that there are four general elements to activities that contribute to resource replenishment (Sonnentag & Fritz, 2007).

Psychological detachment involves distancing oneself and disconnecting from work-related demands (Sonnentag, Binnewies, et al., 2010; Sonnentag & Fritz, 2015). In-line with the Stressor-Detachment model, psychological detachment may help workers recover from negative effects of high work-related demands and generally be good for worker well-being (Sonnentag & Bayer, 2005; Sonnentag, Binnewies, et al., 2010). Lack of detachment (or inability to detach) also seems to be particularly detrimental to overall well-being (Sonnentag & Fritz, 2007). *Mastery experiences* are a second element to recovery that can facilitate detachment from work, as these experiences are generally linked with activities that require intensive focus, challenge, and learning (e.g., engaging in competitive sport, learning a new language, accomplishing a personal nonwork goal; Sonnentag & Fritz, 2007). Mastery experiences facilitate development of new competencies, self-efficacy, and other resources (Bandura, 1997; Hobfoll & Lilly, 1993; Sonnentag & Fritz, 2007). Both mastery and detachment experiences do not happen on their own; workers typically have to invest some resources to make these experiences possible – this is a main reason for a major paradox associated with recovery: Those who need recovery most often are least able to optimally achieve it.

Control is a third main element to recovery of resources and involves workers having the ability to exercise some degree of choice over what they do and when they do it (Sonnentag & Fritz, 2007). The positive psychological effects of control are well-understood in general (Skinner, 1996) and within work environments (e.g., Melamed et al., 1991; Schat & Kelloway, 2000). Finally, the fourth main element to recovery is *relaxation*, which is generally seen as an experience made possible by engaging in a recovery activity that the individual controls to some degree and in which there are minimal demands of any kind imposed upon them (Sonnentag & Fritz, 2007; Tinsley & Eldredge, 1995). Relaxation is important to recovery, because it seems to boost positive affect and reduce the effects of negative emotions that may be tied to work-related stress experiences (Fredrickson et al., 2000). In terms of the effort-recovery model noted earlier, relaxation is essential to facilitating workers' return to their pre-stress homeostasis state.

As our knowledge about work-related stress and recovery continues to develop, it is also important to challenge and expand the way we think about resource replenishing (i.e., recovery) experiences. To illustrate, consider the essential recovery element of detachment. Although we typically think of detachment as the result of a decision to separate from work, it can also be facilitated by purposeful and strategic exposure to restorative environments or stimuli. In particular, exposure to nature has been shown to be impactful (Bowler et al., 2010; Gritzka et al., 2020). *Attention restoration theory* (Berto, 2014; Kaplan, 1995) helps explain how and why this is true, expanding on early ideas from James (1892) regarding the concept of *involuntary attention*, which occurs when we experience moments of fascination or awe (e.g., surrounded by nature, standing in a cathedral). In these moments, no matter how resource-depleted we may be, we automatically find ourselves focusing on where and how we are. This form of attention does not require the expenditure of additional attentional resources and could be as simple as looking at images of something beautiful or taking a walk in a park. These moments can inspire and help us to detach, find calm, and restore depleted resources. Imagine what could happen if we all experienced more daily moments of fascination?

Understanding Demands and Resources in the Work Environment

Many theoretical perspectives guide OHP research and practice pertaining to work-related stress and recovery. These perspectives generally emphasize the importance of alignment between the demands imposed upon workers and the resources that are available to meet these demands. Various contextual and personal factors are also considered in many of the most dominant stress and recovery theories, which we summarize in this section.

Regulating Demand and Resource Alignment

Responding to demands requires workers to leverage their available resources. This makes understanding the concept of a balanced demand-resource relationship essential to protecting and improving WHSWB. As workers deplete their available resources, they become less able to respond effectively to future demands and they develop a need for recovery (e.g., Sonnentag, Kuttler, et al., 2010). When such needs are not met, there can be negative effects on worker behavior and well-being (Sonnentag, 2001; Sonnentag & Fritz, 2015). It is easy to feel daunted by the complex and person-specific nature of stress and recovery. We often find it helpful in our own research and practice efforts to approach stress and recovery challenges from a resource alignment perspective (represented in Figure 6.2). Focusing on specific work demands makes it possible to help workers and organizations identify and obtain specific and corresponding resources that are needed to meet those demands.

Work-related resources can take a variety of forms. One resource framework that corresponds well with most dominant theories in this domain, positions resources in terms of their source (personal or contextual) and longevity (short-term or long-lasting; Hobfoll, 2002; Ten Brummelhuis & Bakker, 2012). *Personal resources* (e.g., cognitive abilities, energy, traits) emanate from and are maintained within the person, while *contextual resources* (e.g., financial stability, social support, adequate shelter) are influenced by external factors. In terms of longevity, short-term or *volatile resources* (e.g., time, energy, attention) are

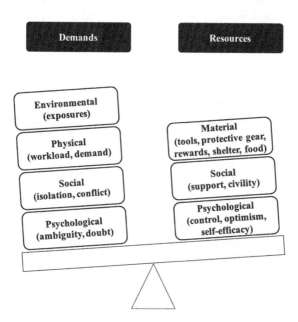

Figure 6.2 Conceptual Representation of Demand-Resource Relationship

easily and quickly used up or depleted and cannot be directly replenished. In contrast, longer-lasting or *structural resources* (e.g., shelter, social connections, technology) are more stable and can be replenished and used again (Ten Brummelhuis & Bakker, 2012). There is likely no single optimal blend of specific resources or even types of resources that ensure resilience for all workers. Instead, workers need access to a broad foundational set of psychological, social, and material resources to respond effectively to day-to-day work and nonwork demands. This assertion is clarified and developed further by several major theories in this domain, which we discuss in the following subsections.

Job Demands and Control

A simple, yet powerful perspective on demands and resources comes from Karasek's (1979) *job demands and control (D-C) model* (see also Wall et al., 1996). In our own work, the core concepts of this model have repeatedly provided a framework and structure that makes sense even to individuals who have no background or interest in applied psychology. The main D-C model focuses attention on the relatively easy-to-understand factors of how workers perceive and experience (a) work-related demands and (b) control or latitude to decide how they will address those demands. According to this model, workers are better able to handle demands when they have at least some degree of control over how they can meet those demands. In contrast, workers struggle in situations characterized by high work demands and low control.

Research involving this D-C model has identified factors that condition or moderate the relationship between work demands and the resource of control. The two most notable conditioning factors are presence of needed social support (e.g., Brough et al., 2018; Van Yperen & Hagedoorn, 2003) and the extent to which workers believe they are capable of doing what is necessary to meet demands (e.g., self-efficacy; Salanova et al., 2002; Schaubroeck & Merritt, 1997). These and potentially other conditioning factors can help to explain why more control and fewer demands is not always the optimal blend for all workers in all work contexts. Setting the stage for the next two theories we discuss, workers who are generally more resource rich may be able to maintain a stronger sense of control even in high demand situations than workers who are resource poor.

Conservation of Resources Theory

The *conservation of resources (COR) theory* (Hobfoll, 1989) provides a broadly applicable set of tenets and corollaries that help us understand how people experience stress as a function of real or impending resource loss, as well as how recovery occurs through replenishing depleted resources. Within this theory, resources can take many forms, including objects, personal characteristics, conditions, and energies (Hobfoll, 1989). The underlying logic of COR is closely aligned with Lazarus and Folkman's (1984) definition of stress as a psychological and physiological reaction to the overtaxing of one's available resources and with many other resource-related models that have been posited and tested over the years (see Hobfoll, 2002 for a summary). The utility of COR theory is not limited to work environments, making it very popular with OHP professionals working to understand and address domain spanning WHSWB issues involving stress and recovery.

COR can also help to explain individual differences in resource availability and resource-related resilience, through the concept of *resource caravans* (i.e., the notion that resources build on and exist with each other). An implication here is that those who have more will generally continue to have more, while those who have less, will continue to have less (Hobfoll, 2002, 2011). As an example, when we secure a good job that provides sufficient income, this often also engenders a sense of confidence and positive emotions. In contrast, losing employment tends to trigger multiple negative consequences (i.e., *loss spirals* in COR) linked to your inability to meet financial obligations, a loss of personal confidence, and difficulty maintaining positive emotions.

Job Demands and Resources Theory

The *job demands and resources (JD-R) theory* (Bakker & Demerouti, 2007) has many elements that are similar to COR, but its specific focus is the work domain. The JD-R theory highlights, very importantly, that worker ability to manage demands comes from resources that are not just person-specific, but also job-specific (e.g., support, time, technology, materials, healthy environment). Both JD-R and COR illustrate how such resources can neither come only from within the worker, nor solely from the organization. The JD-R model has been effectively leveraged to help explain how and when workers experience positive states, like motivation and engagement, and negative outcomes, such as worker burnout (Bakker, 2015; Bakker & Demerouti, 2017; Bakker et al., 2014; Lesener et al., 2018). JD-R theory also has relevance to work-related stress and recovery (e.g., Sonnentag et al., 2012; Tadić et al., 2015). COR and JD-R theory clearly emphasize that protecting WHSWB requires ensuring workers have the resources they need to meet their work-related demands.

The Importance of Appraisals and Context

If you have completed an introductory psychology course, you probably learned that most stimuli only take on meaning when they are perceived and cognitively appraised or evaluated. This is a dominant psychological perspective on how the work stress process operates (Harris & Daniels, 2007; Kinman & Jones, 2007). One of the earliest and most widely cited general stress-related theories along these lines is the transactional theory of Lazarus and colleagues (Folkman et al., 1986; Lazarus & Folkman, 1984). This theory outlines a process by which exposure to stimuli may lead to a stress response through a two-step appraisal process, through which we perceive a stressor and first evaluate the extent to which we are threatened (i.e., primary appraisal), and then second evaluate and determine what we can do in response (i.e., secondary appraisal). Applying this framework, a work demand is stressful when we see it as threatening our resources and when we do not feel we have sufficient resources to adequately respond.

Over time, this concept of appraisal has been revisited and expanded to include the appraisal of stressors in both positive and negative ways (Gerich, 2017; Podsakoff et al., 2007; Tuckey et al., 2015). It is intriguing to consider the possibility that some stressors may be less bad, maybe even positively perceived at times. However, it is important to recognize that our body's response to stress is nonspecific and relatively uniform; physiologically speaking, there is no healthy form of stress (French et al., 2019; Mazzola & Disselhorst, 2019; Rosen et al., 2020).

How we perceive and appraise work-related demands, and how we ultimately respond to them is also linked to social norms around stress in our work environments. There is increasing recognition of the role that social environments, norms, and shared social identities play in determining the stressors we perceive and appraise, and the stress experiences we ultimately have (Haslam & Reicher, 2006; Haslam & van Dick, 2011). Related to all of this is the normalization of stress and even the elevation of stress into something to be prized or treated as a status symbol. We have both observed (and researched) this tendency among some workers to try to frame work-related stress as a sort of badge of honor (Black & Britt, 2018). It is also true that some workers may acknowledge a need for recovery, but believe that they do not have the opportunity or ability to respond to this need. The weight of their work-related stress may even make some workers feel guilty for taking time to attempt to recover (i.e., relaxation remorse), further negatively impacting their health (Black & Britt, 2018). This is an important area for more research and practice, given the power of social norms over how we perceive and respond to work demands.

An Integrated Stress and Recovery Framework

Summarizing the information presented up to this point, the work-related stress and recovery process can be seen as an expanded stimulus-response process similar to what we outlined in Chapter 2. Figure 6.3 provides a high-level conceptual representation of a broad OHP perspective on work-related stress and recovery as a single, cyclical process. Workers are exposed to stressors or demands (stimuli). Some stressors have a direct influence, while others are processed cognitively and emotionally before triggering our psychological and physiological stress response. This figure also highlights ways in which work demands can impact our personal and contextual resources, and also our perceptions and responses to further work demands. This type of broad conceptual model can help to connect many of the seemingly disparate elements of the vast OHP literature pertaining to stress and recovery. We believe this type of model can also support practical thinking about ways to address workers' stress and recovery needs.

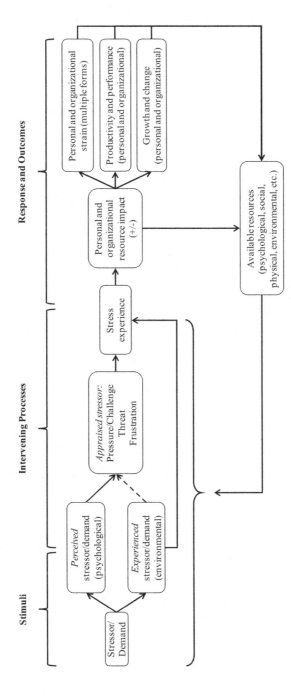

Figure 6.3 Expanded Conceptual Model of Work-Related Stress and Recovery Process

Why Work-Related Stress and Recovery Matter

There are so many reasons why work-related stress and recovery need to be understood and managed well by workers and within organizations. In the following subsections, we discuss a couple of the most compelling general reasons. Additional reasons pertaining to specific forms of demands and resources are explored in subsequent chapters.

Self-Care is Essential

Understanding the work-related stress and recovery cycle enables OHP professionals to help workers protect and replenish the resources needed to meet work-related demands. This is accomplished through *self-care* or recovery practices. For most adults, working and work environments constitute the main, controllable source of demands, as well as a major source of opportunities to build and replenish essential resources. Surveys of populations of working adults routinely indicate now that most workers are regularly experiencing stress associated with their work (American Psychological Association, 2020). Even more troubling, is that one in five working adults may never engage in activities to manage this stress (American Psychological Association, 2015).

Absent or insufficient self-care can have implications that extend far beyond any individual worker. In many occupations, the consequences of not performing at optimal levels can be critical. In healthcare, for example, suboptimal performance could even lead to someone's death. This reality is what makes self-care an absolute responsibility and necessity as outlined by many professions' ethics codes, graduate education requirements, and licensing laws. Unfortunately, no such guidance exists for workers and organizations in more general business and manufacturing contexts. This means that it is often up to organizational leaders to emphasize the importance of adequate self-care as a personal and occupational necessity for their workers.

Business Implications

As noted earlier, the omnipresence of demands at work should not surprise or necessarily concern us – if most workers reported experiencing no demands and no stress at work, we would expect to see numerous other serious problems emerge with worker motivation, engagement, and other important indicators of positive worker and organization alignment. The challenge here is ensuring that workers' normal state of existence at work is not one of chronic psychological and physiological stress arousal. Workers' personal demand-resource equation can only balance with personal and organizational efforts to ensure all necessary resources are available when needed.

This all becomes a business-level concern due to cost-related ripple effects triggered by consequences of workers' chronic or acute stress experiences, which are exacerbated by the absence of necessary resources and ineffective or nonexistent recovery. Organizations can experience costs associated with work-related stress that are both indirect and direct in form. Direct costs are typically associated with negative worker attitudes and performance, and increased numbers of accidents and healthcare-related expenditures. Indirect costs often include less tangible effects on workers (e.g., reduced energy, poor communication, negative relationship quality; Hargrove et al., 2011; Macik-Frey et al., 2007). Ensuring workers have the resources needed to address work-related demands is not just about cost-avoidance; it is really a matter of good human resource and talent management.

Methodological Considerations and Practical Recommendations

Work-related stress and recovery are person-level phenomena that can also impact groups and organizations. When researching or attempting to intervene to address stress and recovery challenges, OHP professionals must make several methodological and measurement decisions. These decisions are influenced by the types of stressors, demands, resources, and outcomes involved in a particular research or intervention scenario. Given this reality, we save our discussion of more specific methodologies and intervention strategies for our discussion of specific demands and resources in the following chapters. Here, however, we present a few general recommendations for working with stress and recovery.

Measuring and Monitoring Work-Related Stress and Recovery

Work-related stress and recovery are not singular constructs that can be easily measured or monitored with a single scale or measure at a single point in time. With respect to stress, a determination has to be made about whether the measurement target is the stress arousal or experience itself, or rather the extent to which exposure is happening or is perceived as happening. Alternatively, maybe the focus is on how workers are appraising and choosing to respond to specific stressors. When it comes to recovery, direct measurement of the extent to which a person is "recovered" is really difficult and typically indirect. Generally, what is measured is some form of need for recovery (e.g., de Croon et al., 2006) or the extent to which a person perceives that they have sufficiently replenished depleted resources.

The personal nature of both of these phenomena is what makes most measurement strategies pertaining to work-related stress and recovery focus on the self-reported perceptions of feelings and energy levels (Giumetti et al., 2013; Zacher et al., 2014), among other related phenomena. Typically such data are gathered using cross-sectional surveys, though if the goal is to more fully capture stress and recovery as part of the cyclical process described in this chapter, a repeated measurement design or longitudinal data measurement and monitoring methodology can be employed (e.g., diary studies, experience sampling; Sonnentag et al., 2013). There may also be opportunities to branch out a bit from strict self-report and engage in some form of more observational data gathering when trying to understand stress and recovery. This is particularly true now that wearable tracking technology is now readily available that can monitor physiological arousal (as we discussed in Chapter 2). A related measurement strategy is to gather physiological and biological indicators of stress as a way of corroborating self-reported experiences and obtaining more objective data regarding stress levels in a group or organization. Such efforts might include measurement and tracking of cortisol, hormones, neurotransmitters, etc. that are known to increase (or decrease) when individuals are experiencing stress or in need of recovery (Carrasco & Van de Kar, 2003; Ganster & Rosen, 2013; Heckenberg et al., 2020).

Intervening to Address Work-Related Stress and Recovery

Although there is still much to learn about work-related stress and recovery, OHP professionals are not prevented from leveraging the broad theory and knowledge we do have to positively impact WHSWB with evidence-based intervention efforts. In the following subsections, we outline a few generally important points to remember when intervening in this area at the person, group, leader, and organization levels.

Strategies for Individuals

When intervening at this level, it is important to note that implementing an intervention can itself introduce a new work-related stressor or set of demands. Related to the dual appraisal concept of stress and basic underlying physiological realities of stress discussed earlier in this chapter, Boyatzis and colleagues have put forth *the intentional change theory* (e.g., Boyatzis & McKee, 2006; Boyatzis & Jack, 2018; Jack et al., 2013). Leveraging this theory, we can minimize the risk of triggering a counterproductive stress reaction among workers by not leading into an intervention with lots of data, a strong problem statement, or heavy educational material that demand workers' time and energy (resources that are already likely seriously depleted). Instead, we can approach workers with a coaching mentality (e.g., McGonagle et al., 2014), initiating an intervention with discussion about what workers value and what their health-related or more general life goals may be. This approach is more likely to trigger positive emotions and help workers remain open to the possibility of making adjustments that can help them move toward better alignment between their work experiences and their personal values and goals. More concretely, self-care and stressor/demand management are two general intervention strategies that can help individual workers better manage their work-related stress experiences and recovery needs.

IMPROVING SELF-CARE AND RESOURCE MANAGEMENT

Even if all of a worker's current work-related demands could be resolved, more will arise tomorrow. Because demands will always exist, workers need pragmatic strategies for recognizing and addressing their personal resource recovery needs. Similar to how workers in industrial settings are often taught how to not lift more than they safely can, all workers need to learn their personal resource limits. The problem is that many workers lack self-awareness regarding their personal demand and resource imbalance. This is one reason why mindfulness-based interventions are so common in this area (Howarth et al., 2019; Kabat-Zinn, 2003), for improving workers' self-awareness pertaining to what our minds and bodies are doing (or not doing).

However, it is one thing to improve workers' awareness of demands and resources, and another thing to develop personally relevant strategies for reducing exposure and enhancing resource replenishment. Mindfulness alone does not address both of these goals. There is also the possibility that trying to combat stress by increasing self-awareness may open workers up to perceiving more demands and also recognizing that they have even fewer resources than originally thought. Thus, mindfulness by itself may not always lead to the results we are seeking with OHP interventions. Instead, such techniques may need to be partnered with more specific strategies for building and actively managing resources.

Thankfully, our evidence base pertaining to this is growing rapidly, with some consistent findings about the qualities or elements of recovery and the types of environments that can be particularly restorative. As we also noted earlier in this chapter, recovery does not have to occur only in demand-free contexts or exclusively outside of working time. Responding to demands generally depletes workers' resources and can lead to strain, but at times it may also provide opportunities to replenish or build new resources (as illustrated in Figure 6.3). This is especially true if our responses to demands lead us to grow or change in a positive way, or to achieve or experience mastery in a particular domain or over a particular task. Thus, there is value in helping workers identify and take advantage of opportunities for recovery that may be present in the workplace, not just during nonwork time (Cranley et al., 2015; Henning et al., 1997; Hunter & Wu, 2016; Trougakos et al., 2013; Zacher et al., 2014). Opportunities for recovery both during and after the workday have the potential to support psychological health (Sonnentag et al., 2017). Research developments continue along these lines that may eventually make it possible to identify and target specific recovery experiences and strategies to replenish specific resources needed to meet specific work demands (e.g., Bennett et al., 2016; Tuckey et al., 2017).

STRESSOR/DEMAND MANAGEMENT AND STRESS COPING STRATEGIES

A second general form of intervention for workers can involve empowering workers to manage stressors and stress experiences that they can control. To be clear, what we are describing here really is more about stressor management and stress coping strategies, and less about managing our feelings of stress that happen in response to exposure to a stressor. The theory and evidence base in OHP supports a more targeted approach to managing stress than what is typically taught in high-level stress management workshops. Targeted approaches can help workers prepare themselves to handle anticipated and unanticipated stressor exposures.

Extending from the transactional model of stress noted earlier (Folkman et al., 1986; Lazarus & Folkman, 1984), personal coping responses to stress can vary. Active or problem-focused coping strategies enable workers to exert at least some control and potentially gain mastery over a particular demand. Sometimes, though, other forms of coping may be needed, even if just to temporarily reduce an emotional response to stress. For example, if directly addressing a stressor requires more resources than a person has available, it may be effective for the individual to temporarily pursue some form of resource recovery. A risk here is that pursuing recovery activities can turn into a form of avoidance coping. Socializing with friends, exercising, going for a walk are all potentially good recovery activities, but their overall value is reduced if doing these things is a way of not dealing with a demand that needs to be addressed. Also, there is the risk that avoidance may turn into withdrawal or more seriously negative and counterproductive psychological states if workers are chronically exposed to demands that exceed their resources (e.g., Podsakoff et al., 2007; Sliter et al., 2012). Workers can be helped by learning how to avoid getting to this point. There is also evidence that workers can learn to build resources and resilience through a process of self-reflection regarding how and why they respond as they do to specific stressors (Crane et al., 2019).

Strategies for Groups, Leaders, and Organizat-ions

Workers' stress and recovery experiences are often influenced by social norms pertaining to how work-related demands are perceived, appraised, and responded to within a particular work context. Interventions delivered at the workgroup, leadership, and broader organization level can all help to adjust and reinforce healthy social norms along these lines. In addition, there are many forms of intervention within organizational control that can directly improve workers' access to resources needed to address work-related demands. We explore three general forms of such interventions in this section.

MANAGE DEMANDS RELATIVE TO RESOURCES

Organizational leaders can help to ensure workers have access to the resources they need to meet work demands. This is an extension of an idea known as the "match hypothesis" outlined by Cohen and Wills (1985) when discussing the role that social support can play in protecting workers from the negative effects of work stress. The majority of OHP intervention efforts have focused on stress and recovery as person-level phenomena. At a more macro or organizational level, however, controlling demands or otherwise minimizing workers' exposures can be a direct and effective way of addressing the challenges outlined in this chapter. Sometimes these types of adjustments may involve taking steps that are controversial to different stakeholders at first, but potentially helpful over time (e.g., limiting work hours, restricting demands on workers outside of work, rotating worker exposure to different workloads).

FACILITATE JOB CRAFTING

There is growing evidence that workers are often better able to respond to work demands when they have some say or control over how their jobs operate and/or how they respond to such demands. This type of design-related worker autonomy is commonly known as *job crafting* (Wrzesniewski & Dutton, 2001). One reason that job crafting works for individuals and organizations is that it puts into practice what we understand about stress from the D-C model discussed earlier (Karasek, 1979). Specifically, when work demands are high, workers are more likely to handle them well if they at least perceive that they have some control over how they go about meeting those demands. Unfortunately, many managers regularly create imbalanced demand-control situations by forcing employees to do things that they are not prepared to do, that they are not supported to do, and they have no latitude to modify or craft into a form that works for them. A direct way to address these types of situations is to intervene at the leader and workgroup level to enable workers (when possible) to engage in job crafting and allow workers to process alternative ways of getting the work done. There is no single way to go about crafting a job, but key elements are to ensure that workers are supported in identifying their resources, strengths, values, and motivations, and then encouraged to consider these elements as they determine how they will approach meeting work-related demands going forward (e.g., Niessen et al., 2016). Job crafting has been shown to positively impact workers' ability to manage work demands and sustain positive well-being (e.g., Tims et al., 2013).

SUPPORTIVE POLICIES AND PRACTICES

Addressing work-related stress and recovery may also require interventions in the form of organization-wide policy and practice changes. Often, we think about this in terms of trying to grant workers more flexibility over their work schedules, encouraging supervisors to provide more support for utilizing such flexibility and taking time to recover. Keep in mind, however, that organizations can also implement policies and practices that are more extreme. Remember that the concept of an eight-hour workday is an organizational invention, from Henry Ford's well-known efforts to balance production against risk of injury in his assembly line factories in the early 1900s. It is important to keep in mind that limiting work demands, workload, and even work hours and generally supporting WHSWB are all strategies that can be managed through policies and practices. These strategies also can help to provide workers with at least the opportunity to engage in resource recovery.

Evaluating Work-Related Stress and Recovery Interventions

Methods and strategies for evaluating work-related stress and recovery interventions can take many forms and need to be designed to fit a specific intervention context and focus. We encourage you to revisit Chapter 2 for an overall discussion of generally relevant information along these lines. We also explore these evaluation challenges with respect to specific forms of OHP-related intervention efforts over the next few chapters in which we discuss specific work-related demands.

Concluding Thoughts and Reality Check

As discussed in this and the following chapters, a major WHSWB concern for OHP researchers and practitioners is how workers manage the complex and continuous balancing of work demands with the personal and work-related resources needed to meet those demands. Stress associated with working has become part of our common language and experience. We can see this so clearly in our everyday interactions, when a question of "How is work going?" is so often met with a response of, "Busy!" or "Stressful." At some level, all workers experience demands and therefore are likely to experience moments of stress arousal. What is less clear is whether such stress needs to be a chronic or in some cases continuous feature of workers' realities, or whether it is an unfortunate and at least somewhat preventable consequence of poor work design, poor management, misaligned organizational policies and practices, and poor person-environment fit.

Interrupting workers' stress cycle is difficult. Efforts to improve workers' stress and recovery experiences must be multi-faceted and multi-level, focused on re-balancing the demands-resources relationship using a combination of person, group, leader, and organization level strategies. These issues cannot be resolved just by increasing workers' personal awareness and resilience. Also required are modifications and improvements to actual work environments, to ensure workers have access to necessary resources and appropriate levels of demands.

The linkage between chronic stress arousal and a wide variety of psychological and physiological consequences elevates these concerns about work-related stress and recovery to the level of a global public health concern. Stress is not a new phenomenon in our modern world, but the types and frequency of our demands are continuously changing and there is increasing competition for our limited attention and time. This is especially true in work environments, where technology and social norms combine to keep workers nearly continuously engaged in demanding, resource-depleting activities. There is a tremendous opportunity for OHP research and practice to inform a new way of working and living, in which workers, organizations, and society are strengthened by maintaining a healthier demand and resource balance. Imagine the possibilities.

Media Resources

- Business-focused article exploring the link between work stress and worker mental health and well-being:
  https://www.entrepreneur.com/article/321451
- Article translating research that highlights the importance of detachment after work to the quality of sleep and health:
  https://www.forbes.com/sites/daviddisalvo/2018/04/26/science-just-served-up-a-great-reason-to-do-something-fun-after-work/#7fd2e7987b61
- Translational article summarizing research that suggests long-term effects of work stress on workers' mental health
  https://www.psychologytoday.com/us/blog/the-new-resilience/201811/midlife-work-stress-may-hurt-long-term-mental-health

---

Discussion Questions

1) What is stress and recovery, in the context of work?
2) Why and how does it make sense to conceptualize and work with stress and recovery as elements to the same broader process?
3) What are some work environment and task characteristics that likely influence the stress and recovery processes of workers?
4) Why does it make sense to talk about work stress and recovery in terms of demands and resources?
5) Is it an organizational responsibility to minimize worker stress or facilitate worker recovery? If not, how far does an organization's responsibility extend along those lines?

Professional Profile: M. Gloria Gonzalez-Morales, Ph.D.

Country/region: USA

Current position title: Associate Professor

Background: I have been working to improve the health, safety, and/or well-being of workers for 18 years. I got my Psychology Degree at Universidad de La Laguna (Canary Islands, Spain). I completed my dissertation, awarded with the European Ph.D. (Doctor Europaea), in 2006 at the University of Valencia (Spain) in the Work and Organizational Psychology Interuniversity Doctoral Program. In 2007, I moved to Virginia (USA) to do research as a Fulbright Visiting Scholar at George Mason University for two years. I was a postdoctoral researcher at the University of Delaware for a year before starting at the University of Guelph in May 2010. I recently moved to California as Associate Professor in Claremont Graduate University. I currently belong to the Society for Occupational Health Psychology, the Society for Industrial and Organizational Psychology, and am a member of the American Psychological Association.

I started my M.A./Ph.D. in Organizational Psychology thinking that I would end up being a Human Resources consultant. As I was working on my research as a graduate student I realized that as a Psychologist my main focus was to improve well-being, and I decided to do my research on work stress and gender. While finishing up my dissertation I was involved in a project with an insurance company to understand absenteeism. That project helped me understand the diversity of views about occupational health from management, HR professionals, health professionals, unions, and employees. It was clear to me that there was a need to integrate those views using Organizational Psychology and OHP.

I am a feminist university professor and I teach undergraduate and graduate students on I-O and OHP courses. I have also been the Executive Director of OMS (Organization & Management Solutions) that is a consulting group at the University of Guelph I-O Psychology program used to train graduate students in I-O Psychology Practice. As part of that role, I have been consulting for organizations during the last 10 years. This has helped me understand better my role as a scientist-practitioner, especially in the area of Occupational Health and Well-Being. I am the director of my lab, the Occupational Health and Positive Psychology lab, where I do research with graduate students and other colleagues on the disciplines of OHP and positive organizational psychology.

How my work impacts WHSWB: My research focuses on stress, work-life issues, workplace victimization, harassment and incivility, workplace diversity and positive organizational interventions to enhance well-being and performance. The end goal is to figure out ways to create and maintain healthy organizations in which workers are not only safe and healthy, but can also flourish and grow at work. I wish my research studies directly impacted more people (over and above the samples of participants who volunteer), but I like to think that my research papers are contributing to the pool of knowledge and resources that practitioners can use to impact workers. That is why most of my research is guided with practice in mind. OHP is an applied discipline, and as such, I do applied research that can be translated to the real world.

I recently received the best paper award from the *Journal of Occupational Health Psychology* for my paper (with Kernan, Becker, and Eisenberger, published in 2018, https://doi.org/10.1037/ocp0000061) on training supervisors to support subordinates and curbing abusive supervision. I wished I had had more opportunities like that one to start a research-practice project. We collaborated with a restaurant chain and they were very supportive of the research and the training we wanted to implement. For example, before starting the project, they allowed us to collect data from workers using a critical incident technique to understand the specific situations of that organization. We were able to use all that data to design a training based on examples that were customized to the organization.

We designed the training program from scratch using theory and previous empirical evidence to guide both the content, the learning outcomes and the learning activities. We delivered the training to supervisors and I learned so much by doing so, not only about the restaurant business, but also about the key role of supervisors in workers' well-being and how to support them to understand how important they are. I enjoyed every part of that project, especially because it was a hands-on experience on applying all my knowledge not only to the content but the delivery and the research aspects. The fact that the organization was on board was crucial and that speaks to the importance of buy-in from upper management. Without it, it is hard to make an impact.

My motivation: I am a Psychologist, and as such, well-being is my priority. As an Organizational Psychologist, I am motivated to promote well-being in the context of work. Work is what we devote most of our awake hours to. We should work to live, not live to work, and we should work in decent, safe, and healthy conditions. Moreover, we should be able to develop and grow during the activity that we invest most of our time and energy to.

Chapter References

Allen, T. A., & Armstrong, J. (2006). Further examination of the link between work-family conflict and physical health: The role of health-related behaviors. *American Behavioral Scientist, 49*(9), 1204–1221. https://doi.org/10.1177/0002764206286386

American Psychological Association. (2015). *Stress in America: Paying with our health.* https://www.apa.org/news/press/releases/stress/2014/stress-report.pdf

Association, A. P. (2020). *Stress in America TM 2020: A national mental health crisis.* https://www.apa.org/news/press/releases/stress/2020/sia-mental-health-crisis.pdf

Bakker, A. B. (2015). A Job Demands-Resources approach to public service motivation. *Public Administration Review, 75*(5), 723–732. https://doi.org/10.1111/puar.12388

Bakker, A. B., & Demerouti, E. (2007). The job demands-resources model: State of the art. *Journal of Managerial Psychology, 22*(3), 309–328. https://doi.org/10.1108/02683940710733115

Bakker, A. B., & Demerouti, E. (2017). Job demands-resources theory: Taking stock and looking forward. *Journal of Occupational Health Psychology, 22*(3), 273–285. https://doi.org/10.1037/ocp0000056

Bakker, A. B., Demerouti, E., & Sanz-Vergel, A. I. (2014). Burnout and Work Engagement: The JD-R Approach. *Annual Review of Organizational Psychology and Organizational Behavior, 1*(1), 389–411. https://doi.org/10.1146/annurev-orgpsych-031413-091235

Bandura, A. (1997). *Self-efficacy: The exercise of control.* New York: Freeman.

Bennett, A. A., Bakker, A. B., & Field, J. G. (2018). Recovery from work-related effort: A meta-analysis. *Journal of Organizational Behavior, 39*(3), 262–275. https://doi.org/10.1002/job.2217

Bennett, A. A., Gabriel, A. S., Calderwood, C., Dahling, J. J., & Trougakos, J. P. (2016). Better together? Examining profiles of employee recovery experiences. *Journal of Applied Psychology, 101*(12), 1635–1654. https://doi.org/10.1037/apl0000157

Berto, R. (2014). The role of nature in coping with psycho-physiological stress: A literature review on restorativeness. *Behavioral Sciences, 4*(4), 394–409. https://doi.org/10.3390/bs4040394

Black, K. J. & Britt, T.W. (2018, April). *Stress as a badge of honor: Relationships with health and performance.* In K. J. Black (Chair), Good for work, bad for life: Individual characteristics with divergent effects. 33rd Annual Conference of the Society for Industrial and Organizational Psychology, Chicago, IL.

Bowler, D. E., Buyung-Ali, L. M., Knight, T. M., & Pullin, A. S. (2010). A systematic review of evidence for the added benefits to health of exposure to natural environments. *BMC Public Health, 10,* 456. https://doi.org/10.1186/1471-2458-10-456

Boyatzis, R. E., & Jack, A. I. (2018). The neuroscience of coaching. *Consulting Psychology Journal: Practice and Research, 70*(1), 11–27. https://doi.org/10.1037/cpb0000095

Boyatzis, R., & McKee, A. (2006). Intentional change. *Journal of Organizational Excellence, 25*(3), 49–60. https://doi.org/10.1002/joe.20100

Brough, P., Drummond, S., & Biggs, A. (2018). Job support, coping, and control: Assessment of simultaneous impacts within the occupational stress process. *Journal of Occupational Health Psychology, 23*(2), 188–197. https://doi.org/10.1037/ocp0000074

Carrasco, G. A., & Van de Kar, L. D. (2003). Neuroendocrine pharmacology of stress. *European Journal of Pharmacology, 463*(1–3), 235–272. https://doi.org/10.1016/s0014-2999(03)01285-8

Clark, M. M., Warren, B. A., Hagen, P. T., Johnson, B. D., Jenkins, S. M., Werneburg, B. L., & Olsen, K. D. (2011). Stress level, health behaviors, and quality of life in employees joining a wellness center. *American Journal of Health Promotion*, 26(1), 21–25. https://doi.org/10.4278/ajhp.090821-QUAN-272

Clarke, S. (2012). The effect of challenge and hindrance stressors on safety behavior and safety outcomes: A meta-analysis. *Journal of Occupational Health Psychology*, 17(4), 387–397. https://doi.org/10.1037/a0029817

Cohen, S., & Wills, T. A. (1985). Stress, social support, and the buffering hypothesis. *Psychological Bulletin*, 98(2), 310–357. https://doi.org/10.1037/0033-2909.98.2.310

Crane, M. F., Searle, B. J., Kangas, M., & Nwiran, Y. (2019). How resilience is strengthened by exposure to stressors: The systematic self-reflection model of resilience strengthening. *Anxiety, Stress, & Coping*, 32(1), 1–17. https://doi.org/10.1080/10615806.2018.1506640

Cranley, N. M., Cunningham, C. J. L., & Panda, M. (2015). Understanding time use, stress and recovery practices among early career physicians: An exploratory study. *Psychology, Health & Medicine*, 21(3), 362–367. https://doi.org/10.1080/13548506.2015.1061675

de Croon, E. M., Sluiter, J. K., & Frings-Dresen, M. H. (2006). Psychometric properties of the Need for Recovery after work scale: test-retest reliability and sensitivity to detect change. *Occupational and Environmental Medicine*, 63(3), 202–206. https://doi.org/10.1136/oem.2004.018275

Edwards, J. R. (2008). Person-environment fit in organizations: An assessment of theoretical progress. *The Academy of Management Annals*, 2(1), 167–230. https://doi.org/10.1080/19416520802211503

Folkman, S., Lazarus, R. S., Dunkel-Schetter, C., DeLongis, A., & Gruen, R. J. (1986). Dynamics of a stressful encounter: cognitive appraisal, coping, and encounter outcomes. *Journal of Personality and Social Psychology*, 50(5), 992–1003. https://doi.org/10.1037//0022-3514.50.5.992

Ford, M. T., Matthews, R. A., Wooldridge, J. D., Mishra, V., Kakar, U. M., & Strahan, S. R. (2014). How do occupational stressor-strain effects vary with time? A review and meta-analysis of the relevance of time lags in longitudinal studies. *Work & Stress*, 28(1), 9–30. https://doi.org/10.1080/02678373.2013.877096

Fredrickson, B. L., Mancuso, R. A., Branigan, C., & Tugade, M. M. (2000). The undoing effect of positive emotions. *Motivation and Emotion*, 24(4), 237–258. https://doi.org/10.1023/A:1010796329158

French, K. A., Allen, T. D., & Henderson, T. G. (2019). Challenge and hindrance stressors and metabolic risk factors. *Journal of Occupational Health Psychology*, 24(3), 307–321. https://doi.org/10.1037/ocp0000138

Fritz, C., & Ellis, A. M. (2015). *A marathon, not a sprint: The benefits of taking time to recover from work demands* [White paper]. Society for Industrial and Organizational Psychology. https://www.siop.org/Portals/84/docs/White%20Papers/AMarathonNotASprint.pdf

Ganster, D. C., & Rosen, C. C. (2013). Work stress and employee health: A multidisciplinary review. *Journal of Management*, 39(5), 1085–1122. https://doi.org/10.1177/0149206313475815

Gerich, J. (2017). The relevance of challenge and hindrance appraisals of working conditions for employees' health. *International Journal of Stress Management*, 24(3), 270–292. https://doi.org/10.1037/str0000038

Giumetti, G. W., Hatfield, A. L., Scisco, J. L., Schroeder, A. N., Muth, E. R., & Kowalski, R. M. (2013). What a rude e-mail! Examining the differential effects of incivility versus support on mood, energy, engagement, and performance in an online context. *Journal of Occupational Health Psychology, 18*(3), 297–309. https://doi.org/10.1037/a0032851

Godin, I., Kittel, F., Coppieters, Y., & Siegrist, J. (2005). A prospective study of cumulative job stress in relation to mental health. *BMC Public Health, 5,* 67. https://doi.org/10.1186/1471-2458-5-67

Greenhalgh, L., & Rosenblatt, Z. (1984). Job insecurity: Toward conceptual clarity. *Academy of Management Review, 9*(3), 438–448. https://doi.org/10.5465/amr.1984.4279673

Gritzka, S., MacIntyre, T. E., Dorfel, D., Baker-Blanc, J. L., & Calogiuri, G. (2020). The effects of workplace nature-based interventions on the mental health and well-being of employees: A systematic review. *Frontiers in Psychiatry, 11,* 323. https://doi.org/10.3389/fpsyt.2020.00323

Hargrove, M. B., Quick, J. C., Nelson, D. L., & Quick, J. D. (2011). The theory of preventive stress management: A 33-year review and evaluation. *Stress and Health, 27,* 182–193. https://doi.org/10.1002/smi.1417

Harris, C., & Daniels, K. (2007). The role of appraisal-related beliefs in psychological well-being and physical symptom reporting. *European Journal of Work and Organizational Psychology, 16*(4), 407–431. https://doi.org/10.1080/13594320701506054

Haslam, S. A., & Reicher, S. (2006). Stressing the group: Social identity and the unfolding dynamics of responses to stress. *Journal of Applied Psychology, 91*(5), 1037–1052. https://doi.org/10.1037/0021-9010.91.5.1037

Haslam, S. A., & van Dick, R. (2011). A social identity approach to workplace stress. In D. De Cremer, R. van Dick, & J. K. Murnighan (Eds.), *Organization and management series: Social psychology and organizations* (pp. 325–352). Routledge/Taylor & Francis Group.

Heckenberg, R. A., Hale, M. W., Kent, S., & Wright, B. J. (2020). Empathy and job resources buffer the effect of higher job demands on increased salivary alpha amylase awakening responses in direct-care workers. *Behavioural Brain Research, 394,* 112826. https://doi.org/10.1016/j.bbr.2020.112826

Henning, R. A., Jacques, P., Kissel, G. V., Sullivan, A. B., & Alteras-Webb, S. M. (1997). Frequent short rest breaks from computer work: Effects on productivity and well-being at two field sites. *Ergonomics, 40*(1), 78–91. https://doi.org/10.1080/001401397188396

Hobfoll, S. E. (1989). Conservation of resources: A new attempt at conceptualizing stress. *American Psychologist, 44*(3), 513–524. https://doi.org/10.1037/0003-066x.44.3.513

Hobfoll, S. E. (2002). Social and psychological resources and adaptation. *Review of General Psychology, 6*(4), 307–324. https://doi.org/10.1037//1089-2680.6.4.307

Hobfoll, S. E. (2011). Conservation of resource caravans and engaged settings. *Journal of Occupational and Organizational Psychology, 84*(1), 116. https://doi.org/10.1111/j.2044-8325.2010.02016.x

Hobfoll, S. E., & Lilly, R. S. (1993). Resource conservation as a strategy for community psychology. *Journal of Community Psychology, 21*(2), 128–148. https://doi.org/10.1002/1520-6629(199304)21:2%3C128::AID-JCOP2290210206%3E3.0.CO;2-5

Hoffman, B. J., & Woehr, D. J. (2006). A quantitative review of the relationship between person–organization fit and behavioral outcomes. *Journal of Vocational Behavior, 68*(3), 389–399. https://doi.org/10.1016/j.jvb.2005.08.003

Houdmont, J., Leka, S., & Bulger, C. A. (2008). The definition of curriculum areas in occupational health psychology. In *Occupational Health Psychology: European Perspectives on Research, Education and Practice*. Nottingham University Press.

Howarth, A., Smith, J. G., Perkins-Porras, L., & Ussher, M. (2019). Effects of brief mindfulness-based interventions on health-related outcomes: A systematic review. *Mindfulness*. https://doi.org/10.1007/s12671-019-01163-1

Hunter, E. M., & Wu, C. (2016). Give me a better break: Choosing workday break activities to maximize resource recovery. *Journal of Applied Psychology, 101*(2), 302–311. https://doi.org/10.1037/apl0000045

Hurrell, J. J. Jr., & McLaney, M. A. (1988). Exposure to job stress-a new psychometric instrument. *Scandinavian Journal of Work, Environment & Health, 14*, 27–28.

Ilies, R., Aw, S. S. Y., & Lim, V. K. G. (2016). A naturalistic multilevel framework for studying transient and chronic effects of psychosocial work stressors on employee health and well-being. *Applied Psychology, 65*(2), 223–258. https://doi.org/10.1111/apps.12069

Jack, A. I., Boyatzis, R. E., Khawaja, M. S., Passarelli, A. M., & Leckie, R. L. (2013). Visioning in the brain: An fMRI study of inspirational coaching and mentoring. *Social Neuroscience, 8*(4), 369–384. https://doi.org/10.1080/17470919.2013.808259

James, W. (1982) *Psychology-briefer course*. New York: Henry Holt and Company

Jarczok, M. N., Jarczok, M., Mauss, D., Koenig, J., Li, J., Herr, R. M., & Thayer, J. F. (2013). Autonomic nervous system activity and workplace stressors-a systematic review. *Neuroscience and Biobehavioral Reviews, 37*(8), 1810–1823. https://doi.org/10.1016/j.neubiorev.2013.07.004

Juster, R. P., McEwen, B. S., & Lupien, S. J. (2010). Allostatic load biomarkers of chronic stress and impact on health and cognition. *Neuroscience & Biobehavioral Reviews, 35*(1), 2–16. https://doi.org/10.1016/j.neubiorev.2009.10.002

Kabat-Zinn, J. (2003). Mindfulness-based interventions in context: Past, present and future. *Clinical Psychology: Science and Practice, 12*(2), 144–156. https://doi.org/10.1093/clipsy.bpg016

Kahn, R. L., Wolfe, D. M., Quinn, R. P., Snoek, J. D., & Rosenthal, R. A. (1964). *Organizational stress: Studies in role conflict and ambiguity*. John Wiley.

Kaplan, S. (1995). The restorative benefits of nature: Toward an integrative framework. *Journal of Environmental Psychology, 15*(3), 169–182. https://doi.org/10.1016/0272-4944(95)90001-2

Karasek, R. A., Jr. (1979). Job demands, job decision latitude, and mental strain: Implications for job redesign. *Administrative Science Quarterly, 24*(2), 285–308. https://doi.org/10.2307/2392498

King, L. A., & King, D. W. (1990). Role conflict and role ambiguity: A critical assessment of construct validity. *Psychological Bulletin, 107*, 48–64. https://doi.org/10.1037/0033-2909.107.1.48

Kinman, G., & Jones, F. (2007). Lay representations of workplace stress: What do people really mean when they say they are stressed? *Work & Stress, 19*(2), 101–120. https://doi.org/10.1080/02678370500144831

Kristof-Brown, A. L., Zimmerman, R. D., & Johnson, E. C. (2005). Consequences of individuals' fit at work: A meta-analysis of person-job, person-organization, person-group, and person-supervisor fit. *Personnel Psychology, 58*(2), 281–342. https://doi.org/10.1111/j.1744-6570.2005.00672.x

Lamb, S., & Kwok, K. C. (2016). A longitudinal investigation of work environment stressors on the performance and wellbeing of office workers. *Applied Ergonomics, 52*, 104–111. https://doi.org/10.1016/j.apergo.2015.07.010

Lazarus, R. S., & Folkman, S. (1984). *Stress, appraisal, and coping.* Springer Publishing Company, Inc.

LePine, J. A., LePine, M. A., & Jackson, C. L. (2004). Challenge and hindrance stress: Relationships with exhaustion, motivation to learn, and learning performance. *Journal of Applied Psychology, 89*(5), 883–891. https://doi.org/10.1037/0021-9010.89.5.883

Lesener, T., Gusy, B., & Wolter, C. (2018). The job demands-resources model: A meta-analytic review of longitudinal studies. *Work & Stress, 33*(1), 76–103. https://doi.org/10.1080/02678373.2018.1529065

Liu, Y., Prati, L. M., Perrewe, P. L., & Ferris, G. R. (2008). The relationship between emotional resources and emotional labor: An exploratory study. *Journal of Applied Social Psychology, 38*(10), 2410–2439. https://doi.org/10.1111/j.1559-1816.2008.00398.x

Lucas, L. R., Celen, Z., Tamashiro, K. L., Blanchard, R. J., Blanchard, D. C., Markham, C., Sakai, R. R., & McEwen, B. S. (2004). Repeated exposure to social stress has long-term effects on indirect markers of dopaminergic activity in brain regions associated with motivated behavior. *Neuroscience, 124*(2), 449–457. https://doi.org/10.1016/j.neuroscience.2003.12.009

Macik-Frey, M., Quick, J. C., & Nelson, D. L. (2007). Advances in occupational health: From a stressful beginning to a positive future. *Journal of Management, 33*(6), 809–840. https://doi.org/10.1177/0149206307307634

Mazzola, J. J., & Disselhorst, R. (2019). Should we be "challenging" employees?: A critical review and meta-analysis of the challenge-hindrance model of stress. *Journal of Organizational Behavior, 40*(8), 949–961. https://doi.org/10.1002/job.2412

McBain, W. N. (1961). Noise, the arousal hypothesis, and monotonous work. *Journal of Applied Psychology, 45*(5), 309–317. https://doi.org/10.1037/h0049015

McEwen, B. S. (1998). Stress, adaptation, and disease: Allostasis and allostatic load. *Annals of the New York Academy of Sciences, 840*(1), 33–44. https://doi.org/10.1111/j.1749-6632.1998.tb09546.x

McEwen, B. S. (2000). The neurobiology of stress: From serendipity to clinical relevance. *Brain Research, 886*(1–2), 172–189. https://doi.org/10.1016/s0006-8993(00)02950-4

McGonagle, A. K., Beatty, J. E., & Joffe, R. (2014). Coaching for workers with chronic illness: Evaluating an intervention. *Journal of Occupational Health Psychology, 19*(3), 385–398. https://doi.org/10.1037/a0036601

McVicar, A., Ravalier, J. M., & Greenwood, C. (2014). Biology of stress revisited: intracellular mechanisms and the conceptualization of stress. *Stress and Health, 30*(4), 272–279. https://doi.org/10.1002/smi.2508

Meijman, T. F., & Mulder, G. (1998). Psychological aspects of workload. In P. J. D. Drength, H. Thierry, & C. J. de Wolff (Eds.), *Handbook of work and organizational psychology* (Vol. 2, pp. 5–33). Psychology Press.

Melamed, S., Kushnir, T., & Meir, E. I. (1991). Attenuating the impact of job demands: Additive and interactive effects of perceived control and social support. *Journal of Vocational Behavior, 39*(1), 40–53. https://doi.org/10.1016/0001-8791(91)90003-5

Merriam-Webster. (n.d. a). Demand. In *Merriam-Webster.com dictionary.* Retrieved October 29, 2020, from https://www.merriam-webster.com/dictionary/demand

Merriam-Webster. (n.d. b). Resource. In *Merriam-Webster.com dictionary.* Retrieved October 29, 2020, from https://www.merriam-webster.com/dictionary/resource

Michl, L. C., McLaughlin, K. A., Shepherd, K., & Nolen-Hoeksema, S. (2013). Rumination as a mechanism linking stressful life events to symptoms of depression and anxiety: Longitudinal evidence in early adolescents and adults. *Journal of Abnormal Psychology*, 122(2), 339–352. https://doi.org/10.1037/a0031994

Niessen, C., Weseler, D., & Kostova, P. (2016). When and why do individuals craft their jobs? The role of individual motivation and work characteristics for job crafting. *Human Relations*, 69(6), 1287–1313. https://doi.org/10.1177/0018726715610642

Nixon, A. E., Mazzola, J. J., Bauer, J., Krueger, J. R., & Spector, P. E. (2011). Can work make you sick? A meta-analysis of the relationships between job stressors and physical symptoms. *Work & Stress*, 25(1), 1–22. https://doi.org/10.1080/02678373.2011.569175

Park, H. I., Beehr, T. A., Han, K., & Grebner, S. I. (2012). Demands-abilities fit and psychological strain: Moderating effects of personality. *International Journal of Stress Management*, 19(1), 1–33. https://doi.org/10.1037/a0026852

Pejtersen, J. H., Kristensen, T. S., Borg, V., & Bjorner, J. B. (2010). The second version of the Copenhagen Psychosocial Questionnaire. *Scandinavian Journal of Public Health*, 38(3 Suppl), 8–24. https://doi.org/10.1177/1403494809349858

Pilcher, J. J., Nadler, E., & Busch, C. (2002). Effects of hot and cold temperature exposure on performance: A meta-analytic review. *Ergonomics*, 45(10), 682–698. https://doi.org/10.1080/00140130210158419

Pindek, S., & Spector, P. E. (2016). Organizational constraints: A meta-analysis of a major stressor. *Work & Stress*, 30(1), 7–25. https://doi.org/10.1080/02678373.2015.1137376

Podsakoff, N. P., LePine, J. A., & LePine, M. A. (2007). Differential challenge stressor-hindrance stressor relationships with job attitudes, turnover intentions, turnover, and withdrawal behavior: A meta-analysis. *Journal of Applied Psychology*, 92(2), 438–454. https://doi.org/10.1037/0021-9010.92.2.438

Prinet, J. C., & Sarter, N. B. (2016). The effects of high stress on attention. *Proceedings of the Human Factors and Ergonomics Society Annual Meeting*, 59(1), 1530–1534. https://doi.org/10.1177/1541931215591331

Raio, C. M., Orederu, T. A., Palazzolo, L., Shurick, A. A., & Phelps, E. A. (2013). Cognitive emotion regulation fails the stress test. *Proceedings of the National Academies of Sciences*, 110(37), 15139–15144. https://doi.org/10.1073/pnas.1305706110

Rook, J. W., & Zijlstra, F. R. H. (2006). The contribution of various types of activities to recovery. *European Journal of Work and Organizational Psychology*, 15(2), 218–240. https://doi.org/10.1080/13594320500513962

Rosen, C. C., Dimotakis, N., Cole, M. S., Taylor, S. G., Simon, L. S., Smith, T. A., & Reina, C. S. (2020). When challenges hinder: An investigation of when and how challenge stressors impact employee outcomes. *Journal of Applied Psychology*, 105(10), 1181–1206. https://doi.org/10.1037/apl0000483

Salanova, M., Peiró, J. M., & Schaufeli, W. B. (2002). Self-efficacy specificity and burnout among information technology workers: An extension of the job demand-control model. *European Journal of Work and Organizational Psychology*, 11(1), 1–25. https://doi.org/10.1080/13594320143000735

Sapolsky, R. M. (1998). *Why zebras don't get ulcers: An updated guide to stress, stress-related diseases, and coping*. W. H. Freeman and Company.

Schat, A. C., & Kelloway, E. K. (2000). Effects of perceived control on the outcomes of workplace aggression and violence. *Journal of Occupational Health Psychology*, 5(3), 386–402. https://doi.org/10.1037/1076-8998.5.3.386

Schaubroeck, J. M., & Merritt, D. E. (1997). Divergent effects of job control on coping with work stressors: The key role of self-efficacy. *The Academy of Management Journal*, 40(3), 738–754.

Schwarzer, R., & Hallum, S. (2008). Perceived teacher self-efficacy as a predictor of job stress and burnout: Mediation analyses. *Applied Psychology*, 57(s1), 152–171. https://doi.org/10.1111/j.1464-0597.2008.00359.x

Siegrist, J. (2002). Effort-reward imbalance at work and health. *Historical and Current Perspectives on Stress and Health*, 2, 261–291. https://doi.org/10.1016/S1479-3555(02)02007-3

Skinner, E. A. (1996). A guide to constructs of control. *Journal of Personality and Social Psychology: Personality Processes and Individual Differences*, 71(3), 549–570. https://doi.org/10.1037/0022-3514.71.3.549

Sliter, M., Sliter, K., & Jex, S. (2012). The employee as a punching bag: The effect of multiple sources of incivility on employee withdrawal behavior and sales performance. *Journal of Organizational Behavior*, 33(1), 121–139. https://doi.org/10.1002/job.767

Sonnentag, S. (2001). Work, recovery activities, and individual well-being: A diary study. *International Journal of Psychology*, 35(3–4), 196–210. https://doi.org/10.1037/1076-8998.6.3.196

Sonnentag, S. (2003). Recovery, work engagement, and proactive behavior: A new look at the interface between nonwork and work. *Journal of Applied Psychology*, 88(3), 518–528. https://doi.org/10.1037/0021-9010.88.3.518

Sonnentag, S., & Bayer, U.-V. (2005). Switching off mentally: Predictors and consequences of psychological detachment from work during off-job time. *Journal of Occupational Health Psychology*, 10(4), 393–414. https://doi.org/10.1037/1076-8998.10.4.393

Sonnentag, S., & Frese, M. (2012). Stress in organizations. In I. B. Weiner, N. W. Schmitt, & S. Highhouse (Eds.), *Handbook of psychology* (2nd ed., Vol. 12, pp. 560–592). Wiley.

Sonnentag, S., & Fritz, C. (2007). The Recovery Experience Questionnaire: development and validation of a measure for assessing recuperation and unwinding from work. *Journal of Occupational Health Psychology*, 12(3), 204–221. https://doi.org/10.1037/1076-8998.12.3.204

Sonnentag, S., & Fritz, C. (2015). Recovery from job stress: The stressor-detachment model as an integrative framework. *Journal of Organizational Behavior*, 36, S72–S103. https://doi.org/10.1002/job.1924

Sonnentag, S., & Jelden, S. (2009). Job stressors and the pursuit of sport activities: A day-level perspective. *Journal of Occupational Health Psychology*, 14(2), 165–181. https://doi.org/10.1037/a0014953

Sonnentag, S., Binnewies, C., & Mojza, E. J. (2010). Staying well and engaged when demands are high: The role of psychological detachment. *Journal of Applied Psychology*, 95(5), 965–976. https://doi.org/10.1037/a0020032

Sonnentag, S., Binnewies, C., & Ohly, S. (2013). Event-sampling methods in occupational health psychology. In R. R. Sinclair, M. Wang, & L. E. Tetrick (Eds.), *Research methods in occupational health psychology: Measurement, design, and data analysis* (pp. 208–228). Routledge/Taylor & Francis Group.

Sonnentag, S., Kuttler, I., & Fritz, C. (2010). Job stressors, emotional exhaustion, and need for recovery: A multi-source study on the benefits of psychological detachment. *Journal of Vocational Behavior*, 76(3), 355–365. https://doi.org/10.1016/j.jvb.2009.06.005

Sonnentag, S., Mojza, E. J., Demerouti, E., & Bakker, A. B. (2012). Reciprocal relations between recovery and work engagement: the moderating role of job stressors. *Journal of Applied Psychology, 97*(4), 842–853. https://doi.org/10.1037/a0028292

Sonnentag, S., Venz, L., & Casper, A. (2017). Advances in recovery research: What have we learned? What should be done next? *Journal of Occupational Health Psychology, 22*(3), 365–380. https://doi.org/10.1037/ocp0000079

Spector, P. E. (1986). Perceived control by employees: A meta-analysis of studies concerning autonomy and participation at work. *Human Relations, 39*(11), 1005–1016. https://doi.org/10.1177/001872678603901104

Spector, P. E., & Jex, S. M. (1998). Development of four self-report measures of job stressors and strain Interpersonal Conflict at Work Scale, Organizational Constraints Scale, Quantitative Workload Inventory, and Physical Symptoms Inventory. *Journal of Occupational Health Psychology, 3*(4), 356–367. https://doi.org/10.1037/1076-8998.3.4.356

Tadić, M., Bakker, A. B., & Oerlemans, W. G. M. (2015). Challenge versus hindrance job demands and well-being: A diary study on the moderating role of job resources. *Journal of Occupational and Organizational Psychology, 88*(4), 702–725. https://doi.org/10.1111/joop.12094

Ten Brummelhuis, L. L., & Bakker, A. B. (2012). A resource perspective on the work–home interface. *American Psychologist, 67*(7), 545–556. https://doi.org/10.1037/a0027974

Tims, M., Bakker, A. B., & Derks, D. (2013). The impact of job crafting on job demands, job resources, and well-being. *Journal of Occupational Health Psychology, 18*(2), 230–240. https://doi.org/10.1037/a0032141

Tinsley, H. E. A., & Eldredge, B. D. (1995). Psychological benefits of leisure participation: A taxonomy of leisure activities based on their need-gratifying properties. *Journal of Counseling Psychology, 42*(2), 123–132. https://doi.org/10.1037/0022-0167.42.2.123

Trougakos, J. P., Hideg, I., Cheng, B. H., & Beal, D. J. (2013). Lunch breaks unpacked: The role of autonomy as a moderator of recovery during lunch. *Academy of Management Journal.* https://doi.org/10.5465/amj.2011.1072

Tuckey, M. R., Boyd, C. M., Winefield, H. R., Bohm, A., Winefield, A. H., Lindsay, A., & Black, Q. (2017). Understanding stress in retail work: Considering different types of job demands and diverse applications of job resources. *International Journal of Stress Management, 24*(4), 368–391. https://doi.org/10.1037/str0000032

Tuckey, M. R., Searle, B. J., Boyd, C. M., Winefield, A. H., & Winefield, H. R. (2015). Hindrances are not threats: Advancing the multidimensionality of work stress. *Journal of Occupational Health Psychology, 20*(2), 131–147. https://doi.org/10.1037/a0038280

Tugade, M. M., & Fredrickson, B. L. (2004). Resilient individuals use positive emotions to bounce back from negative emotional experiences. *Journal of Personality and Social Psychology, 86*(2), 320–333. https://doi.org/10.1037/0022-3514.86.2.320

van Hooff, M. L. M., Geurts, S. A. E., Beckers, D. G. J., & Kompier, M. A. J. (2011). Daily recovery from work: The role of activities, effort and pleasure. *Work & Stress, 25*(1), 55–74. https://doi.org/10.1080/02678373.2011.570941

Van Yperen, N. W., & Hagedoorn, M. (2003). Do high job demands increase intrinsic motivation or fatigue or both? The role of job control and job social support. *Academy of Management Journal, 46*(3), 339–348.

Wall, T. D., Jackson, P. R., Mullarkey, S., & Parker, S. K. (1996). The demands-control model of job strain: A more specific test. *Journal of Occupational and Organizational Psychology, 69*, 153–166. https://doi.org/10.1111/j.2044-8325.1996.tb00607.x

Wargocki, P., & Wyon, D. P. (2017). Ten questions concerning thermal and indoor air quality effects on the performance of office work and schoolwork. *Building and Environment, 112*, 359–366. https://doi.org/10.1016/j.buildenv.2016.11.020

Wichary, S., Mata, R., & Rieskamp, J. (2016). Probabilistic inferences under emotional stress: How arousal affects decision processes. *Journal of Behavioral Decision Making, 29*(5), 525–538. https://doi.org/10.1002/bdm.1896

Wilkinson, H., Whittington, R., Perry, L., & Eames, C. (2017). Examining the relationship between burnout and empathy in healthcare professionals: A systematic review. *Burnout Research, 6*, 18–29. https://doi.org/10.1016/j.burn.2017.06.003

Wrzesniewski, A., & Dutton, J. E. (2001). Crafting a job: Revisioning employees as active crafters of their work. *The Academy of Management Review, 26*(2), 179–201. https://doi.org/10.5465/amr.2001.4378011

Zacher, H., Brailsford, H. A., & Parker, S. L. (2014). Micro-breaks matter: A diary study on the effects of energy management strategies on occupational well-being. *Journal of Vocational Behavior, 85*(3), 287–297. https://doi.org/10.1016/j.jvb.2014.08.005

7

PSYCHOLOGICAL AND SOCIAL DEMANDS AND RESOURCES

Christopher J. L. Cunningham and Kristen Jennings Black

Psychological and social (i.e., psychosocial) demands and resources are present in every work situation. They can vary in form and intensity, from the sustained vigilance required by air traffic controllers, to the emotional regulation required by retail or service personnel. Psychosocial demands can negatively impact worker health, safety, and well-being (WHSWB) if workers do not have access to the resources needed to meet such demands. Although few psychosocial demands can be entirely removed from our work experiences, they can often be alleviated or better managed with the help of occupational health psychology (OHP) theories, research, and interventions.

> **When you are finished reading this chapter, you should be able to:**
>
> LO 7.1: Define and provide examples of psychosocial work demands and resources present in most occupational contexts.
> LO 7.2: Explain how/why all workers are not uniformly affected by similar psychosocial demands.
> LO 7.3: Describe an optimal balancing of psychosocial demands and resources for a specific occupation.
> LO 7.4: Identify and evaluate the effectiveness of interventions to address psychosocial work demands.

Psychosocial Demands and Resources at Work

Many psychological and social aspects of work jointly influence or affect each other, creating complex demands and resources that can have strong and lasting effects on WHSWB. Sometimes these influences are positive, such as when perceived job control helps to weaken the relationship between specific work demands and unhealthy worker outcomes. Sometimes these shared influences are negative, such as when workers experience boredom at work, which is made worse by lack of control over how tasks are accomplished.

Psychosocial demands and resources are difficult to manage or control in work settings because they often operate invisibly within and between workers. The effects of such demands can also be exacerbated by the presence of more visible physical and environmental work demands (Winwood & Lushington, 2006), which we discuss in Chapter 10. The theories we reviewed in Chapter 6 provide a starting point for understanding psychosocial demands and resources. Here we dig deeper into this essential area of OHP research and practice, beginning with a review of the main types of psychosocial demands and resources.

Cognitive Demands and Resources

Extending from the expanded stimulus-response model we presented in Chapter 2, recall that most of the intervening processes linking stimuli to responses involve some level of cognitive processing. These cognitive processes can involve perception, appraisal, and response planning (e.g., Hockey, 1997). More negative forms of appraisal (e.g., demands as hindrances or threats) are associated with negative and resource-draining personal responses and work-related outcomes, while more positive appraisals (i.e., challenge) are associated with opportunities for achievement, growth, and resource replenishment (e.g., Boswell et al., 2004; Crane & Searle, 2016; also, note that, regardless of how demands are cognitively appraised, their physiological effects are consistent; Mazzola & Disselhorst, 2019). In this section we explore several essential forms of psychosocial demand and resource phenomena linked to workers' cognitive processing of work-related stimuli.

Perceived Workload

A commonly studied and strongly influential psychosocial demand is workers' perception of workload, or the amount and difficulty one's work (i.e., its quantitative and qualitative aspects; Bowling & Kirkendall, 2012). Research has shown that perceived workload is significantly associated with multiple other forms of work-related demands and general stress reactions, and can be detrimental to WHSWB in a variety of ways (e.g., Bowling et al., 2015; Karasek et al., 1988). Perceived workload can be an excellent, broad-spectrum indicator of overall work demands. However, it is important in OHP research and practice to clearly differentiate between workload as a demand and workload as an indication of strain.

Perceived Constraints

Paradoxically, one of the most common psychosocial demands is the perception of having inadequate resources to address demands. This is the essence of perceived resource, situational, or organizational *constraints* (e.g., Peters & O'Connor, 1980; Villanova & Roman, 1993). Constraints can also be real or perceived impediments or barriers that prevent workers from accessing resources that are needed. Addressing constraints directly is more likely to be effective than broadly increasing general resources; sometimes the solution to a constraint is not more resources, but better access to existing resources. For example, if it takes a complex form and 48 hours to access the talents of your organization's IT department, the fix for this constraint is simply streamlining the process of accessing the resources that already exist. The point to note here is perceived constraints can emerge from missing resources and difficulty accessing available resources. We agree with others who have argued that these types of constraints need more attention (Pindek & Spector, 2016), given that they are a major factor connected to many WHSWB phenomena.

Perceived Control and Self-Efficacy

Perceived control (or lack thereof) when responding to a work demand can be a resource when present or an additional form of cognitive demand when it is absent. This complex psychosocial resource (Spector, 1998), and sometimes demand, is perhaps most directly aligned with the Demands-Control theory (Karasek, 1979), but also with the resource theories we outlined in Chapter 6. Recall that workers tend to manage demands better when they have more latitude or control over how they respond to those demand. Research from a Job Demands-Resources perspective has also shown that control, as a resource at work, can protect workers from the negative effects of job demands (Bakker & Demerouti, 2007; Schaufeli & Taris, 2014). In a related fashion, control over work breaks and time outside of work can help workers more easily replenish resources (e.g., Sonnentag et al., 2017).

Often linked to actual or perceived control over one's work, is a worker's own sense of confidence in their abilities to be able to meet work demands. This cognitive belief is often studied as self-efficacy (Bandura, 1977). Heavy work demands are daunting, even when perceived control is high, unless the worker is confident in their ability to address those demands (e.g., Meier et al., 2008). In essence, then, the presence of self-efficacy enables workers to access or make use of the resource of available control. Additional research has further shown that those with higher levels of self-efficacy are also more likely to seek and find ways to exercise control over their work, sometimes in the form of job crafting, especially in resource poor work environments (Tims et al., 2014; Xanthopoulou et al., 2009). As an interesting reciprocal effect, engaging in job crafting also appears to strengthen workers' self-efficacy and perception of other resources more generally (e.g., van Wingerden et al., 2017).

Perceived (In)security and (In)justice

Another type of cognitively oriented psychosocial demand that is also a potential resource is the extent to which workers generally perceive and experience (in)security and (in)justice while working. Insecurity regarding one's working and corresponding financial situation is a form of psychological demand that can negatively affect workers' attitudes (Ashford et al., 1989), lead to emotional exhaustion (De Cuyper et al., 2014), and weigh heavily on workers' minds, potentially distracting or impairing general cognitive and behavioral functioning (Sverke et al., 2002). This particular psychological demand also can have serious consequences at the level of an organization, including increased rates of turnover and reductions in worker performance (cf., Greenhalgh & Rosenblatt, 1984).

The psychological demand of insecurity is often explored in OHP research with respect to job and income insecurity (e.g., Ashford et al., 1989; Strazdins et al., 2004), but there are deeper issues here that are only beginning to be explored. Perceptions of insecurity at work are likely to correlate with experiences of injustice, which may be due to workers' minority status, poverty, education level, citizenship, or any number of other factors in work settings (e.g., Landsbergis et al., 2014). Research jointly considering insecurity and injustice has identified additive effects on psychological strain (Francis & Barling, 2005) and sometimes complex conditioning effects, such that these demands mediate or moderate the effects of each other on various health-related outcomes (e.g., Piccoli & De Witte, 2015). Insecurity and perceptions of injustice in and about our work create additional interrelated demands, over and above those associated with normal work responsibilities. These types of demands also force workers to question the security and stability of other important resources (e.g., relationships, shelter and food, pay and benefits).

Emotional Demands and Resources

Another major class of psychosocial demands and resources are those experienced emotionally. Two commonly studied examples of these phenomena are emotional labor and empathy.

Emotional Labor

Workers in many occupations must manage their emotional displays so as not to upset the people they serve. Although we all regulate our emotions to some degree in social interactions (e.g., Beal et al., 2013), so-called *emotional display rules* (Diefendorff & Richard, 2003; Gosserand & Diefendorff, 2005) are essentially added demands on workers to demonstrate and exhibit emotions that are contrary to what they may actually be feeling. This type of forced emotional labor is a major psychosocial demand that especially affects service-oriented workers.

Researchers have distinguished between *surface acting* and *deep acting* as distinct forms of emotional labor (i.e., faking versus experiencing emotions to meet display rules), with some evidence suggesting that former may be more detrimental to workers than the latter (Brotheridge & Grandey, 2002). While researchers continue to tease apart that nuance, the more essential WHSWB reality here is that covering up what we are really feeling and displaying an entirely different set of opposing feelings drains important resources (Grandey et al., 2005).

It is most certainly true that controlling our emotions is often essential to controlling our behaviors in sometimes challenging work situations – emotional labor is not wrong or unnecessary; it is a very real and often challenging component to many occupations. Thankfully, developing research suggests that certain types of resources (e.g., emotional intelligence, general affectivity) may support workers' efforts to manage these demands (e.g., Liu et al., 2008). There are also some trainable skills for regulating one's emotions that can help workers who often encounter these demands (Buruck et al., 2016; Edelman & van Knippenberg, 2016; Hulsheger et al., 2015).

Empathy

Whereas emotional labor is clearly a demand on workers' psychosocial resources, empathy is more complex. Empathy is, "the action of understanding, being aware of, being sensitive to, and vicariously experiencing the feelings, thoughts, and experience of another…", often without knowledge of the other person's complete circumstances (Merriam-Webster, n.d.). In research and practice, empathy is often treated as an emotional resource (with cognitive elements), though demonstrating empathy, especially over extended periods of time or in otherwise very demanding situations, can be quite resource depleting – ask any social worker, nurse, physician, teacher, etc.

Research regarding empathy is limited in OHP, but interest is growing, especially given the broad impact empathy can have on workers and the people they serve. For example, Heckenberg et al. (2020) found that job resources and empathy (a personal resource) seemed to protect social service care workers (as shown by effects on various physiological indicators of stress and health). This is consistent with the findings of a systematic review by Wilkinson et al. (2017), which highlighted the negative linkage between empathy and burnout among healthcare professionals. Interestingly, although practicing empathy is a form of emotional labor, existing research suggests that outcomes associated with empathy are more positive than negative (e.g., Pohl et al., 2015). Findings like these support intervention efforts to improve workers' capacity for empathy (e.g., Krasner et al., 2009). The effectiveness of such interventions may be improved with corresponding efforts to provide workers with the resources they need to demonstrate empathy on the job (e.g., coworker and supervisor support).

Social Demands and Resources

A third general class of psychosocial work demands and resources pertain to the nature of social interactions within the work environment. Such interactions can absolutely serve as a resource when they are supportive, but they can also quickly become major demands if they turn negative or uncivil. Workers also struggle when the social expectations and requirements of work-related role(s) are unclear or interfere with workers' other life role demands. We explore in more detail issues of interpersonal and role-related demands and resources in Chapters 8 and 9, respectively. Here, we examine a couple of more general and essential forms of social demands and resources.

Norms and Expectations

Social pressures to conform and match general behavioral expectations at work are major demands that can manifest in various ways. Sometimes these social demands take the form of *stigmas* in the workplace or concerns over being treated differently because of a socially undesirable attribute or behavior. In OHP, stigma have been studied pertaining to psychological health (Corrigan et al., 2015), physical health (McGonagle & Barnes-Farrell, 2014), and involvement in safety incidents (Black et al., 2019). Social norms and expectations can be particularly difficult for newcomers to manage, especially in organizations that do not engage in high-volume hiring and where a new hire may continue to be the "new person" for some time. At this level, we tend to think about social pressure as negative demands, but social norms and expectations also provide needed clarity and structure for workers, particularly those in new situations. Social norms also help to establish appropriate and inappropriate forms of social interaction among people in particular work environments.

Social Support

Another work-related social factor that can be both demand and resource is social support. Lack of social support is an influential demand (e.g., Blanton & Morris, 1999), while the presence of social support can be seen as a valuable resource. Such support may come from coworkers or supervisors (Halbesleben, 2006), and from sources outside the organization who encourage and facilitate workers' success (e.g., Russo et al., 2016). The large body of OHP-related research on the topic of social support highlights the power of our connections with others to directly and indirectly improve workers' abilities to manage effects of demands (e.g., Cohen & Wills, 1985; Frese, 1999). Sometimes, however, social support is present, but not helpful because it is insufficient or not properly matching a specific need. Social support can even "hurt" workers if it draws attention to insecurity, such as a boss trying to help with a project, but a worker feeling like the boss does not trust in their abilities (Beehr et al., 2010; Tucker et al., 2018).

Support also functions at a number of levels within organizations. Perhaps the best examples of this come from the research on perceived organizational support (Eisenberger et al., 2016; Rhoades & Eisenberger, 2002) and supportive supervisors who can help workers manage a variety of WHSWB issues (e.g., Russo et al., 2018; Sianoja et al., 2020). Consistent support has emerged for a positive link between organizational support and worker attitudes and well-being (Kurtessis et al., 2015). Although there are nuances in these findings, the main takeaway is that supportive connections at work are broadly useful and valuable resources for workers.

Why Psychosocial Demands and Resources Matter

Psychosocial demands and the resources needed to meet them are typically experienced at a personal level, hidden from general view. This is an important area of OHP research and practice because every organization is a combination of individuals managing these issues. In this section, we explore several more specific reasons psychosocial demands and resources need our attention.

Worker Health and Resilience

Responding to demands of any form requires workers to expend or use up available resources. Chronic exposure to psychosocial work demands without sufficient resources to meet those demands can be especially debilitating over time (e.g., Elovainio et al., 2015). This is largely because these types of demands are very difficult to "shift gears" away from or accommodate compared to physical and environmental work demands (see Chapter 10). For example, when physical work demands (e.g., lifting a heavy load) are met, the demand stops requiring resources to manage. In contrast, many psychosocial demands do not automatically turn off or leave us alone even when we have physically detached from our work environments. As discussed more in Chapter 6, an inability to detach psychologically and socially from work-related demands can be particularly detrimental to workers' ability to recover depleted resources and generally maintain positive psychological and physical health and generally demonstrate resilience (e.g., Meier & Cho, 2019; Sonnentag et al., 2017). Ensuring that workers have the necessary psychosocial resources to meet work-related demands (through high-quality recovery and self-care practices) is an essential component to protecting and promoting WHSWB (Leka & Jain, 2010; Niedhammer et al., 2015).

Business and Societal Reasons

Our work environments are major sources of psychosocial demands *and* resources. Although psychosocial demands may be personally experienced, there are many ways in which organizations can help workers to manage these challenges with adjustments to work assignments, and features of the work environment and culture. For some individuals, work roles provide more consistent and controlled opportunities for psychosocial resource replenishment than nonwork roles.

Helping workers to manage psychosocial demands and resources is a legitimate business concern. Organizations suffer when workers regularly must manage high levels of psychosocial work demands without the necessary and corresponding resources. Work-related psychosocial demands and resources can also have a broader impact on society, due to their tendency to have lasting spillover effects that impact workers, and their families and communities. As just one example, consider how the effects of chronically imbalanced psychosocial demands and resources can negatively affect the functioning of *essential* workers in all segments of society (e.g., healthcare professionals, police officers, social workers, teachers).

Part of the complexity involving psychosocial demands and resources is that their effects are often intertwined and difficult to tease apart. A positive implication of this is that focusing on one or two essential psychosocial demands and resources can have multiple positive ripple effects at the organizational and societal level. For example, worker performance and perceived ability to be productive suffer when perceived resources like control and support are absent, but increase when such resources are present (e.g., Dollard et al., 2000; Madrid et al., 2017; Park et al., 2004). In a particularly strong intervention study, Nielsen et al. (2002) demonstrated how attention to psychosocial risk factors can even help organizations reduce turnover and absenteeism, and improve worker health. These types of positive effects can also transcend organizations and impact society, as shown by Stansfeld et al. (2013), who found that support and control at work are connected to population-level well-being.

Methodological Considerations and Practical Recommendations

Given the personal nature of psychosocial demands and resources, most OHP research and practice in this area leverages self-reported measures and individual-level interventions. Effective use of these methodologies requires consideration of several important details, as outlined in this section.

Measuring and Monitoring Psychosocial Demands and Resources

Efforts to understand and intervene to address psychosocial demands and resources will be more effective if informed by high quality data. When relying on self-report methods, such data are best obtained using *mixed method* (i.e., quantitative and qualitative) data collection efforts that also ideally involve multiple rounds of data collection over time (to permit observation of trends and trajectories of change within individual workers and groups). Primary methodologies used when studying psychosocial demands and resources are surveys and interviews or focus groups (e.g., Gondim & Borges-Andrade, 2009; Montgomery et al., 2013). Other common methods include experience sampling and diary studies to monitor and track how these types of demands and resources develop and are managed, and social network analyses to illustrate social interaction patterns and interdependencies (e.g., Almeida et al., 2016; Ferrin et al., 2006).

With respect to survey techniques, most OHP professionals will design and build surveys to fit specific research and intervention evaluation situations. These surveys typically gather data that workers report about themselves and are often composed of a mix of established measures of psychosocial demands (e.g., Karasek et al., 1998; Spector & Jex, 1998). It is also possible to gather self-reported survey data on psychosocial resources using measures targeting specific constructs, such as control or social support, or even respondents' more generalized sense of resource availability (Hobfoll et al., 1992). When using less structured approaches, like interviews and focus groups, questions are typically tailored to specific research or intervention evaluation purposes. The same is true for longitudinal measurement efforts involving experience sampling and diary study techniques, which often require adapting and creating measures that are brief and easy to administer in a repeated fashion (e.g., single-item and limited response choice items).

Intervening to Improve Psychosocial Demand and Resource Management

Workers' ability to manage psychosocial demands and resources can be improved through interventions in work settings. Here, we highlight several such strategies at the individual, group, leader, and organizational level.

Strategies for Individuals

At the level of individual workers, interventions that facilitate personal job crafting and emotional regulation are particularly appropriate options for facilitating management of psychosocial work demands and resources. As introduced in Chapter 6, *job crafting* is a technique by which employees drive change in some aspect of their work environment or process (e.g., Wrzesniewski & Dutton, 2001). Facilitating job crafting for individual workers is a way of granting workers more control over at least some aspect of their work, including psychosocial (and other) forms of demands and resources. In allowing and encouraging workers to do this, organizations and leaders also signal trust, respect, and support for workers. There is no single way to go about crafting a job, but it is essential to ensure that workers are supported in identifying their resources, strengths, values, and motivations, and then encouraged to consider these elements as they determine how they will approach meeting work-related demands going forward. Employees who feel good about their work tend to be more likely to engage in job crafting (e.g., Clegg & Spencer, 2007), so some management-driven effort to recognize good work and promote worker self-efficacy may be a necessary foundation before employees feel empowered to craft their own work.

To address challenges associated with emotion-oriented work demands, individual workers may benefit from *emotion regulation* interventions designed to improve their abilities to recognize and manage emotions. This type of educational and skills-based intervention can be focused internally (within the worker) and externally (when interacting with others). Specific techniques that have been helpful in this type of intervention include mindfulness and meditation (Hulsheger et al., 2015), relaxation (Jain et al., 2007), and prayer (e.g., McCulloch & Parks-Stamm, 2020).

Strategies for Groups, Leaders, and Organizations

Management of psychosocial work demands and resources can also be facilitated by interventions that operate at a higher level than individual workers. We examine three major forms of such interventions.

CLARIFYING NORMS AND FACILITATING SUPPORT

Intervention efforts to improve general awareness of psychosocial demands and resources can be valuable for workers and their supervisors. Shared knowledge of these factors that influence WHSWB can facilitate the setting and maintenance of healthy social norms and expectations pertaining to what demands are made, how they are framed, and how they are evaluated. Supervisor awareness of subordinates' abilities and available resources may also lead to more reasonable demands. Two general and common techniques for doing this in organizations are through new-hire onboarding or *socialization* (e.g., Lapointe et al., 2014) and through mentoring relationships (Allen et al., 2017; Baranik et al., 2010).

Socialization efforts can provide new members to an organization with important information and clarity regarding norms and expectations for responding to common psychosocial demands within a particular work environment. Longer-term, ongoing mentorship relationships can also help to build and sustain organizational cultures in which workers feel supported and are likely to communicate better with others to manage psychosocial work-related demands. There is even some evidence that the benefits of supervisor support can emerge when workers do not actively utilize it (Munc et al., 2017).

REDESIGNING WORK FOR MEANING AND CONTROL

Work redesign is a common strategy for addressing an imbalance between psychosocial (and other forms of) demands and resources. At the individual level, this can happen through job crafting. At a broader group or even organizational level, redesign can occur through enriching, enlarging, or otherwise changing workers' experiences in a way that provides a greater sense of meaning and control related to demands. *Job enrichment* is a job redesign technique supported by elements of *job characteristics theory* (Hackman &

Oldham, 1976), which suggests that workers will be more motivated and perform better if their daily work includes skill variety, task identity, task significance, autonomy, and feedback. These characteristics in jobs are also good for WHSWB. Straightforward applications of this theory help to reduce repetition, monotony, and boredom, and increase overall richness of one's work experiences. More specifically, organizations can increase task identity by creating ways for workers to be involved in a complete work process (which can often also increase task variety). Similarly, task significance can be generated by helping employees to see the importance of their work, either in relation to the functioning of the organization or the service provided to customers, clients, or patients. Autonomy and control are enhanced when a worker has some freedom to choose their tasks or at least the order or manner in which tasks are completed. Feedback helps workers to know how they are doing and can provide room to either recognize their good work or improve their work.

Another intervention approach involves applying principles of the Demand-Control theory (discussed more in Chapter 6). Conti et al. (2006) used this model to explain how an organizational effort to implement a lean production system may improve process efficiency, but lead to high levels of worker stress. Many organizations have experimented with similar workforce optimization and minimization efforts. From a psychosocial demands and resources perspective, lean can become mean very quickly. Organizations operating with no excess personnel create work environments in which there is no room for workers to have a down day, or for someone to be sick, take time off to care for a loved one, or generally to be human. An important point here is that organizational interventions to improve worker control must do more than adjust workers' perceptions of control; workers also need the opportunity to exercise such control. Many studies have documented the power of control within work settings; as one example, Cendales-Ayala et al. (2017) demonstrated how providing control over one's work could improve physiological and psychological health-related outcomes.

ENHANCING SECURITY AND STABILITY

At an organizational level, there is value in exploring ways to enhance workers' feelings of security and stability at work. While the specific approach will vary by organization, typically this will involve improvements to transparency and communication throughout an organization's hierarchy. Organizations can also enhance workers' sense of security and stability by fostering a *growth mindset* (e.g., Dweck & Yeager, 2019) that encourages workers to continue developing their skills and learning new things. This strategy builds adaptability and value in workers, not just within their current organization, but in a more generalizable sense if there is ever a future period of insecurity and workers need to consider other employment options.

Related to this, proper management of talent resources facilitates better management of psychosocial demands and resources. This requires developing and validating consistent and defensible processes for recruiting, selecting, developing, and promoting talent throughout the organization. This also involves realistic *workforce and succession planning* efforts that factor in the need to ensure workers opportunities to recover and perhaps also avoid absurdly high psychosocial demands altogether. Good workforce planning and cross-training within a workforce can also enable organizations to avoid large-scale terminations during periods of economic instability, by using furloughs and other flexible work assignments to provide workers with economic security and stability when it is most needed.

Evaluating Psychosocial Demand and Resource Interventions

Given the complex nature of psychosocial demands and resources, interventions in this area of OHP may generate some short-term benefits, but are more likely to show their full effects only over an extended period of time and effort. Initial attempts to address one psychosocial demand (e.g., lack of support) may end up developing into more complex interventions once other connected demands are identified (e.g., perceived workload limiting socializing opportunities). Similarly, it will take time for workers to adopt new ways of working (e.g., job crafting), especially if they do not feel supported by their supervisors and leaders. We recommend that psychosocial demand and resource intervention evaluations involve multiple time points of data collection, to address these concerns and enable modeling of trends and trajectories of change, and to inform process-related adjustments to an intervention strategy if needed.

Also, because of the personal nature of psychosocial demands and resources, it is recommended that intervention evaluations in this area include a variety of question formats and a combination of quantitative and qualitative data collection strategies. This will help to limit the effects of single/common source bias, and greatly improve your ability to understand and interpret what the evaluation data indicates. Especially valuable are open-ended remarks from workers participating in such interventions, as this information can help us understand what we may otherwise miss in quantitative evaluation data by itself. Similarly, measuring key individual differences among intervention participants (e.g., nonwork demands, demographics) can help explain why an intervention works better for some workers than others, and why this might be.

Concluding Thoughts and Reality Check

Psychosocial demands and resources may be difficult to see, but their impact is real. On the demand side, we have all witnessed this in the exhausted faces of healthcare professionals or restaurant servers who have had to regulate their emotions or deal with work-related constraints. You may have experienced this yourself after navigating a particularly challenging day with coworkers or customers. In terms of psychosocial resources, we all know even the most complex work demands are easier to manage with some control and social support. While research continues in this area of OHP, enough is known at this point to put our theories, knowledge, and methods to work to improve workers' and organizations' management of psychosocial demands and resources. By partnering with organizations and helping workers directly, OHP professionals can improve workers' opportunities to have meaningful and manageable psychological and social experiences at work every day.

Media Resources

- Science-related blog post regarding use of organizational support to reduce worker stress:
  https://blogs.cdc.gov/niosh-science-blog/2020/07/29/org_support_hwd/
- Independent blog post regarding "imposter syndrome" as a stress-inducing self-perception:
  https://forge.medium.com/feel-like-a-fraud-lately-yeah-its-going-around-6ab62f5893a1
- Article about how parents' experiences with psychosocial demands can impact kids:
  https://www.frontiersin.org/articles/10.3389/fpsyg.2020.01713/full

---

Discussion Questions

1) What are examples of psychosocial work demands and corresponding or unique psychosocial resources?
2) What would characteristics of an "optimal psychosocial work experience" be? How might these differ between a manufacturing environment and an office environment?
3) Are all workers affected similarly by psychosocial work demands? What are some factors that might explain differences?
4) What general approaches to intervention can reduce psychosocial demands and increase psychosocial resources in organizations?

Professional Profile: Jeanie Nigam, M.S.

Country/region: USA, Ohio
Current position title: I am a Research Psychologist in the Division of Science Integration with the National Institute for Occupational Safety and Health (NIOSH), in the Centers for Disease Control and Prevention (CDC). I also am a Co-Coordinator of the Healthy Work Design and Well-Being Cross Sector, and a Program Advisor to the NIOSH Total Worker Health® Program.

Background: I have been employed in my current role for 18 years, but prior to that, I worked in Human Resources and management roles where I made every effort to consider employee needs and execute organizational practices in ways that supported both the organization's mission and the employees' work-nonwork fit. I completed my Bachelor's degree in Psychology at the University of Central Florida and I have been completing my graduate work at Wright State University in Dayton, Ohio in a program that focuses both on Industrial-Organizational and Human Factors Psychology. Specifically, I have been studying occupational stress, coping, and how employment is a fundamental resource that supports well-being. I earned a Master of Science degree in Industrial-Organizational and Human Factors Psychology in December 2001, and have since completed all additional coursework and the Ph.D. qualifying exam toward my Ph.D. in the near future. I am a member of the American Psychological Association (APA), the Society for Industrial and Organizational Psychology, the Society for Occupational Health Psychology, the Society for Human Resource Management, and the European Academy for OHP.

When I entered graduate school, I envisioned myself working in an applied position either within an organization or a consulting firm. During my study of occupational stress, I came across a series of books published by the APA that were edited by Gwendolyn Keita at the APA and scholars at NIOSH including Steven Sauter, Larry Murphy, and Joe Hurrell. The books sparked my interest in learning more about NIOSH, which led me to approach them about employment opportunities. I was drawn to NIOSH because of their emphasis on public health and doing research of practical value – research designed to solve occupational problems and improve the lives of workers. The recommendations NIOSH makes are based on rigorous research and grounded in scientific evidence. The opportunity to do practical research that could really make a difference in workers' lives is what ultimately led me to this role as a research psychologist.

In my current position, I propose and conduct research to learn more about how work organization relates to worker health and well-being. I am specifically interested in reducing occupational stress and improving both mental and physical well-being through comprehensive approaches that consider the whole worker. I help coordinate the NIOSH Healthy Work Design and Well-Being Cross Sector, and the Total Worker Health® program. I help shape and manage the programs' research portfolios and institute's efforts to advance science and application in these OHP-related areas. I respond to requests for information about our areas of expertise, provide technical input as needed, and collaborate with internal partners and external stakeholders. I am also on the core planning committee for the Work, Stress, and Health Conference, which involves working on the scientific program for the conference, recruiting speakers, and planning special sessions, among other tasks.

How my work impacts WHSWB: One of my current projects is designed to learn more about the way work organization affects the safety, health, and well-being of local/short-haul commercial truck drivers, and the other is developing a short tool for assessing aspects of work that cause stress and affect safety and health in manufacturing. Both of these projects address a research gap and will provide knowledge about how work relates to health and safety in industries where workers are at increased risk for injury and poor health. These studies will help point to aspects of work organization that can be improved, with the ultimate goal of reducing worker stress and improving their safety, health, and well-being. Each takes a very comprehensive view of factors at work that can influence well-being – which encompasses both physical and mental health – and considers both work and nonwork outcomes so has the potential to be wide-reaching in the benefits for workers.

From a programmatic perspective, the programs I work with have developed National Occupational Research Agendas (NORA) that identify research needs and priorities for the nation. These agendas bring attention to critical occupational health topics and emphasize the need to apply our scientific knowledge to solve real-world problems to promote public health. These NORAs drive research and are often cited by external researchers as justification for devoting resources to addressing a particular occupational health topic. Evaluating where we are and where the field needs to go through developing these agendas helps improve worker safety, health, and well-being by inspiring researchers and practitioners to work together to address important long-standing and emerging risks that threaten the safety and health of our nation's workers.

My motivation: I always viewed the field of psychology as a tool for understanding people and helping them be the very best versions of themselves. We know that work is a key determinant of well-being and that most people spend the majority of their waking hours involved in work-related activities – including time spent commuting to and from work, working in remote locations, or at a physical worksite. Fundamentally, work should always be safe. No worker should fear losing their life doing their job, and no family should worry that their loved ones will return home injured, or not return home at all. Work should also not induce injury or illness that impairs quality of life or impedes people's abilities to live life to its fullest. I believe that many work-related occupational injuries and illnesses can be prevented through healthy work design and proper support of workers. I also see opportunity for work to support and enhance well-being – for healthy, rewarding work to be a tool for preventing mental illness and supporting people's efforts to advance and achieve their goals on and off the job. When I hear that employers learn something from our efforts, that they are spreading the message about the need to modify work, to approach safety and health from a comprehensive perspective, and to recognize the wholeness of workers, I am inspired to keep doing this important work. No matter how small the change, it makes a difference over time as more people gain a better understanding of how work affects health, and as more workplaces adapt integrated approaches and strive to keep workers safe and well.

Chapter References

Allen, T. D., Eby, L. T., Chao, G. T., & Bauer, T. N. (2017). Taking stock of two relational aspects of organizational life: Tracing the history and shaping the future of socialization and mentoring research. *Journal of Applied Psychology, 102*(3), 324–337. https://doi.org/10.1037/apl0000086

Almeida, D. M., Davis, K. D., Lee, S., Lawson, K. M., Walter, K., & Moen, P. (2016). Supervisor support buffers daily psychological and physiological reactivity to work-to-family conflict. *Journal of Marriage and Family, 78*(1), 165–179. https://doi.org/10.1111/jomf.12252

Ashford, S. J., Lee, C., & Bobko, P. (1989). Content, causes, and consequences of job insecurity: A theory-based measure and substantive test. *Academy of Management Journal, 32*(4), 803–829.

Bakker, A. B., & Demerouti, E. (2007). The Job Demands-Resources model: State of the art. *Journal of Managerial Psychology, 22*(3), 309–328. https://doi.org/10.1108/02683940710733115

Bandura, A. (1977). Self-efficacy: Toward a unifying theory of behavioral change. *Psychological Review, 84*(2), 191–215. https://doi.org/10.1037/0033-295X.84.2.191

Baranik, L., Roling, E. A., & Eby, L. T. (2010). Why does mentoring work? The role of perceived organizational support. *Journal of Vocational Behavior, 76*(3), 366–373. https://doi.org/10.1016/j.jvb.2009.07.004

Beal, D. J., Trougakos, J. P., Weiss, H. M., & Dalal, R. S. (2013). Affect spin and the emotion regulation process at work. *Journal of Applied Psychology, 98*(4), 593–605. https://doi.org/10.1037/a0032559

Beehr, T. A., Bowling, N. A., & Bennett, M. M. (2010). Occupational stress and failures of social support: When helping hurts. *Journal of Occupational Health Psychology, 15*(1), 45–59. https://doi.org/10.1037/a0018234

Black, K. J., Munc, A., Sinclair, R. R., & Cheung, J. H. (2019). Stigma at work: The psychological costs and benefits of the pressure to work safely. *Journal of Safety Research, 70*, 181–191. https://doi.org/10.1016/j.jsr.2019.07.007

Blanton, P. W., & Morris, M. L. (1999). Work-related predictors of physical symptoms and emotional well-being among clergy and spouses. *Review of Religious Research, 40*(4), 331–348. https://doi.org/10.2307/3512120

Boswell, W. R., Olson-Buchanan, J. B., & LePine, M. A. (2004). Relations between stress and work outcomes: The role of felt challenge, job control, and psychological strain. *Journal of Vocational Behavior, 64*(1), 165–181. https://doi.org/10.1016/S0001-8791(03)00049-6

Bowling, N. A., & Kirkendall, C. (2012). Workload: A review of causes, consequences, and potential interventions. In J. Houdmount, S. Leka, & R. R. Sinclair (Eds.), *Contemporary occupational health psychology: Global perspectives on research and practice* (Vol. 2, pp. 221–238). John Wiley & Sons, Ltd. https://doi.org/10.1002/9781119942849.ch13

Bowling, N. A., Alarcon, G. M., Bragg, C. B., & Hartman, M. J. (2015). A meta-analytic examination of the potential correlates and consequences of workload. *Work & Stress, 29*(2), 95–113. https://doi.org/10.1080/02678373.2015.1033037

Brotheridge, C. M., & Grandey, A. A. (2002). Emotional labor and burnout: Comparing two perspectives of "people work". *Journal of Vocational Behavior, 60*(1), 17–39. https://doi.org/10.1006/jvbe.2001.1815

Buruck, G., Dorfel, D., Kugler, J., & Brom, S. S. (2016). Enhancing well-being at work: The role of emotion regulation skills as personal resources. *Journal of Occupational Health Psychology, 21*(4), 480–493. https://doi.org/10.1037/ocp0000023

Cendales-Ayala, B., Useche, S. A., Gomez-Ortiz, V., & Bocarejo, J. P. (2017). Bus operators' responses to job strain: An experimental test of the job demand-control model. *Journal of Occupational Health Psychology, 22*(4), 518–527. https://doi.org/10.1037/ocp0000040

Clegg, C., & Spencer, C. (2007). A circular and dynamic model of the process of job design. *Journal of Occupational and Organizational Psychology, 80,* 321–339. https://doi.org/10.1348/096317906X113211

Cohen, S., & Wills, T. A. (1985). Stress, social support, and the buffering hypothesis. *Psychological Bulletin, 98*(2), 310–357. https://doi.org/10.1037/0033-2909.98.2.310

Conti, R., Angelis, J., Cooper, C., Faragher, B., & Gill, C. (2006). The effects of lean production on worker job stress. *International Journal of Operations & Production Management, 26*(9), 1013–1038. https://doi.org/10.1108/01443570610682616

Corrigan, P. W., Bink, A. B., Fokuo, J. K., & Schmidt, A. (2015). The public stigma of mental illness means a difference between you and me. *Psychiatry Research, 226*(1), 186–191. https://doi.org/10.1016/j.psychres.2014.12.047

Crane, M. F., & Searle, B. J. (2016). Building resilience through exposure to stressors: The effects of challenges versus hindrances. *Journal of Occupational Health Psychology, 21*(4), 468–479. https://doi.org/10.1037/a0040064

De Cuyper, N., Schreurs, B., Vander Elst, T., Baillien, E., & De Witte, H. (2014). Exemplification and Perceived Job Insecurity. *Journal of Personnel Psychology, 13*(1), 1–10. https://doi.org/10.1027/1866-5888/a000099

Diefendorff, J. M., & Richard, E. M. (2003). Antecedents and consequences of emotional display rule perceptions. *Journal of Applied Psychology, 88*(2), 284–294. https://doi.org/10.1037/0021-9010.88.2.284

Dollard, M. F., Winefield, H. R., Winefield, A. H., & de Jonge, J. (2000). Psychosocial job strain and productivity in human service workers: A test of the demand-control-support model. *Journal of Occupational and Organizational Psychology, 73,* 501–510. https://doi.org/10.1348/096317900167182

Dweck, C. S., & Yeager, D. S. (2019). Mindsets: a view from two eras. *Perspectives on Psychological Science, 14*(3), 481–496. https://doi.org/10.1177/1745691618804166

Edelman, P. J., & van Knippenberg, D. (2016). Training leader emotion regulation and leadership effectiveness. *Journal of Business and Psychology, 32*(6), 747–757. https://doi.org/10.1007/s10869-016-9471-8

Eisenberger, R., Malone, G. P., & Presson, W. D. (2016). *Optimizing perceived organizational support to enhance employee engagement* [White paper]. Society for Human Resource Managment and Society for Industrial and Organizational Psychology Science of HR Series. https://www.shrm.org/hr-today/trends-and-forecasting/special-reports-and-expert-views/Documents/SHRM-SIOP%20Perceived%20Organizational%20Support.pdf

Elovainio, M., Heponiemi, T., Jokela, M., Hakulinen, C., Presseau, J., Aalto, A. M., & Kivimaki, M. (2015). Stressful work environment and wellbeing: What comes first? *Journal of Occupational Health Psychology, 20*(3), 289–300. https://doi.org/10.1037/a0038684

Ferrin, D. L., Dirks, K. T., & Shah, P. P. (2006). Direct and indirect effects of third-party relationships on interpersonal trust. *Journal of Applied Psychology, 91*(4), 870–883. https://doi.org/10.1037/0021-9010.91.4.870

Francis, L., & Barling, J. (2005). Organizational injustice and psychological strain. *Canadian Journal of Behavioural Science, 37*(4), 250–261. https://doi.org/10.1037/h0087260

Frese, M. (1999). Social support as a moderator of the relationship between work stressors and psychological dysfunctioning: a longitudinal study with objective measures. *Journal of Occupational Health Psychology, 4*(3), 179–192. https://doi.org/10.1037/1076-8998.4.3.179

Gondim, S. M. G., & Borges-Andrade, J. E. (2009). Emotional regulation at work: A case study after air disaster. *PSICOLOGIA CIÊNCIA E PROFISSÃO, 29*(3), 512–533. https://doi.org/10.1590/S1414-98932009000300007

Gosserand, R. H., & Diefendorff, J. M. (2005). Emotional display rules and emotional labor: The moderating role of commitment. *Journal of Applied Psychology, 90*(6), 1256–1264. https://doi.org/10.1037/0021-9010.90.6.1256

Grandey, A. A., Fisk, G. M., & Steiner, D. D. (2005). Must "service with a smile" be stressful? The moderating role of personal control for American and French employees. *Journal of Applied Psychology, 90*(5), 893–904. https://doi.org/10.1037/0021-9010.90.5.893

Greenhalgh, L., & Rosenblatt, Z. (1984). Job insecurity: toward conceptual clarity. *Academy of Management Review, 9*(3), 438–448. https://doi.org/10.5465/amr.1984.4279673

Hackman, J. R., & Oldham, G. R. (1976). Motivation through the design of work: Test of a theory. *Organizational Behavior and Human Performance, 16*, 250–279. https://doi.org/10.1016/0030-5073(76)90016-7

Halbesleben, J. R. (2006). Sources of social support and burnout: a meta-analytic test of the conservation of resources model. *Journal of Applied Psychology, 91*(5), 1134–1145. https://doi.org/10.1037/0021-9010.91.5.1134

Heckenberg, R. A., Hale, M. W., Kent, S., & Wright, B. J. (2020). Empathy and job resources buffer the effect of higher job demands on increased salivary alpha amylase awakening responses in direct-care workers. *Behavioural Brain Research, 394*, 112826. https://doi.org/10.1016/j.bbr.2020.112826

Hobfoll, S. E., Lilly, R. S., & Jackson, A. P. (1992). Conservation of social resources and the self. In H. O. F. Veiel & U. Baumann (Eds.), *The series in clinical and community psychology. The meaning and measurement of social support* (pp. 125–141). Hemisphere Publishing Corp.

Hockey, G. R. J. (1997). Compensatory control in the regulation of human performance under stress and high workload: A cognitive-energetical framework. *Biological Psychology, 45*, 73–93. https://doi.org/10.1016/S0301-0511(96)05223-4

Hulsheger, U. R., Lang, J. W., Schewe, A. F., & Zijlstra, F. R. (2015). When regulating emotions at work pays off: A diary and an intervention study on emotion regulation and customer tips in service jobs. *Journal of Applied Psychology, 100*(2), 263–277. https://doi.org/10.1037/a0038229

Jain, S., Shapiro, S. L., Swanick, S., Roesch, S. C., Mills, P. J., Bell, I., & Schwartz, G. E. R. (2007). A randomized controlled trial of mindfulness meditation versus relaxation training: Effects on distress, positive states of mind, rumination, and distraction. *Annals of Behavioral Medicine, 33*(1), 11–21. https://doi.org/10.1207/s15324796abm3301_2

Karasek, R. A., Jr. (1979). Job demands, job decision latitude, and mental strain: Implications for job redesign. *Administrative Science Quarterly, 24*(2), 285–308. https://doi.org/10.2307/2392498

Karasek, R., Brisson, C., Kawakami, N., Houtman, I., Bongers, P., & Amick, B. (1998). The Job Content Questionnaire (JCQ): An instrument for internationally comparative assessments of psychosocial job characteristics. *Journal of Occupational Health Psychology*, *3*(4), 322–355. https://doi.org/10.1037//1076-8998.3.4.322

Karasek, R. A., Theorell, T., Schwartz, J. E., Schnall, P. L., Pieper, C. F., & Michela, J. L. (1988). Job characteristics in relation to the prevalence of myocardial infarction in the US Health Examination Survey (HES) and the Health and Nutrition Examination Survey (HANES). *American Journal of Public Health*, *78*(8), 910–918. https://doi.org/10.2105/ajph.78.8.910

Krasner, M. S., Epstein, R. M., Beckman, H., Suchman, A. L., Chapman, B., Mooney, C. J., & Quill, T. E. (2009). Association of an educational program in mindful communication with burnout, empathy, and attitudes among primary care physicians. *Journal of the American Medical Association*, *302*(12), 1284–1293. https://doi.org/10.1001/jama.2009.1384

Kurtessis, J. N., Eisenberger, R., Ford, M. T., Buffardi, L. C., Stewart, K. A., & Adis, C. S. (2015). Perceived organizational support: A meta-analytic evaluation of Organizational Support Theory. *Journal of Management*, *43*(6), 1854–1884. https://doi.org/10.1177/0149206315575554

Landsbergis, P. A., Grzywacz, J. G., & LaMontagne, A. D. (2014). Work organization, job insecurity, and occupational health disparities. *American Journal of Industrial Medicine*, *57*(5), 495–515. https://doi.org/10.1002/ajim.22126

Lapointe, É., Vandenberghe, C., & Boudrias, J.-S. (2014). Organizational socialization tactics and newcomer adjustment: The mediating role of role clarity and affect-based trust relationships. *Journal of Occupational and Organizational Psychology*, *87*(3), 599–624. https://doi.org/10.1111/joop.12065

Leka, S., Jain, A., & World Health Organization (2010). *Health impact of psychosocial hazards at work: An overview*. World Health Organization. https://apps.who.int/iris/handle/10665/44428

Liu, Y., Prati, L. M., Perrewe, P. L., & Ferris, G. R. (2008). The relationship between emotional resources and emotional labor: An exploratory study. *Journal of Applied Social Psychology*, *38*(10), 2410–2439. https://doi.org/10.1111/j.1559-1816.2008.00398.x

Madrid, H. P., Diaz, M. T., Leka, S., Leiva, P. I., & Barros, E. (2017). A finer grained approach to psychological capital and work performance. *Journal of Business and Psychology*, *33*(4), 461–477. https://doi.org/10.1007/s10869-017-9503-z

Mazzola, J. J., & Disselhorst, R. (2019). Should we be "challenging" employees?: A critical review and meta-analysis of the challenge-hindrance model of stress. *Journal of Organizational Behavior*, *40*(8), 949–961. https://doi.org/10.1002/job.2412

McCulloch, K. C., & Parks-Stamm, E. J. (2020). Reaching resolution: The effect of prayer on psychological perspective and emotional acceptance. *Psychology of Religion and Spirituality*, *12*(2), 254–259. https://doi.org/10.1037/rel0000234

McGonagle, A. K., & Barnes-Farrell, J. L. (2014). Chronic illness in the workplace: Stigma, identity threat and strain. *Stress and Health*, *30*(4), 310–321. https://doi.org/10.1002/smi.2518

Meier, L. L., & Cho, E. (2019). Work stressors and partner social undermining: Comparing negative affect and psychological detachment as mechanisms. *Journal of Occupational Health Psychology*, *24*(3), 359–372. https://doi.org/10.1037/ocp0000120

Meier, L. L., Semmer, N. K., Elfering, A., & Jacobshagen, N. (2008). The double meaning of control: Three-way interactions between internal resources, job control, and stressors at work. *Journal of Occupational Health Psychology*, *13*(3), 244–258. https://doi.org/10.1037/1076-8998.13.3.244

Merriam-Webster. (n.d.). Empathy. In *Merriam-Webster.com dictionary*. Retrieved November 13, 2020, from https://www.merriam-webster.com/dictionary/empathy

Montgomery, A., Todorova, I., Baban, A., & Panagopoulou, E. (2013). Improving quality and safety in the hospital: The link between organizational culture, burnout, and quality of care. *British Journal of Health Psychology*, *18*(3), 656–662. https://doi.org/10.1111/bjhp.12045

Munc, A., Eschleman, K., & Donnelly, J. (2017). The importance of provision and utilization of supervisor support. *Stress and Health*, *33*(4), 348–357. https://doi.org/10.1002/smi.2716

Niedhammer, I., Lesuffleur, T., Algava, E., & Chastang, J. F. (2015). Classic and emergent psychosocial work factors and mental health. *Occupational Medicine*, *65*(2), 126–134. https://doi.org/10.1093/occmed/kqu173

Nielsen, M. L., Kristensen, T. S., & Smith-Hansen, L. (2002). The Intervention Project on Absence and Well-being (IPAW): Design and results from the baseline of a 5-year study. *Work & Stress*, *16*(3), 191–206. https://doi.org/10.1080/02678370210164003

Park, K., Wilson, M. G., & Lee, M. S. (2004). Effects of social support at work on depression and organizational productivity. *American Journal of Health Behavior*, *28*(5), 444–455. https://doi.org/10.5993/AJHB.28.5.7

Peters, L. H., & O'Connor, E. J. (1980). Situational constraints and work outcomes: The influences of a frequently overlooked construct. *Academy of Management Review*, *5*(3), 391–397. https://doi.org/10.5465/amr.1980.4288856

Piccoli, B., & De Witte, H. (2015). Job insecurity and emotional exhaustion: Testing psychological contract breach versus distributive injustice as indicators of lack of reciprocity. *Work & Stress*, *29*(3), 246–263. https://doi.org/10.1080/02678373.2015.1075624

Pindek, S., & Spector, P. E. (2016). Organizational constraints: A meta-analysis of a major stressor. *Work & Stress*, *30*(1), 7–25. https://doi.org/10.1080/02678373.2015.1137376

Pohl, S., Dal Santo, L., & Battistelli, A. (2015). Empathy and emotional dissonance: Impact on organizational citizenship behaviors. *European Review of Applied Psychology*, *65*(6), 295–300. https://doi.org/10.1016/j.erap.2015.10.001

Rhoades, L., & Eisenberger, R. (2002). Perceived organizational support: A review of the literature. *Journal of Applied Psychology*, *87*(4), 698–714. https://doi.org/10.1037//0021-9010.87.4.698

Russo, M., Buonocore, F., Carmeli, A., & Guo, L. (2018). When family supportive supervisors meet employees' need for caring: Implications for work–family enrichment and thriving. *Journal of Management*, *44*(4), 1678–1702. https://doi.org/10.1177/0149206315618013

Russo, M., Shteigman, A., & Carmeli, A. (2016). Workplace and family support and work–life balance: Implications for individual psychological availability and energy at work. *The Journal of Positive Psychology*, *11*(2), 173––188. https://doi.org/10.1080/17439760.2015.1025424

Schaufeli, W. B., & Taris, T. W. (2014). A critical review of the Job Demands-Resources model: Implications for improving work and health. In G. F. Bauer & O. Hämmig (Eds.), *Bridging occupational, organizational and public health* (pp. 43–68). https://doi.org/10.1007/978-94-007-5640-3_4

Sianoja, M., Crain, T. L., Hammer, L. B., Bodner, T., Brockwood, K. J., LoPresti, M., & Shea, S. A. (2020). The relationship between leadership support and employee sleep. *Journal of Occupational Health Psychology*, *25*(3), 187–202. https://doi.org/10.1037/ocp0000173

Sonnentag, S., Venz, L., & Casper, A. (2017). Advances in recovery research: What have we learned? What should be done next? *Journal of Occupational Health Psychology*, *22*(3), 365–380. https://doi.org/10.1037/ocp0000079

Spector, P. E. (1998). A control theory of the job stress process. In C. L. Cooper (Ed.), *Theories of organizational stress* (pp. 153–169). Oxford University Press.

Spector, P. E., & Jex, S. M. (1998). Development of four self-report measures of job stressors and strain Interpersonal Conflict at Work Scale, Organizational Constraints Scale, Quantitative Workload Inventory, and Physical Symptoms Inventory. *Journal of Occupational Health Psychology*, *3*(4), 356–367. https://doi.org/10.1037/1076-8998.3.4.356

Stansfeld, S. A., Shipley, M. J., Head, J., Fuhrer, R., & Kivimaki, M. (2013). Work characteristics and personal social support as determinants of subjective well-being. *PLoS One*, *8*(11), e81115. https://doi.org/10.1371/journal.pone.0081115

Strazdins, L., D'Souza, R. M., Lim, L. L., Broom, D. H., & Rodgers, B. (2004). Job strain, job insecurity, and health: Rethinking the relationship. *Journal of Occupational Health Psychology*, *9*(4), 296–305. https://doi.org/10.1037/1076-8998.9.4.296

Sverke, M., Hellgren, J., & Näswall, K. (2002). No security: A meta-analysis and review of job insecurity and its consequences. *Journal of Occupational Health Psychology*, *7*(3), 242–264. https://doi.org/10.1037/1076-8998.7.3.242

Tims, M., Bakker, A. B., & Derks, D. (2014). Daily job crafting and the self-efficacy – performance relationship. *Journal of Managerial Psychology*, *29*(5), 490–507. https://doi.org/10.1108/jmp-05-2012-0148

Tucker, M. K., Jimmieson, N. L., & Bordia, P. (2018). Supervisor support as a double-edged sword: Supervisor emotion management accounts for the buffering and reverse-buffering effects of supervisor support. *International Journal of Stress Management*, *25*(1), 14–34. https://doi.org/10.1037/str0000046

van Wingerden, J., Bakker, A. B., & Derks, D. (2017). Fostering employee well-being via a job crafting intervention. *Journal of Vocational Behavior*, *100*, 164–174. https://doi.org/10.1016/j.jvb.2017.03.008

Villanova, P., & Roman, M. A. (1993). A meta-analytic review of situational constraints and work-related outcomes: Alternative approaches to conceptualization. *Human Resource Management Review*, *3*(2), 147–175. https://doi.org/10.1016/1053-4822(93)90021-U

Wilkinson, H., Whittington, R., Perry, L., & Eames, C. (2017). Examining the relationship between burnout and empathy in healthcare professionals: A systematic review. *Burnout Research*, *6*, 18–29. https://doi.org/10.1016/j.burn.2017.06.003

Winwood, P. C., & Lushington, K. (2006). Disentangling the effects of psychological and physical work demands on sleep, recovery and maladaptive chronic stress outcomes within a large sample of Australian nurses. *Journal of Advanced Nursing*, *56*(6), 679–689. https://doi.org/10.1111/j.1365-2648.2006.04055.x

Wrzesniewski, A., & Dutton, J. E. (2001). Crafting a job: Revisioning employees as active crafters of their work. *The Academy of Management Review*, *26*(2), 179–201. https://doi.org/10.5465/amr.2001.4378011

Xanthopoulou, D., Bakker, A. B., Demerouti, E., & Schaufeli, W. B. (2009). Reciprocal relationships between job resources, personal resources, and work engagement. *Journal of Vocational Behavior*, *74*(3), 235–244. https://doi.org/10.1016/j.jvb.2008.11.003

8

INTERPERSONAL MISTREATMENT AT WORK

Kristen Jennings Black and Christopher J. L. Cunningham

Workers in most jobs frequently face challenges in maintaining positive, healthy, and safe interpersonal relationships with supervisors and coworkers, as well as others who are in some way served by the work being done. Interpersonal mistreatment can take a variety of forms, ranging from subtle incivilities to more severe and intentional arguments, bullying, and even aggression. In this chapter, we discuss antecedents and outcomes associated with variety of common work-related interpersonal interactions along this spectrum. We pay special attention to some of the challenging aspects of interpersonal mistreatment, like the potential for fairly mild incidents to escalate to more serious problems, as well as the role of social dynamics in creating negative cultural norms. There are many strategies for managing conflict, but they must be tailored to fit the unique situation, considering the type of conflict, the source of the conflict (e.g., supervisor, customer, coworker), and the broader organizational and cultural context. We provide several examples of methods for managing and discouraging interpersonal mistreatment, including preventive policy (e.g., zero-tolerance), positive organizational norms for interactions, and direct responses to incidents.

When you are finished reading this chapter, you should be able to:

LO 8.1: Describe the major forms of interpersonal mistreatment at work and the factors that distinguish different forms.

LO 8.2: Describe personal or environmental characteristics that can make interpersonal mistreatment more likely to occur.

LO 8.3: Explain why and in what ways different forms of interpersonal mistreatment are harmful for organizations and their members.

LO 8.4: Propose proactive strategies for reducing the occurrence of interpersonal mistreatment, as well as strategies for responding when mistreatment does occur.

Perspectives on Interpersonal Mistreatment at Work

All too often, we see news headlines involving aggression in the workplace, ranging from angry customers harassing fast food workers to troubling incidents of worker-on-worker violence or the many sexual harassment allegations that surfaced in the #MeToo movement that grew rapidly beginning in 2017 (e.g., Chicago Tribune, 2020). These examples remind us that a main source of stressors in modern workplaces are our social and interpersonal interactions with other people. Sometimes these stressors are acute (e.g., interpersonal violence), but more often they are subtle, chronic, and ambiguous (e.g., discourteous or unpleasant interactions).

How can conflict or rudeness at work be so harmful to worker health, safety, and well-being (WHSWB)? We can at least partially understand interpersonal mistreatment at work through the lens of work-related stress theories (discussed in Chapter 6); any type of conflict constitutes a demand that could threaten or exhaust personal resources. Expanding on this theoretical orientation, conflict between persons is particularly difficult to manage given our innate human desires and needs for affiliation and belongingness (e.g., Baumeister & Leary, 1995). These needs influence our daily motivations, behaviors, emotions, and thoughts, causing us to seek out positive interpersonal interactions and feel unsettled by negative ones. At work, our needs for meaningful interactions can be achieved or frustrated, making interpersonal mistreatment an important area of study and practice for occupational health psychology (OHP) professionals.

We can more fully understand the occurrence, impact, and motivations for interpersonal mistreatment at work if we consider this phenomenon as a form *of counterproductive workplace behavior* (CWB) or workplace deviance. CWB is traditionally seen as one of three forms of work-related performance outcomes; the other two are actual job task performance (i.e., what is expected of employees) and organizational citizenship behavior (OCB) or contextual performance (i.e., voluntary behaviors that benefit the company or its members; Motowidlo & Van Scotter, 1994). CWB encompasses behaviors that violate workplace norms and pose some threat to the well-being of the organization and/or its members (Gruys & Sackett, 2003; Robinson & Bennett, 1995). Negative interpersonal dynamics are captured in commonly used typologies of CWB, with behaviors ranging from minor (e.g., favoritism, gossip) to severe (e.g., sexual harassment, aggression) violations of acceptable interpersonal norms in a work environment (Gruys & Sackett, 2003; Robinson & Bennett, 1995).

The CWB literature has developed quickly over the past couple of decades and now offers additional theoretical explanations for why workers may engage in interpersonal mistreatment and other such behaviors. One perspective is that positive emotions tend to propel workers to engage in OCB, while negative emotions might lead a worker to engage in CWB (Spector & Fox, 2002). This perspective makes logical sense, in that we feel more like helping others if we are in a good mood, but would be more likely to be rude to someone if we are in a bad mood. Alternatively, a reciprocity perspective also makes sense, in that we return bad treatment in exchange for received bad treatment as a way of maintaining balance in our social relationships (Helm et al., 1972). These and other theoretical perspectives help us understand the complex psychology that drives and sustains interpersonal mistreatment at work. This background also guides OHP professionals working to address various forms of interpersonal mistreatment, as we discuss later in this chapter.

Types of Interpersonal Mistreatment at Work

To understand interpersonal mistreatment at work, we need to start by defining the wide range of experiences that fit within this domain. These experiences vary extensively in terms of their visibility, severity, and content (Neuman & Baron, 1998). Before going further, please note that *not all conflict is bad*. In fact, conflict in work teams that is directed at a task can potentially benefit performance (e.g., de Wit et al., 2012). The focus of this chapter is person-oriented conflict that is known to have more negative effects for organizations and workers. In the next few subsections, we review types of mistreatment, moving from mild to severe forms.

Incivility and Interpersonal Conflict

Workplace *incivility* includes, "low-intensity deviant behavior with ambiguous intent to harm the target, in violation of workplace norms for mutual respect" (Andersson & Pearson, 1999, p. 457). Examples of such behaviors include condescending comments, demeaning remarks, unprofessional language, jokes at another's expense, or purposefully excluding a colleague from social interactions (Cortina et al., 2001; Matthews & Ritter, 2016). These behaviors can occur face-to-face or through virtual modes of communication (Giumetti et al., 2012). Incivilities like these can be pervasive because they are a "low risk" way to mistreat someone, with the ambiguity in intent making these behaviors difficult to recognize and address (i.e., the same act could be meant as "just a joke" by one individual, but as a malicious remark by another).

A major concern with incivility is that "low-intensity" acts can and often do escalate into more serious offenses. This phenomenon is captured in Andersson and Pearson's (1999) concept of an *incivility spiral*, which develops when the target of an uncivil act reciprocates potentially at a level that is stronger than the original act (Greco et al., 2019). After a few back-and-forth rounds of interaction, acts that were originally low-intensity can become intentionally harmful. Research does indeed show that retaliatory behavior often increases in severity from interaction to interaction (Greco et al., 2019). Related to this, incivility can be "contagious" within workgroups (Foulk et al., 2016). Andersson and Pearson (1999) proposed that secondary spirals of incivility can occur when an uncivil exchange occurs between individuals who were not the original target and instigator (i.e., the primary spiral), perhaps when the original target acts out toward others or when others begin to mimic observed negative interactional patterns.

In contrast to incivility, *interpersonal conflict* tends to be more overt. This form of interpersonal mistreatment is often defined as a "negative interpersonal encounter characterized by a contentious exchange, hostility, or aggression" (Ilies et al., 2015, p. 2). Some aspects of interpersonal conflict overlap with incivility (e.g., disrespectful behavior), but interpersonal conflict can also take place in a manner that is respectful, albeit involving opposing views. These interactions can occur among coworkers, as well as in interactions with customers or clients. Service workers are particularly vulnerable to customer-related conflicts (Grandey et al., 2004; Sliter & Jones, 2016). Some studies have even found that interpersonal conflict with customers demonstrates a stronger relationship with burnout and performance, when compared to coworker conflict (Sliter et al., 2011).

Harassment, Bullying, and Physical Violence

Some forms of interpersonal mistreatment are intentional, with the goal of harming someone psychologically, socially, or physically (Anderson & Bushman, 2002). The International Labour Organization (2019) defines violence and harassment at work as "a range of unacceptable behaviors and practices, or threats thereof, whether in a single occurrence or repeated, that aim at, result in, or are likely to result in physical, psychological, sexual, or economic harm..." (Definitions, section a). For instigators, such behaviors may be instrumental (i.e., to achieve some goal) or impulsive, often driven by emotion (Anderson & Bushman, 2002).

Workplace *harassment* can be defined several ways, with legal definitions often involving negative behaviors targeted at someone based on race, color, sex, religion, or disability differences. Many governing bodies consider these behaviors to be illegal when enduring negative treatment becomes a condition of employment or when the unwanted conduct creates a work environment that is hostile, degrading, intimidating, or offensive (e.g., European Commission, 2000; U.S. EEOC, n.d.). Harassment is, unfortunately, very common, with 28,000 harassment allegations in 2015 just in the United States (Feldblum & Lipnic, 2016). Even more concerning, victims of harassment often do not report their experiences, and deny, downplay, or try to ignore the situation. Many victims realistically fear retaliation if they report harassment; one study found 66% of employees who did voice concerns about mistreatment experienced negative social and work-related consequences (e.g., shunning, gossip, denied training/promotion opportunities, undesirable task assignments; Cortina & Magley, 2003).

Bullying occurs when an individual is frequently the target of negative actions for which they have difficulty defending themselves (Einarsen &

Skogstad, 1996). When multiple perpetrators are involved, this form of mistreatment is sometimes referred to as *mobbing* (Leymann, 1996). Bullying behaviors can include isolating someone, manipulating information given to them, creating uncomfortable working conditions, engaging in actual emotional abuse, or discrediting or devaluing their professional abilities (Escartín et al., 2009). The most harmful effects of bullying develop and worsen over time as these behaviors tend to be repeated (Nielsen & Einarsen, 2012). Workplace bullying does not just occur in-person or among colleagues. For example, school bus drivers have been noted to be victim to bullying from students (Goodboy et al., 2016), which can take a major toll on psychological health and even their ability to focus on working safely. As electronic and virtual communication increases, attention to workplace cyberbullying is also growing (Kowalski et al., 2018).

Abusive supervision is often considered a type of bullying that comes from one's supervisor. Tepper (2000) describes abusive supervision as, "sustained displays of hostile verbal and nonverbal behaviors, excluding physical contact" (p. 178). Key features of abusive supervision are that it is enduring and unlikely to end unless either the target or perpetrator ends the relationship or the perpetrator decides to change their behavior. Drawing from broader theories on abuse, Tepper describes that abusive supervision can be particularly problematic because the "abuser" holds some level of perceived or real power over the victim. Abusers also tend to be either unaware or unwilling to acknowledge the harmfulness of their behaviors. Because it is difficult to change abusive supervisors' behavior and because such behavior is likely to have consistent, negative effects on WHSWB outcomes (Mackey et al., 2015), organizations are encouraged to adopt a zero-tolerance approach when it comes to abusive supervision (Tepper et al., 2008).

Most forms of mistreatment we have explored so far are social or psychological in form. Sometimes, however, mistreatment can take physically aggressive forms as acts of *violence*. Thankfully, such acts are much less common than other forms of mistreatment in organizations. When worker-on-worker violence does occur, it is often associated with perceived injustice and other forms of social and environmental stressors (Neuman & Baron, 1998). More often, violence experienced at work is committed by organizational outsiders who simply plan to do harm (e.g., robberies) or affiliated individuals like former employees, significant others, customers, or clients (Barling et al., 2009). Although some violent acts are certainly unpredictable, Barling et al. (2009) point out that some occupational features are associated with a greater risk for violence (e.g., regular contact with customers or clients who may be in a heightened state of negative emotions). It may surprise you that healthcare workers are among the most likely to experience work-related aggression or violence, second only to police officers, who experience the highest rates of violence from the public (LeBlanc & Kelloway, 2002). Even the fear or anticipation of violence can be related to poor psychological health effects on workers (Rogers & Kelloway, 1997). Thankfully, there is some evidence that a work climate emphasizing violence prevention can offset some of these effects (Mueller & Tschan, 2011).

Individual and Environmental Predictors of Interpersonal Mistreatment

Earlier, we introduced the idea that emotions and reciprocity motives could influence decisions to engage in CWB in general. Researchers have tried to understand more specifically who tends to be involved in acts of interpersonal mistreatment and the various factors that make these behaviors more likely to occur at work. While certain forms of mistreatment are associated with unique antecedents, we focus in this section on essential and general contributing factors.

Who Tends to Be Involved?

The term *perpetrator* is often used when referring to the person who commits an uncivil act that affects a *target* or *victim*. These labels can lead us to oversimplify interpersonal mistreatment as a phenomenon. It may be more accurate to consider mistreatment as an unfolding relationship, in which an initial perpetrator can become a target and vice versa (Hershcovis & Reich, 2013). This relational perspective makes sense given what is understood about reciprocity motives (mentioned earlier), and is supported by evidence that individuals are prone to retaliate to experienced mistreatment (e.g., Gallus et al., 2014; Hershcovis et al., 2012). Keeping this in mind reinforces the idea that incivility is often an ongoing, dynamic interaction rather than a one-way exchange. This perspective also highlights that addressing incivility at work requires more than simply trying to identify and remove any "bad seeds"; in reality, all workers have the potential at times to be both perpetrator and victim.

That said, those involved in workplace conflict (targets and perpetrators), tend to have underlying dispositions that are characterized by strong negative traits (e.g., neuroticism, negative affect, hostile attribution style) and negative self-perceptions (e.g., low self-esteem; Aquino & Thau, 2009; Hershcovis & Reich, 2013). It is a challenge to clearly determine if traits like negative affect or neuroticism "attract" incivility, or if these traits enhance sensitivity to potentially uncivil acts, given the experiential and perceptual nature of incivility (Aquino & Thau, 2009). For instance, those higher in negative affect tend to be more likely to report experiencing incivility and are likely to have stronger responses to perceived uncivil acts (Penney & Spector, 2005). More specifically with respect to perpetrators, trait anger and impulsivity are fairly consistent predictors of workplace aggression (Glomb et al., 2002). Finally, some researchers suggest that combinations of individual traits can be more likely to create patterns of negative interactions, such as a domineering perpetrator and a submissive target (Aquino & Lamertz, 2004).

Research also suggests that power differentials and task interdependence can impact aggressive exchanges, sometimes in fairly complex ways. For instance, targets of aggression may only retaliate against perpetrators with a higher power status if they do not work together closely (Hershcovis et al., 2012). Somewhat related to this perspective is the finding that several forms of conflict and aggression are more commonly experienced by those who belong to disadvantaged or underrepresented groups. Although research testing this theory does not always yield consistent findings, women, older employees, and ethnic minorities are often more likely to be targets of mistreatment at work (Aquino & Thau, 2009). Because of this, the concept of *intersectionality* is particularly important to also consider when working to address this WHSWB issue. For example, black women may be more likely to be the target of interpersonal mistreatment as members of two common, but minority groups (based on race and sex) and this can take a substantial toll on their health (e.g., Buchanan & Fitzgerald, 2008).

When and Where is Interpersonal Mistreatment Likely?

Many forms of mistreatment are explained less by individual characteristics and more by situational characteristics. Studies suggest that rates of interpersonal conflict vary by occupational groups (e.g., Mikkelsen & Einarsen, 2010), again highlighting the influence of task and contextual characteristics on such behaviors. It is worth repeating that some work environments are particularly prone to fostering incivility from customers or clients, especially if the work being done involves customer service, working with people with heightened emotions, or direct access to cash or valued goods (Barling et al., 2009; Sliter & Jones, 2016). This connection between work environment and interpersonal mistreatment can also be understood with the help of Folger and Skarlicki's (1998) popcorn metaphor – as popcorn kernels are heated in oil, they are increasingly likely to pop. Some kernels are quicker to pop than others, but no kernels will pop if they are not placed in oil. Organizations can proactively create environments that are "oil-free" or at least "oil-light", and therefore not primed to promote incivility, aggression, or violence.

Across a number of work environments, experiences of injustice, poor leadership, high levels of work stressors, and even boredom in some circumstances, are all related to interpersonal mistreatment (Aquino & Thau, 2009; Bowling & Beehr, 2006; Folger & Skarlicki, 1998). These types of negative work experiences may also result in worker aggression because they create a strong emotional response, which for some can lead to impulsive anger, or because these acts negatively affect worker attitudes, which over time can motivate workers toward interpersonal deviance (Glomb et al., 2002). Together, these findings and theoretical perspectives imply that efforts to provide sufficient resources to offset job demands could make a difference in limiting the likelihood of mistreatment among workers.

An organization's culture can also strongly influence interpersonal mistreatment at work (e.g., Glomb et al., 2002). As emphasized throughout this book and in the theory of planned behavior (Ajzen, 1991), worker behaviors are in influenced by perceived interpersonal and social norms. With respect to interpersonal mistreatment, workers are more likely to act in an aggressive way if they perceive that others would either not care or might even approve of such behavior. Some organizational cultures and workgroup climates may generally support aggressive behavior as acceptable (Glomb et al., 2002). Studies find that workgroup climates around incivility affect important outcomes like group cohesion and worker satisfaction (Paulin & Griffin, 2015). Broadly speaking, competitive or very informal climates at work can also promote conflict because of the "no time to be nice" mentality that is often present in competitive environments (Pearson & Porath, 2005) or ambiguity regarding what behaviors would be considered inappropriate with informal climates (Andersson & Pearson, 1999).

Why Interpersonal Mistreatment Matters

Minimizing interpersonal mistreatment at work needs to be a goal for all of us, if for no other reason than that it is a human right to be treated with dignity and respect (e.g., https://www.un.org/en/universal-declaration-human-rights/index.html). In this section, we highlight three additional reasons why this WHSWB topic needs attention.

Prevalence and Impact on Workers

Only about five percent of workers experience severe forms of interpersonal mistreatment such as violence or bullying (European Agency for Safety and Health at Work, 2010). However, these statistics are less encouraging when considering less severe forms of mistreatment: 50% or more of workers may experience some form of harassment (Feldblum & Lipnic, 2016) and potentially 70% of workers experience incivility at work (Cortina et al., 2001). Even with milder forms of mistreatment, "the conduct is subtle, the consequences are not" (Cortina et al., 2017, p. 299). It is therefore unsurprising that all forms of interpersonal mistreatment are negatively associated with a variety of health-related outcomes. Physical harm certainly can occur as a result of workplace violence (Barling et al., 2009), with devastating costs associated with physical injury or even loss of life. More commonly experienced outcomes, however, involve psychological harm. Meta-analyses and detailed reviews of the literature find that harassment (Bowling & Beehr, 2006), incivility (Cortina et al., 2017), and bullying (Nielsen & Einarsen, 2012) all relate negatively to psychological health outcomes and individuals' ability to enjoy their work.

Ripple Effects Beyond the Workplace

Effects of interpersonal mistreatment at work are not constrained to a single worker or specific work environment. Workplace aggression can also affect victimized workers' significant others. For example, experiences with abusive supervision have been positively correlated with work-family conflict (Tepper, 2000) and experiences with workplace aggression are linked to psychological distress in one's partner (Haines et al., 2006). Coping with harassment or abuse is difficult, and some workers may fall into maladaptive routines that include use of alcohol, tobacco, and other drugs (e.g., Richman et al., 2002). Related to and compounding this, instances of incivility can harm workers' abilities to psychologically detach during the evenings after work (Park et al., 2015), affecting their ability to replenish personal resources. Mistreatment experienced at work has negative effects that transcend the workplace; this is why addressing interpersonal mistreatment at work directly is so critical.

Bad for Business

Interpersonal mistreatment can negatively affect organizations in many ways. Severe cases of interpersonal mistreatment can lead to lawsuits and associated legal fees and settlements (Feldblum & Lipnic, 2016; Lieber, 2010). Subtler forms of mistreatment can also be costly to organizations in terms of time spent by managers to respond to incivility when it occurs (Porath & Pearson, 2013) and reduced worker productivity, either due to distraction or withdrawal due to a mistreatment incident (Bowling & Beehr, 2006; Porath & Pearson, 2013). These costs can be substantial. As an example from a well-regarded organization, Cisco estimated that incivility among its workers cost the company around $12 million a year. To address this, Cisco started its global workplace civility program (Porath & Pearson, 2013).

Also bad for business are the CWBs that often occur in response to interpersonal mistreatment at work (Bowling & Beehr, 2006; Tepper et al., 2008). Enacting CWB may be a method for coping or "rebalancing" for those who experience work-related incivility or other forms of mistreatment (Krischer et al., 2010). In this way, CWB can contribute to the incivility spirals mentioned earlier (Andersson & Pearson, 1999) and ignoring persistent incivility among workers may facilitate its spread within workgroups (Foulk et al., 2016). The business-related impacts of this within an organization are obvious, but these effects can also transcend a workforce. For example, mistreatment among coworkers makes customers uncomfortable and can even make them less likely to purchase an organization's goods or services (Porath & Pearson, 2013).

Finally, interpersonal mistreatment can impact the recruitment and retention of talented workers. Interpersonal mistreatment is related to reduced worker empowerment and organizational commitment, and increased levels of work withdrawal and turnover intentions (Kabat-Farr et al., 2018; LeBlanc & Kelloway, 2002). Mistreatment at work also weakens efforts to create inclusive and diverse workforces, as incivilities can sometimes become a subtle, modern form of discrimination, particularly impacting women and racial minorities (Cortina et al., 2011). These issues are particularly concerning when combined with the reality that many workers would rather leave an uncivil work environment (especially if the perpetrator has power or authority), than report a problem and work to resolve it (Zapf & Gross, 2010).

Methodological Considerations and Practical Recommendations

Efforts to prevent and manage interpersonal mistreatment at work have to start with a clear assessment of the state of civility and prevalence of mistreatment within an organization. Building on this information, there are several promising strategies that OHP professionals can use to prevent or reduce mistreatment.

Measuring and Monitoring Interpersonal Mistreatment

Assessment or evaluation is a particularly important step when it comes to addressing interpersonal mistreatment at work (Glomb et al., 2002), allowing organizations to take an "honest look in the mirror" regarding their state of civility (Pearson & Porath, 2005). This could be done through examining filed complaints (though there are legitimate concerns about underreporting; e.g., Feldblum & Lipnic, 2016), intentional observations of interpersonal interactions in the workplace, anonymous surveys, and even exit interview questions about the interpersonal climate. Engaging external observers may help to identify incivility that may not seem problematic (e.g., inappropriate jokes; passive aggressive tendencies), but could be detrimental. Observers could also examine aspects of the physical work environment and review processes and procedures to identify risks for violence or aggression (for an example of review guidelines for a healthcare setting see Occupational Safety and Health Administration, 2016).

When gathering self-report data, a variety of formats can be used (e.g., measuring the frequency, source, duration, and intensity of interpersonal mistreatment). Several established measures have been developed to assess incivility at work (Cortina et al., 2001; Matthews & Ritter, 2016), interpersonal conflict at work (Spector & Jex, 1998), workplace violence (Rogers & Kelloway, 1997), or abusive supervision (Tepper, 2000). You may also want to evaluate the overall climate toward aggression, incivility, or violence prevention in a work environment (e.g., Mueller & Tschan, 2011; Paulin & Griffin, 2015; Spector et al., 2007) to understand whether mistreatment is socially normative or accepted. If problems with interpersonal mistreatment are evident, more granular assessment of critical incidents and worker interactions (perhaps with daily diary studies) may be helpful (Hershcovis & Reich, 2013).

When gathering mistreatment-related data of any form, workers have to be able to trust that the experiences, feelings, and perspectives they share are confidential and secure. Unfortunately, retaliation against workers who report mistreatment is not uncommon (Cortina & Magley, 2003), so these types of data collection efforts must be managed well. Workers also must believe their reports will be truly considered and acted upon as valid. Organizations that conduct general employee attitude or specific civility surveys, uncover issues with mistreatment, but then do nothing in response, are signaling that mistreatment is not a real concern and implicitly allowing it to continue. When it comes to interpersonal mistreatment, asking and then doing nothing in response to reports could do more damage than not asking in the first place.

Intervening to Improve Interpersonal Treatment

Efforts to address incivility and mistreatment at work necessarily target reduction and prevention of such behaviors at the individual worker, group, and broader organizational level. We examine a number of these strategies in this section. For readers wanting more, Yamada (2020) summarizes an extensive list of resources pertaining especially to workplace bullying.

Strategies for Individuals

Despite careful selection efforts and a generally positive workplace culture, interpersonal mistreatment will likely occur in work settings. A first step toward reducing incivility at work is to educate all workers about behaviors that are acceptable and unacceptable, and what to do when any form of interpersonal mistreatment is observed or experienced. For example, sexual harassment trainings can help workers to understand what is considered an inappropriate behavior and what an appropriate response to such behaviors would be (Buchanan et al., 2014). Also helpful may be training to improve effective communication strategies in typical and stressful situations (e.g., Howard & Embree, 2020). Simple misunderstandings and unclear communication can unfortunately be common triggers for incivility. Workers can also be taught emotion regulation techniques (Glomb et al., 2002), emotional expression skills (e.g., Kirk et al., 2011), and positive-refocusing strategies (Grandey et al., 2004) to effectively manage reactions that might otherwise lead to incivility or other forms of mistreatment. Feelings of control and empowerment can be bolstered by these types of trainings, and function as particularly important resources when workers confront difficult interpersonal situations with coworkers or customers (e.g., Ben-Zur & Yagil, 2005).

Another approach to intervention at the individual worker level involves training workers how to reduce negative reactions or harm caused by others' incivility. *De-escalation training* is an example of this approach that can be useful for workers who are at risk for negatively charged interpersonal interactions, particularly with clients, customers, or patients. Such trainings are common elements to violence prevention initiatives for law enforcement professionals (Engel et al., 2020) and healthcare workers (Arbury et al., 2017). Typically, these trainings focus on teaching verbal communication skills to "talk down" individuals in a heightened negative state and behavioral response techniques that can be employed if physical safety is threatened. Related to this, there may also be value in helping workers to identify and more civilly respond when personal interactions begin to turn uncivil (Andersson & Pearson, 1999; Cortina et al., 2017). Although this may seem simplistic, the reality is many of us can get caught up in interpersonal disputes and have a hard time recognizing when we should simply be apologizing, rather than defending or trying to get even.

Strategies for Groups, Leaders, and Organizations

We know that incivility can be "contagious" (Foulk et al., 2016), so there is often a need to address civility in workgroups. The Civility, Respect, and Engagement in the Workplace (CREW) intervention, which was developed and administered among workers at Veteran's Health Administration Hospitals, is one example of this type of group intervention (Osatuke et al., 2009). The concept of this intervention is simple: give work groups regular (i.e., weekly) time to talk about their strengths and weaknesses around civility. When originally evaluated, this intervention lasted six months, starting with sharing baseline data on civility in workgroups and allowing participating workers to set goals for what aspects of civility they wanted to address. This intervention was intentionally flexible, providing structured meeting times that were used to address unique needs, with guidance from consultants as needed to provide educational resources and even information sharing on what other workgroups were finding to be effective. Through regular conversations and support from the organization (i.e., time and encouragement to engage in these conversations), participating workgroups reported higher civility compared to comparison groups. Additional tests of this CREW intervention among healthcare workers showed positive and lasting effects on job satisfaction and trust in management (Leiter et al., 2012; Leiter et al., 2011).

Workgroups, as well as supervisors, can also be powerful sources of support that can help to limit interpersonal mistreatment. Strong perceptions of shared responsibility within a workgroup can increase the likelihood of a coworker intervening if they witness some form of interpersonal mistreatment (e.g., Hershcovis et al., 2017). This is the logic driving efforts to empower workers to stand up for one another, leveraging their own personal resources (e.g., power, status, confidence) to protect other coworkers. A specific form of this type of approach is found in *bystander training interventions* trainings, which encourage witnesses of mistreatment to speak up. Bystander interventions are gaining more attention as a strategy with potential to prevent sexual harassment (McDonald et al., 2015), as well as workplace intimate partner violence and workplace bullying (Lassiter et al., 2018). Despite their potential, these sorts of interventions can only achieve maximum effect if reinforced by the culture and leadership throughout an organization (e.g., Meyer & Zelin, 2019). Going beyond training bystanders to recognize and report mistreatment incidents, organizations can also consider allyship initiatives to encourage those with advantaged status (e.g., based on role, race, gender) to proactively advocate for those who may be disadvantaged in some way (see Edwards, 2006 for a conceptual review).

Shifting more completely to the level of leaders and the overall organization, it is worth noting that, "Leadership practices and organisational culture, as well as the misuse of power are always, somehow, related to the onset, as well as the prevention of, harassment at work" (European Agency for Safety and Health at Work, 2010, p. 112). To reduce interpersonal mistreatment, leaders and organizations have two general responsibilities: (1) try to stop negative treatment from occurring in the first place (prevention), and (2) respond quickly and appropriately when mistreatment is observed or reported (response). These prevention and response strategies can take a number of forms.

Although there is no single profile that describes all perpetrators of incivility, thorough background and reference checks, resume reviews, and interviews can help to reveal subtle warning signs and competencies that are predictive of civil or uncivil behavior, such as ability to handle stress and control negative emotions (Glomb et al., 2002; Pearson & Porath, 2005). Similarly, organizations can modify some aspects of the work environment to make interpersonal mistreatment less likely. Because workplace stress can be a strong driver of negative emotion and conflict (Bowling & Beehr, 2006; Glomb et al., 2002), efforts to manage psychosocial and physical work demands (discussed more in Chapters 7 and 10) are particularly warranted.

Establishing a clear and consistent response to interpersonal mistreatment is another way organizations can protect and promote WHSWB. This message can be incorporated into formal policy, connected to the company's overall mission and values (e.g., that all workers are treated with respect), and included in onboarding and regular training. Emphasizing expectations for civility and mutual respect to newcomers, both formally and informally, can enhance efforts to foster positive work relationships (Pearson & Porath, 2005). Formal values of civility, along with reinforcing messages by leadership and peers, are the foundation for a positive organizational climate that minimizes harassment, aggression, and other forms of incivility (e.g., Buchanan et al., 2014).

Correspondingly, when mistreatment incidents do occur, victims need opportunities to voice concerns safely (i.e., without risk of retaliation), allowing them to regain a sense of control and experience fewer negative personal outcomes (Cortina & Magley, 2003). Formal policies and practices should be clear regarding how incidents of incivility will be handled, and consistently applied, regardless of who the perpetrator(s) or target(s) may be. Failing to respond to concerns can leave the reporting victim feeling unheard and set a precedent that there are not real consequences to uncivil behavior.

Finally, organizations may also want to consider larger-scale interventions or trainings focused on unlikely, but costly incidents of violence. Increasingly, organizations are preparing for the rare, but terrible instances of violence with active-shooter trainings and other preventative measures (McGraw, 2016). Proactively, organizations can create climates where violence prevention is a key value (Mueller & Tschan, 2011), helping workers to feel more secure in their work roles, even when these risks exist.

Evaluating Interpersonal Mistreatment Interventions

In addition to our overall intervention evaluation recommendations in Chapter 2, there are a few specific concepts to keep in mind when addressing interpersonal mistreatment. First, as is true with interventions to address several WHSWB challenges, some indicators associated with interpersonal mistreatment may get worse before they get better. For instance, recently refined company policies and procedures for reporting harassment, may initially lead to an influx of new reports. These metrics may not mean that these changes are causing more mistreatment. Instead, this increase in reporting is a sign that workers feel more comfortable and empowered to come forward. Second, and related to the first point, evaluating incivility-related interventions is likely to require an extended time period of observation before real change is evident. You will want to see if reports of harassment and related mistreatment do diminish over time, as evidence that the organization's new policies and procedures are working as intended.

It is also important to ensure that the benefits of efforts to improve the social climate within work groups are lasting. Think of team-building type interventions that create a sense of trust and solidarity that shatters when a stressful event occurs. Workgroups may not master civil interactions after a single training and they may need reminders/refreshers over time as they transfer their skills or discussions into their everyday work environments. Lastly, we cannot emphasize enough that evaluating interpersonal mistreatment interventions requires gathering data from many stakeholders with potentially many different perspectives, paying careful attention to confidentiality. When it comes to incivility and other forms of mistreatment, those with little power can be prone to keeping quiet if they do not feel safe voicing their opinions. The social climate at work cannot be improved unless we know workers' honest perceptions and concerns.

Concluding Thoughts and Reality Check

The workplace provides an opportunity for individuals to feel a real sense of connection to one another through positive interpersonal relationships. Working closely with others can also create opportunities for interpersonal mistreatment, which can take many different forms, ranging from mild instances of rudeness to persistent and severe forms of psychological abuse and potentially physical violence. These sorts of experiences can often arise from customers or clients experiencing negative emotions, as well as among colleagues who are experiencing stress and working in an organizational culture in which there is "no time to be nice". Even the most subtle acts of mistreatment can take a negative toll on workers' well-being, job attitudes, and commitment to the organization. This is why interpersonal mistreatment is a topic that cannot be ignored by OHP professionals or organization leaders.

Work environments in which interpersonal mistreatment is normalized and accepted are simply incompatible with values for human dignity, inclusion, and diversity. This is especially true given that subtle forms of mistreatment often affect disadvantaged groups at a higher rate (i.e., *selective incivility*, a form of discrimination; Cortina et al., 2011). Current global trends toward diversity in society and work organizations means that there will be more potential for differing perspectives and viewpoints, and unfortunately, interpersonal mistreatment. OHP professionals can help organizations to prioritize civility and increase the prevalence of proper and positive interpersonal treatment at work.

Media Resources

- Lawsuit related to customer violence:
  https://www.npr.org/2019/11/21/781236230/mcdonalds-failed-to-protect-workers-against-violent-customers-lawsuit-says
- Workplace incivility after COVID-19:
  https://newworkplace.wordpress.com/2020/05/17/coronavirus-what-can-we-expect-in-terms-of-workplace-bullying-incivility-and-conflict-as-we-reopen-our-physical-workspaces/
- How COVID and anti-racism efforts may improve accountability:
  https://www.cnbc.com/2020/08/19/diversity-equity-and-inclusion-expert-what-leaders-must-do-to-commit-to-anti-racism-in-workplace.html

---

Discussion Questions

1) What is incivility and how does it relate to and/or differ from interpersonal conflict, harassment, bullying, and abusive supervision?
2) In different types of work environments, how would you expect interpersonal mistreatment to manifest and what makes it more likely?
3) Is it possible for organizations to reduce or prevent interpersonal treatment in a work environment? If so, how?
4) What strategies can individual workers use to reduce interpersonal mistreatment and its negative consequences?

Professional Profile: Martina Carroll-Garrison, DM

Country/region: USA, Georgia
Current position title: Owner of Dr. Tina Talks Work (consultant to U.S. federal agencies)
Background: I have 20 years of experience working to improve the health, safety, and well-being of workers. I earned my Doctor of Management (DM) in Organizational Leadership from the University of Phoenix Arizona. I also hold a M.S. in Resource Strategy from the National Defense University at the Industrial College of the Armed Forces and an M.A. in Global Strategy from Excelsior College. I completed my undergraduate studies at Excelsior, earning a B.S. in Liberal Studies. I also have a Higher National Diploma in Civil Engineering/Construction Management from Limerick Institute of Technology in Ireland. Finally, I attended the Federal Bureau of Investigation (FBI) Academy for Adjunct Faculty Skills Training. I am also an International Coach Federation (ICF) certified executive leadership coach.

As a leadership practitioner, I began my professional life as an engineer and construction manager. This work experience along with my work as a leader, manager, and influencer within the U.S. Defense and Intelligence community for over three decades is what differentiates me now as a coach, trainer, mentor, and work life expert. I have firsthand experience with the complex challenges faced by today's global executives, including the need for effective leadership skills and responsive organizational structures. For 28 years I supported the Department of the Army's global footprint. I then concluded my federal career by supporting the FBI in Washington, DC. At this point, I pivoted into the next phase of my work life and began sharing my expertise as an executive leadership coach and transformative workplace/work life organizational consultant. I continue to passionately serve as an advocate for the needs and challenges of the federal workforce as they support the process of democracy. I also strongly encourage a work life philosophy built upon a foundation of civility, autonomy, mastery, and purpose. My current work involves executive coaching, organization development, and leadership development.

How my work impacts WHSWB: Workplace civility is very important to me because incivility can suck the joy out of your organization. I help individuals and organizations shape their future at work by learning about and using soft skills to relate more effectively with colleagues, clients, and external stakeholders, therefore reducing workplace incivility. One particular project that I was involved in and which had a meaningful impact was when I was contracted to respond to a poor climate survey, specifically around elements of leadership and supervisory skills. I developed and delivered a series of interactive, executive level workshops for leaders, managers, and supervisors focused on building leadership capacity, improving soft skills, and developing strategies to better manage/teach/influence their subordinates towards more effective interpersonal skills. I am especially proud, based on feedback from participants and the organization, of a one-day workshop titled "Civility in the Workplace and Conflict Resolution". In it, I helped participants explore and understand the costs of workplace incivility, as well as the rewards of civility, within the workplace. Other objectives included learning: (a) practical ways of practicing workplace etiquette, (b) basic styles of conflict resolution and skills in diagnosing the causes of uncivil behavior, (c) about the role of forgiveness and conflict resolution in the creation of a civil working environment, (d) facilitative communication skills such as listening and appreciative inquiry, (e) specific interventions that can be utilized when there is conflict within the workplace, and (f) a recommended procedure for systematizing civil behavior within the workplace.

My motivation: What motivates me so much in my career is that I believe we have forgotten the fundamental WHY of why we work! We work because we are Human Beings… not Human Doings or Human Earnings. Therefore, we should encourage a work life philosophy built upon a foundation of civility, autonomy, mastery, and purpose. I also believe we have forgotten the basics of acceptable human behavior at work, which is why workplace incivility is so prevalent. When workplace civility was first talked about, I got the feeling that no one understood what I was talking about. However, when I would talk about it, I got direct and immediate feedback that everyone has experienced it. I am constantly amazed by the numbers of people who experience incivility in their work life and how organizations ignore such behavior and its impact on performance outcomes, employee well-being, and the bottom-line.

Chapter References

Ajzen, I. (1991). The Theory of Planned Behavior. *Organizational Behavior and Human Decision Processes, 50*, 179–211. https://doi.org/10.1016/0749-5978(91)90020-T

Anderson, C. A., & Bushman, B. J. (2002). Human aggression. *Annual Review of Psychology, 53*, 27–51. https://doi.org/10.1146/annurev.psych.53.100901.135231

Andersson, L. M., & Pearson, C. M. (1999). Tit for tat? The spiraling effect of incivility in the workplace. *Academy of Management Review, 24*, 452–471. https://doi.org/https://doi.org/10.5465/amr.1999.2202131

Aquino, K., & Lamertz, K. (2004). A relational model of workplace victimization: Social roles and patterns of victimization in dyadic relationships. *Journal of Applied Psychology, 89*(6), 1023–1034. https://doi.org/10.1037/0021-9010.89.6.1023

Aquino, K., & Thau, S. (2009). Workplace victimization: Aggression from the target's perspective. *Annual Review of Psychology, 60*, 717–741. https://doi.org/10.1146/annurev.psych.60.110707.163703

Arbury, S., Hodgson, M., Zankowski, D., & Lipscomb, J. (2017). Workplace violence training programs for health care workers: An analysis of program elements. *Workplace Health and Safety, 65*(6), 266–272. https://doi.org/10.1177/2165079916671534

Barling, J., Dupre, K. E., & Kelloway, E. K. (2009). Predicting workplace aggression and violence. *Annual Review of Psychology, 60*, 671–692. https://doi.org/10.1146/annurev.psych.60.110707.163629

Baumeister, R. F., & Leary, M. R. (1995). The need to belong: Desire for interpersonal attachments as a fundamental human motivation. *Psychological Bulletin, 117*(3), 497–529. https://doi.org/10.1037/0033-2909.117.3.497

Ben-Zur, H., & Yagil, D. (2005). The relationship between empowerment, aggressive behaviours of customers, coping, and burnout. *European Journal of Work and Organizational Psychology, 14*(1), 81–99. https://doi.org/10.1080/13594320444000281

Bowling, N. A., & Beehr, T. A. (2006). Workplace harassment from the victim's perspective: a theoretical model and meta-analysis. *Journal of Applied Psychology, 91*(5), 998–1012. https://doi.org/10.1037/0021-9010.91.5.998

Buchanan, N. T., & Fitzgerald, L. F. (2008). Effects of racial and sexual harassment on work and the psychological well-being of African American women. *Journal of Occupational Health Psychology, 13*(2), 137–151. https://doi.org/10.1037/1076-8998.13.2.137

Buchanan, N. T., Settles, I. H., Hall, A. T., & O'Connor, R. C. (2014). A review of organizational strategies for reducing sexual harassment: Insights from the U. S. military. *Journal of Social Issues, 70*(4), 687–702. https://doi.org/10.1111/josi.12086

Chicago Tribune. (2020, August 10). #MeToo: A Timeline of events. *Chicago Tribune.* https://www.chicagotribune.com/lifestyles/ct-me-too-timeline-20171208-htmlstory.html

Cortina, L. M., & Magley, V. J. (2003). Raising voice, risking retaliation: Events following interpersonal mistreatment in the workplace. *Journal of Occupational Health Psychology, 8*(4), 247–265. https://doi.org/10.1037/1076-8998.8.4.247

Cortina, L. M., Kabat-Farr, D., Leskinen, E. A., Huerta, M., & Magley, V. J. (2011). Selective incivility as modern discrimination in organizations. *Journal of Management, 39*(6), 1579–1605. https://doi.org/10.1177/0149206311418835

Cortina, L. M., Kabat-Farr, D., Magley, V. J., & Nelson, K. (2017). Researching rudeness: The past, present, and future of the science of incivility. *Journal of Occupational Health Psychology, 22*(3), 299–313. https://doi.org/10.1037/ocp0000089

Cortina, L. M., Magley, V. J., Williams, J. H., & Langhout, R. D. (2001). Incivility in the workplace: Incidence and impact. *Journal of Occupational Health Psychology, 6*(1), 64–80. https://doi.org/10.1037/1076-8998.6.1.64

de Wit, F. R., Greer, L. L., & Jehn, K. A. (2012). The paradox of intragroup conflict: A meta-analysis. *Journal of Applied Psychology, 97*(2), 360–390. https://doi.org/10.1037/a0024844

Edwards, K. (2006). Aspiring social justice ally identity development: A conceptual model. *NASPA Journal, 43*(4), 39–60. https://doi.org/10.2202/1949-6605.1722

Engel, R. S., McManus, H. D., & Herold, T. D. (2020). Does de-escalation training work? *Criminology & Public Policy.* https://doi.org/10.1111/1745-9133.12467

Escartín, J., Rodríguez-Carballeira, A., Zapf, D., Porrúa, C., & Martín-Peña, J. (2009). Perceived severity of various bullying behaviours at work and the relevance of exposure to bullying. *Work & Stress, 23*(3), 191–205. https://doi.org/10.1080/02678370903289639

European Agency for Safety and Health at Work. (2010). *Workplace Violence and Harassment: A European Picture* (ISSN 1830–5946). https://osha.europa.eu/en/publications/workplace-violence-and-harassment-european-picture

European Commission. (2000). *Council Directive 2000/78/EC of 27 November 2000 establishing a general framework for equal treatment in employment and occupation.* European Commission. https://eur-lex.europa.eu/legal-content/EN/TXT/?uri=celex:32000L0078

Feldblum, C. R., & Lipnic, V. A. (2016). *Select task force on the study of harrassment in the workplace.* https://www.eeoc.gov/select-task-force-study-harassment-workplace

Folger, R., & Skarlicki, D. P. (1998). A popcorn metaphor for employee aggression. In R. W. Griffin, A. O'Leary-Kelly, & J. M. Collins (Eds.), *Monographs in organizational behavior and industrial relations* (Vol. 23, pp. 43–81). Elsevier Science/JAI Press.

Foulk, T., Woolum, A., & Erez, A. (2016). Catching rudeness is like catching a cold: The contagion effects of low-intensity negative behaviors. *Journal of Applied Psychology, 101*(1), 50–67. https://doi.org/10.1037/apl0000037

Gallus, J. A., Bunk, J. A., Matthews, R. A., Barnes-Farrell, J. L., & Magley, V. J. (2014). An eye for an eye? Exploring the relationship between workplace incivility experiences and perpetration. *Journal of Occupational Health Psychology, 19*(2), 143–154. https://doi.org/10.1037/a0035931

Giumetti, G. W., McKibben, E. S., Hatfield, A. L., Schroeder, A. N., & Kowalski, R. M. (2012). Cyber incivility @ work: The new age of interpersonal deviance. *Cyberpsychology, Behavior, and Social Networking, 15*(3), 148–154. https://doi.org/10.1089/cyber.2011.0336

Glomb, T. M., Steel, P. D. G., & Arvey, R. D. (2002). Office sneers, snipes, and stab wounds: Antecedents, consequences, and implications of workplace violence and aggression. In R. G. Lord (Ed.), *Emotions in the workplace: Understanding the structure and role of emotions in organizational behavior* (pp. 227–259). Jossey-Bass.

Goodboy, A. K., Martin, M. M., & Brown, E. (2016). Bullying on the school bus: Deleterious effects on public school bus drivers. *Journal of Applied Communication Research, 44*(4), 434–452. https://doi.org/10.1080/00909882.2016.1225161

Grandey, A. A., Dickter, D. N., & Sin, H.-P. (2004). The customer is not always right: Customer aggression and emotion regulation of service employees. *Journal of Organizational Behavior, 25*(3), 397–418. https://doi.org/10.1002/job.252

Greco, L. M., Whitson, J. A., O'Boyle, E. H., Wang, C. S., & Kim, J. (2019). An eye for an eye? A meta-analysis of negative reciprocity in organizations. *Journal of Applied Psychology, 104*(9), 1117–1143. https://doi.org/10.1037/apl0000396

Gruys, M. L., & Sackett, P. R. (2003). Investigating the dimensionality of counter-productive work behavior. *International Journal of Selection and Assessment, 11*(1), 30–42. https://doi.org/10.1111/1468-2389.00224

Haines, V. Y., 3rd, Marchand, A., & Harvey, S. (2006). Crossover of workplace aggression experiences in dual-earner couples. *Journal of Occupational Health Psychology, 11*(4), 305–314. https://doi.org/10.1037/1076-8998.11.4.305

Helm, B., Bonoma, T. V., & Tedeschi, J. T. (1972). Reciprocity for harm done. *The Journal of Social Psychology, 87*, 89–98. https://doi.org/10.1080/00224545.1972.9918651

Hershcovis, M. S., & Reich, T. C. (2013). Integrating workplace aggression research: Relational, contextual, and method considerations. *Journal of Organizational Behavior, 34*(S1), S26–S42. https://doi.org/10.1002/job.1886

Hershcovis, M. S., Neville, L., Reich, T. C., Christie, A. M., Cortina, L. M., & Shan, J. V. (2017). Witnessing wrongdoing: The effects of observer power on incivility intervention in the workplace. *Organizational Behavior and Human Decision Processes, 142*, 45–57. https://doi.org/10.1016/j.obhdp.2017.07.006

Hershcovis, M. S., Reich, T. C., Parker, S. K., & Bozeman, J. (2012). The relationship between workplace aggression and target deviant behaviour: The moderating roles of power and task interdependence. *Work & Stress, 26*(1), 1–20. https://doi.org/10.1080/02678373.2012.660770

Howard, M. S., & Embree, J. L. (2020). Educational intervention improves communication abilities of nurses encountering workplace incivility. *The Journal of Continuing Education in Nursing, 51*(3), 138–144. https://doi.org/10.3928/00220124-20200216-09

Ilies, R., Aw, S. S. Y., & Pluut, H. (2015). Intraindividual models of employee well-being: What have we learned and where do we go from here? *European Journal of Work and Organizational Psychology, 24*(6), 827–838. https://doi.org/10.1080/1359432x.2015.1071422

International Labor Organization. (2019). *Violence and Harassment Convention, 2019 (No. 190).* https://www.ilo.org/dyn/normlex/en/f?p=NORMLEXPUB:12100:0::NO::P12100_ILO_CODE:C190

Kabat-Farr, D., Cortina, L. M., & Marchiondo, L. A. (2018). The emotional aftermath of incivility: Anger, guilt, and the role of organizational commitment. *International Journal of Stress Management, 25*(2), 109–128. https://doi.org/10.1037/str0000045

Kirk, B. A., Schutte, N. S., & Hine, D. W. (2011). Effect of an expressive-writing intervention for employees on emotional self-efficacy, emotional intelligence, affect, and workplace incivility. *Journal of Applied Social Psychology, 41*(1), 179–195. https://doi.org/10.1111/j.1559-1816.2010.00708.x

Kowalski, R. M., Toth, A., & Morgan, M. (2018). Bullying and cyberbullying in adulthood and the workplace. *The Journal of Social Psychology, 158*(1), 64–81. https://doi.org/10.1080/00224545.2017.1302402

Krischer, M. M., Penney, L. M., & Hunter, E. M. (2010). Can counterproductive work behaviors be productive? CWB as emotion-focused coping. *Journal of Occupational Health Psychology, 15*(2), 154–166. https://doi.org/10.1037/a0018349

Lassiter, B. J., Bostain, N. S., & Lentz, C. (2018). Best practices for early bystander intervention training on workplace intimate partner violence and workplace bullying. *Journal of Interpersonal Violence.* https://doi.org/10.1177/0886260518807907

LeBlanc, M. M., & Kelloway, E. K. (2002). Predictors and outcomes of workplace violence and aggression. *Journal of Applied Psychology, 87*(3), 444–453. https://doi.org/10.1037/0021-9010.87.3.444

Leiter, M. P., Day, A., Oore, D. G., & Spence Laschinger, H. K. (2012). Getting better and staying better: Assessing civility, incivility, distress, and job attitudes one year after a civility intervention. *Journal of Occupational Health Psychology, 17*(4), 425–434. https://doi.org/10.1037/a0029540

Leiter, M. P., Laschinger, H. K. S., Day, A., & Oore, D. G. (2011). The impact of civility interventions on employee social behavior, distress, and attitudes. *Journal of Applied Psychology, 96*(6), 1258–1274. https://doi.org/10.1037/a0024442

Leymann, H. (1996). The content and development of mobbing at work. *European Journal of Work and Organizational Psychology, 5*(2), 165–184. https://doi.org/10.1080/13594329608414853

Lieber, L. D. (2010). How workplace bullying affects the bottom line. *Employment Relations Today, 37*(3), 91–101. https://doi.org/10.1002/ert.20314

Mackey, J. D., Frieder, R. E., Brees, J. R., & Martinko, M. J. (2015). Abusive supervision: A meta-analysis and empirical review. *Journal of Management, 43*(6), 1940–1965. https://doi.org/10.1177/0149206315573997

Matthews, R. A., & Ritter, K. J. (2016). A concise, content valid, gender invariant measure of workplace incivility. *Journal of Occupational Health Psychology, 21*(3), 352–365. https://doi.org/10.1037/ocp0000017

McDonald, P., Charlesworth, S., & Graham, T. (2015). Action or inaction: Bystander intervention in workplace sexual harassment. *The International Journal of Human Resource Management, 27*(5), 548–566. https://doi.org/10.1080/09585192.2015.1023331

McGraw, M. (2016). Equipped for the unthinkable. *Human Resources Executive,* May, 40–41.

Meyer, C., & Zelin, A. I. (2019). Bystander as a band-aid: How organization leaders as active bystanders can influence culture change. *Industrial and Organizational Psychology, 12*(3), 342–344. https://doi.org/10.1017/iop.2019.42

Mikkelsen, E. G., & Einarsen, S. (2010). Bullying in Danish work-life: prevalence and health correlates. *European Journal of Work and Organizational Psychology, 10*(4), 393–413. https://doi.org/10.1080/13594320143000816

Motowidlo, S. J., & Van Scotter, J. R. (1994). Evidence that task performance should be distinguished from contextual performance. *Journal of Applied Psychology, 79*(4), 475–480. https://doi.org/https://doi.org/10.1037/0021-9010.79.4.475

Mueller, S., & Tschan, F. (2011). Consequences of client-initiated workplace violence: The role of fear and perceived prevention. *Journal of Occupational Health Psychology, 16*(2), 217–229. https://doi.org/10.1037/a0021723

Neuman, J. H., & Baron, R. A. (1998). Workplace violence and workplace aggression: Evidence concerning specific forms, potential causes, and preferred targets. *Journal of Management, 24*(3), 391–419. https://doi.org/10.1177/014920639802400305

Nielsen, M. B., & Einarsen, S. (2012). Outcomes of exposure to workplace bullying: A meta-analytic review. *Work & Stress, 26*(4), 309–332. https://doi.org/10.1080/02678373.2012.734709

Occupational Safety and Health Administration. (2016). *Guidelines for preventing workplace violence for healthcare and social service workers.* https://www.osha.gov/Publications/osha3148.pdf

Osatuke, K., Moore, S. C., Ward, C., Dyrenforth, S. R., & Belton, L. (2009). Civility, Respect, Engagement in the Workforce (CREW): Nationwide organization development intervention at veterans health administration. *The Journal of Applied Behavioral Science*, *45*(3), 384–410. https://doi.org/10.1177/0021886309335067

Park, Y., Fritz, C., & Jex, S. M. (2015). Daily cyber incivility and distress: The moderating roles of resources at work and home. *Journal of Management*, *44*(7), 2535–2557. https://doi.org/10.1177/0149206315576796

Paulin, D., & Griffin, B. (2015). Team Incivility Climate Scale: Development and validation of the team-level incivility climate construct. *Group & Organization Management*, *42*(3), 315–345. https://doi.org/10.1177/1059601115622100

Pearson, C. M., & Porath, C. L. (2005). On the nature, consequences, and remedies of workplace incivility: No time for "nice"? Think again. *Academy of Management Executive*, *19*(1), 7–18. https://doi.org/10.5465/AME.2005.15841946

Penney, L. M., & Spector, P. E. (2005). Job stress, incivility, and counterproductive work behavior (CWB): The moderating role of negative affectivity. *Journal of Organizational Behavior*, *26*(7), 777–796. https://doi.org/10.1002/job.336

Porath, C. L., & Pearson, C. M. (2013). The price of incivility. *Harvard Business Review, January–February*. https://hbr.org/2013/01/the-price-of-incivility

Richman, J. A., Shinsako, S. A., Rospenda, K. M., Flaherty, J. A., & Freels, S. (2002). Workplace harassment/abuse and alcohol-related outcomes: The mediating role of psychological distress. *Journal of Studies on Alcohol*, *63*(4), 412–419. https://doi.org/10.15288/jsa.2002.63.412

Robinson, S. L., & Bennett, R. J. (1995). A typology of deviant workplace behaviors – a multidimensional-scaling study. *Academy of Management Journal*, *38*(2), 555–572.

Rogers, K. A., & Kelloway, E. K. (1997). Violence at work: Personal and organizational outcomes. *Journal of Occupational Health Psychology*, *2*(1), 63–71. https://doi.org/10.1037/1076-8998.2.1.63

Sliter, M., & Jones, M. (2016). A qualitative and quantitative examination of the antecedents of customer incivility. *Journal of Occupational Health Psychology*, *21*(2), 208–219. https://doi.org/10.1037/a0039897

Sliter, M., Pui, S. Y., Sliter, K. A., & Jex, S. M. (2011). The differential effects of interpersonal conflict from customers and coworkers: Trait anger as a moderator. *Journal of Occupational Health Psychology*, *16*(4), 424–440. https://doi.org/10.1037/a0023874

Spector, P. E., & Fox, S. (2002). An emotion-centered model of voluntary work behavior: Some parallels between counterproductive work behavior and organizational citizenship behavior. *Human Resource Management Review*, *12*(2), 269–292. https://doi.org/10.1016/S1053-4822(02)00049-9

Spector, P. E., & Jex, S. M. (1998). Four self-report measures of job stressors and strain: Interpersonal conflict at work scale, organizational constraints scale, quantitative workload inventory, and physical symptoms inventory. *Journal of Occupational Health Psychology*, *3*(4), 356–367. https://doi.org/10.1037/1076-8998.3.4.356

Spector, P. E., Coulter, M. L., Stockwell, H. G., & Matz, M. W. (2007). Perceived violence climate: A new construct and its relationship to workplace physical violence and verbal aggression, and their potential consequences. *Work & Stress*, *21*(2), 117–130. https://doi.org/10.1080/02678370701410007

Tepper, B. J. (2000). Consequences of abusive supervision. *Academy of Management Journal*, *43*(2), 178–190.

Tepper, B. J., Henle, C. A., Lambert, L. S., Giacalone, R. A., & Duffy, M. K. (2008). Abusive supervision and subordinates' organization deviance. *Journal of Applied Psychology, 93*(4), 721–732. https://doi.org/10.1037/0021-9010.93.4.721

U.S. EEOC. (n.d.). *Harassment.* https://www.eeoc.gov/harassment

Yamada, D. (2020, July 18). Understanding workplace bullying and mobbing: Some lockdown resources. *Minding the Workplace.* https://newworkplace.wordpress.com/2020/07/18/understanding-workplace-bullying-and-mobbing-some-lockdown-resources/

Zapf, D., & Gross, C. (2010). Conflict escalation and coping with workplace bullying: A replication and extension. *European Journal of Work and Organizational Psychology, 10*(4), 497–522. https://doi.org/10.1080/13594320143000834

9

WORK AND NONWORK ROLE DYNAMICS

Christopher J. L. Cunningham and Kristen Jennings Black

The emphasis in this chapter is on demand- and resource-related dynamics workers experience within and between their work and nonwork roles, and how these dynamics affect worker health, safety, and well-being (WHSWB). We explore essential occupational health psychology (OHP) theory and research from this domain, examining multiple phenomena associated with work and nonwork role dynamics, including conflict and facilitation within and between roles, strategies for managing multiple roles, and intervention concepts addressing these phenomena at the worker and broader organization levels.

When you are finished reading this chapter, you should be able to:

LO 9.1: Explain the concept and importance of "roles" as essential to understanding human behavior in and outside of work.

LO 9.2: Identify and describe several different ways in which role-related demands, resources, and related experiences can transcend one role to impact many others.

LO 9.3: Identify and discuss factors that contribute to interrole conflict and facilitation for working adults.

LO 9.4: Describe ways in which organizations can facilitate healthy work and nonwork role dynamics for workers.

Overview of Work and Nonwork Role Theory and Dynamics

Work and nonwork role dynamics are deeply intertwined with nearly all WHSWB concerns. This is because everything workers perceive, appraise, and do happens within the context of one or more roles. This reality makes role dynamics a prime target for the development and application of models, theories, and interventions to understand and shape worker behavior. There are many different models and perspectives guiding work in this area of OHP. Regardless of which perspective is taken, all workers manage demands and resources

206

associated with at least two general life role domains: work and nonwork. The dynamics associated with how we manage these roles are complex and constitute a very rich area of OHP research and practice that considers many personal, social, and environmental factors (e.g., Allen et al., 2014; Allen et al., 2020). Most OHP-related work in this area focuses on work and family role dynamics, but the phenomena studied in this area of OHP are highly generalizable across different types of roles. For this reason, we have tried to be a bit more inclusive in this overview chapter by considering work and nonwork role domains as the two main subdivisions of our many life roles (cf., Grawitch et al., 2013).

Roles influence workers' behaviors, beliefs, preferences, and even attitudes. The complexity associated with roles is evident in the diversity of general role theories that have been developed and promoted since the early 1930s (see Biddle, 1986 for an insightful discussion of five main theories that have contributed to current thinking on work/nonwork role dynamics). As noted by Biddle (1986), the concept of a "role" within most role-related theories is a "theatrical metaphor" that enables us to make sense of how we are acting in specific social contexts. Along these lines, the most influential role theory impacting OHP research and practice is *organizational role theory*, which is commonly attributed to Kahn et al. (1964). This theory focuses on roles within organizational contexts, which are social systems that develop around clear objectives and generally include some degree of hierarchy of social position (see Biddle, 1986). Organizational role theory describes roles as being generated by normative expectations, which may vary among individuals and the various groups of others with whom interaction occurs within an organizational context. The same concept applies in our nonwork domains, where our norms and important others create expectations for our various other roles. The possibility of encountering and having to make sense of multiple sets of norms, in and outside of work, establishes the basis for role conflict (discussed later in this chapter), which can lead to role strain and ultimately negative consequences for the worker and organization if unaddressed.

Roles often develop and are sustained, however, by more than behavioral norms. Consider how roles involving religious participation are more defined by shared beliefs and attitudes, than by behavioral expectations. Consider also roles we might occupy due to preference, such as hobbies or sports. Finally, some work and nonwork roles form more because of situational factors than from any norm or personal choice. For example, when adult children become caregivers for aging parents or other dependants, this type of role is not as normative in all countries and cultures today as it was in previous generations; instead it develops out of situational necessity. Across these examples, roles are contexts in which we actively respond to demands with the help of various resources.

Role-Related Demands and Resources

Intra- and interrole dynamics have been studied for decades, tracing back to early work on role theory (Kahn et al., 1964; Katz & Kahn, 1966). Extending from our discussions about demands and resources in the preceding chapters and the overarching focus on WHSWB throughout this book, *intrarole dynamics* refer to the ebb and flow of demands and resources, and their effects within a specific role. *Interrole* or *role boundary dynamics* are similar, but focus on how these phenomena may span multiple role domains and how workers manage transitions between these domains (Allen et al., 2014; French et al., 2019).

Extending from our discussion of demands and resources in Chapters 6 and 7, nearly all life roles involve a never-ending ebb and flow of demands and resources. This is a helpful perspective when studying or helping workers to manage their work and nonwork role dynamics. Typically, this involves leveraging either Hobfoll's (1989) Conservation of Resources (COR) or Demerouti et al.'s (2001) Job Demands and Resources (JD-R) theories (e.g., Grandey & Cropanzano, 1999; Halbesleben et al., 2009). There are also some attempts to adapt these more general resource-based frameworks to something a bit more tailored to work and nonwork role contexts (e.g., work-home resources model; ten Brummelhuis & Bakker, 2012).

Research in this area continues, but it is increasingly well-understood that certain resources have cross-domain relevance and can positively impact work attitudes and even work-related performance (Odle-Dusseau et al., 2012). Such resources include support from one's work supervisors (e.g., family supportive supervisor behaviors [FSSB]; Hammer et al., 2009), support from one's organization more broadly (e.g., family supportive organizational perceptions [FSOP]; Allen, 2001), and scheduling flexibility (Baltes et al., 1999; Swanberg et al., 2011).

Managing work and nonwork role demands and resources is typically seen as a personal challenge, so it is an area in which organizations are often reluctant to meddle. In 2020, however, the COVID-19 pandemic thrust the challenges of managing, integrating, and balancing work and nonwork life into a particularly bright spotlight as workers all over the world were forced to simultaneously manage major changes across all their role domains (e.g., Dey et al., 2020). Organizations do not have the luxury of waiting for this major disturbance to pass and will have to coordinate with workers to establish new role-related expectations moving forward.

Role Boundaries

Understanding and helping workers to better manage work and nonwork role dynamics often requires clearly defining or differentiating roles. This is a natural part of how we all manage our various life roles, by establishing "edges", borders, or boundaries to help us separate work from nonwork domains (e.g., Nippert-Eng, 1996). However, as major life events such as the COVID-19 pandemic remind us, our roles and the boundaries between them are subject to revision when social and environmental forces change. Understanding role boundaries helps us understand how different roles are likely to jointly influence one another. Role boundaries are commonly understood in OHP-related work in terms of their clarity, strength, and consistency.

Boundary Clarity

It is a simple, but important point to note that people respect and benefit from boundaries that are clearly marked. You know this if you (like Chris) have ever been hiking in a new area, and suddenly found yourself challenged by a shotgun-wielding, privacy-seeking individual whose property boundary you have unintentionally crossed. It is often very difficult for us to find or establish boundaries around some of our roles. As an example, Chris recently took over the responsibility of managing a rental property that is jointly owned by a large group of extended family members – Is this new role a family, work, or service role? Similarly, both of us are regularly involved in a variety of groups and roles associated with our churches – Are these social, personal, spiritual, or work-related roles? Trying to answer such questions is not simply an academic exercise. The ability to demarcate our roles with clear boundaries is an important element to effectively managing role-related demands and resources (e.g., Lobel, 1991) as well as how we transition between roles (e.g., Kreiner et al., 2009).

Boundary Strength

In OHP research and practice, there is a tendency to conceptualize role boundary strength as ranging from fully open to cross-role transfer and integration of demand and resource experiences (weak) to completely separated or segmented (strong). The degree to which role boundaries are weak or strong is also linked to their *flexibility* (i.e., whether a role can be "occupied" and managed in a variety of times and places) and *permeability* (i.e., whether a role permits a workers to engage psychologically and behaviorally in meeting demands of another role, while the first role is occupied; e.g., Capitano & Greenhaus, 2018; Clark, 2000). Boundary strength can vary within and between persons due to personal, situational, and occupational factors. For instance, a nurse cannot continue patient care (i.e., their work role) when not at a medical facility (though they may continue thinking about their work). In contrast, a professor can easily continue responding to student emails or working on projects from home, but perhaps not while engaging in other nonwork roles (e.g., attending a religious service, coaching a youth sports team).

Some research suggests that stronger role boundaries are associated with better work-life balance, especially for older adults (Spieler et al., 2018). However, while permeability is associated with higher levels of interrole conflict (Bulger et al., 2007; Olson-Buchanan & Boswell, 2006), it may also be associated with opportunities for interrole enhancement or facilitation (e.g., Bulger et al., 2007). Other studies show that perceived role flexibility (associated with weaker boundary strength) may be associated with less interrole conflict (Matthews & Barnes-Farrell, 2010) and greater work and family/nonwork role enhancement (Bulger et al., 2007). A key resource that can help explain these inconsistencies is the presence or absence of control: Workers are likely to benefit most when they can choose how permeable their boundaries are (and when they are permeable) versus having a certain degree of permeability forced on them (e.g., work email being blocked during "nonwork hours").

Boundary Consistency and Symmetry

Closely related to boundary permeability and flexibility is the extent to which role boundaries vary in terms of their consistency and symmetry (Allen et al., 2014). This is most apparent in OHP work that examines the often-imbalanced directionality of work versus nonwork role-related demands. This issue of consistency is also evident in work that has explored how certain qualities of our personal role boundaries may change or fluctuate over time. Specifically, workers' strategies for managing role dynamics may change over time as various factors influence the extent to which they identify most strongly with work or nonwork roles. Of particular relevance here is workers' *identity prominence* and *salience*, which are typically operationalized in a way that helps us understand the extent to which a person identifies with a particular role, making it central or most salient in their lives (Brenner et al., 2014; Stryker & Serpe, 1994). Individuals with strong work identity salience place a stronger emphasis and value on their work-related role obligations. As a result, these individuals may maintain a less flexible and permeable work than nonwork boundary and be open to interruptions from work when at home, compared to individuals who identify more strongly with their nonwork roles (cf., Kossek et al., 2012). It is interesting to consider whether and how changes to our identities over time may also lead to shifts in role boundary consistency and symmetry needs.

Interrole Transitions and Dynamics

Appreciating that our various life roles have boundaries, we must also remember that we rarely find ourselves managing the demands and resources of only one role at a time. For example, while reading this book in your role as a student, you may also be interrupted or distracted because of your role as someone's significant other or parent, and even have to pause your reading to address competing demands from another role domain. Multiple roles can even combine within a single role domain. Consider a recently promoted mid-level manager who still has individual contributor responsibilities, but now also must facilitate subordinates' success. This person may still have friendships with coworkers who are now subordinates or supervisor-level peers. Now, managing multiple role relationships and demands contributes to the typical work demands this individual must manage.

There is no map or physical boundary that helps us know when we need to switch from being someone and doing something in one role to a different domain. This was evident in early theorizing about within-domain transitions at work (e.g., Nicholson, 1984) and is evident in our current understanding that workers manage intra- and interrole transitions using a variety of physical, behavioral, and psychological strategies (Allen et al., 2014; Kreiner et al., 2009; Sturges, 2012). For example, many workers facilitate their transition between work and home through participation in some sort of buffering activity, such as a commute (Clark et al., 2019; Redmond & Mokhtarian, 2001; van Hooff, 2015).

Our experiences in one role have a variety of direct and indirect impacts on what we are able to do, how we do it, and what we ultimately experience in other roles. There are at least eight models that help to explain the ways in which our experiences in one role may influence our experiences in another. As summarized in Table 9.1, the first three of these models are considered non-causal and are meant mainly as conceptual or explanatory aids of how roles are generally related. The other models are more causal, as they explain mechanisms or ways in which a set of experiences or demands in one role can affect a person in another role.

As these various models suggest, there is some diversity of perspective among OHP professionals when it comes to how we understand and intervene to help workers manage role dynamics. Most OHP attention has focused on conflict or interference between work and family/nonwork roles, where it comes from, and how it develops (e.g., Allen et al., 2020). There is also growing recognition that spillover and compensation does not necessarily have to be negative, and that there are situations where involvement in one role leads to resource gains that can be useful in other roles (i.e., interrole facilitation or enhancement; e.g., Hill et al., 2007; Sieber, 1974).

The discussion in the rest of this subsection elaborates on three broad and commonly examined forms of interrole dynamics: conflict, facilitation, and balance.

Conflict

Workers can experience conflict or interference within and between work and nonwork roles. Such conflict tends to develop when workers experience "two or more incompatible expectations" for their behavior (Biddle, 1986, p. 82). This is especially common when demands or pressures of one role cannot be readily managed amidst demands and pressures from other roles (Greenhaus & Beutell, 1985; Kahn et al., 1964). This can also happen when there is poor alignment or fit between worker abilities and values, and what the organization demands or values (Bogaerts et al., 2018; Kreiner, 2006). Within a role, conflict is especially likely to be experienced by workers who perceive high levels of ambiguity about their role expectations and requirements (i.e., role ambiguity; e.g., King & King, 1990; Rizzo et al., 1970). Some instances of

211

Table 9.1 Summary of Role Interface Models

	Model	Brief description
Non-causal/ conceptual models	Segmentation	Work/nonwork role domains are independent and have no influence on each other (e.g., Kreiner, 2006; Olson-Buchanan & Boswell, 2006)
	Congruence	Work/nonwork roles may fit or align with each other, societal expectations, and workers' underlying values (e.g., Perrone et al., 2005; Stuhlmacher & Poitras, 2010)
	Integration	Work/nonwork roles are so entwined that it is impossible to differentiate between them (e.g., Burke, 2004; Kossek et al., 2006)
Causal/ explanatory models	Spillover	Positive and negative experiences in one role may impact workers in one or more other roles; often conceptualized and measured in terms of time, behavior, or strain effects (Hammer et al., 2005; Hanson et al., 2006)
	Crossover	Workers' experiences in one role may have effects that are transmitted to others in other roles (e.g., Bakker et al., 2009; Hammer et al., 1997)
	Resource drain	Resources used to meet demands in one role are then unavailable to workers trying to meet demands in another role (e.g., Grandey & Cropanzano, 1999; Odle-Dusseau et al., 2012)
	Compensation	Workers' negative experiences in one role can be counteracted by positive experiences in another role (e.g., Baltes & Heydens-Gahir, 2003; Kossek et al., 2011)
	Enhancement/ Facilitation	Workers' experiences in one role may facilitate or enhance their functioning in one or more other roles (e.g., Sieber, 1974; Wiese et al., 2010)

interrole conflict may be more accurately labeled *role overload* (i.e., responding to heavy demands in a role hinders a worker's ability to address all role-related duties; e.g., Brown et al., 2005; Kahn et al., 1964) or *role interference* (i.e., when multiple roles require simultanous involvement, which impairs one's ability to meet the demands of either role; e.g., Janssen et al., 2004).

Managing within and between role conflict is complicated because it is often bi- or multi-directional, and because workers differ in terms of which roles they prioritize and see as most salient and therefore influential (e.g., Greenhaus & Powell, 2003). Many factors can contribute to work and nonwork role conflict, including role-specific demands and various demographic factors (Byron, 2005; Gutek et al., 1991; Matthews et al., 2012; Voydanoff, 1988). An implication is that the experience of work and nonwork role conflicts differs by person, underscoring again the need for consideration of individual differences (as discussed in Chapter 3). Despite the complexity, efforts to help workers avoid work and nonwork role conflicts are needed, given that such conflicts can negatively impact workers' psychological and physical health in many ways (e.g., Allen & Armstrong, 2006; Greenhaus et al., 2006).

Enhancement and Facilitation

Although responding to role-related demands can indeed deplete resources, our involvements and achievements in different roles can also (sometimes concurrently) replenish or enrich our resources and improve our chances of success across role domains. Managing multiple role involvements, therefore, is not necessarily a zero-sum game. Occupying and participating in multiple roles (i.e., *role accumulation*; Sieber, 1974) may provide opportunities and mechanisms for building resources that can enhance, facilitate, or otherwise enrich our performance in other role domains (Chen & Powell, 2012; Greenhaus & Powell, 2006; Hanson et al., 2006; Wiese et al., 2010).

Examples of this phenomenon abound and might include how participating in a musical ensemble can improve one's focus in all roles, or how coaching a child's sports team might improve one's ability to manage subordinates at work. Even more simply, happiness experienced with a family member might help you stay upbeat at work. Another, less commonly discussed way that interrole facilitation works, is when occupying one role makes it possible for you to occupy or participate in another. This is often why people join professional organizations or students pursue education through particular institutions and degree programs, to facilitate access to and achievement of desired career-related roles and outcomes.

Balance

Related to the positive interrole dynamics concept of facilitation is the much-popularized notion of interrole balance. This positively framed form of interrole dynamic is defined in various ways, including as: the degree to which a person is equally involved and satisfied with their work and nonwork roles (Greenhaus et al., 2003); when a person is experiencing no/low conflict or interference between roles (Frone, 2003); and when there is stability among work and nonwork demands and the search for daily accomplishment and satisfaction (Reece et al., 2009). A more recent attempt at a comprehensive definition suggests that workers experience balance when they consider their emotional experiences, involvements, and effectiveness in certain roles to be favorable and in alignment with the value they attach to those roles (Casper et al., 2018, p. 18).

An issue to consider regarding this phenomenon is that use of the "balance" label implies that there is some sort of end-state or condition that can be achieved, in which a worker has all the resources necessary to meet present demands across all of their life roles. Unfortunately, few people ever experience this type of cross-role, homeostatic condition. Instead, most workers are constantly juggling available resources to meet various competing role-related demands. The point here is that *balancing* of role demands and resources is an active and ongoing process, rather than a specific occurrence or event (e.g., Bacigalupe, 2002; Hall & Richter, 1988).

Why Work and Nonwork Role Dynamics Matter

Beyond providing fertile ground for interesting research, workers' abilities to manage work and nonwork roles are critically important to maintaining WHSWB and the functioning of organizations and society in general. In this section, we explore several reasons why this is so.

Person-Level Reasons

As noted earlier, the roles we occupy can change how we think, feel, and behave. These effects also are not limited to one role, but accumulate and impact other roles. Consider military service members or police officers trained to be vigilant while working, who then may struggle with hyperawareness even in casual nonwork situations. Awareness of role-prescribed behaviors or mindsets that can affect workers across multiple life roles can help workers, and their families and friends to better manage role boundaries, role demands, and role-related resources.

Work is just one major role domain in our lives. It can affect and be affected by our role-related experiences in all other nonwork roles. There is a delicate ecosystem in which each of us lives and it takes coordination among all roles for this ecosystem to function well. Balancing our various role-related demands and resources in this ecosystem allows our role involvements to facilitate and structure our social engagements and opportunities for goal achievement, meaningfulness, purpose, etc.

Role engagements also grant us access to others who can help us to make sense of what is happening when our work and/or nonwork roles become ambiguous or otherwise difficult to manage. We process information, not just through our own minds, but also from those around us in the various roles we hold (e.g., social information processing and social comparison theory; Gerber et al., 2018; Salancik & Pfeffer, 1978). In addition, facilitating healthy management of work and nonwork roles increases workers' opportunities to maintain a robust source of many essential resources such as affiliation with others, financial security, and a sense of meaning or purpose.

Business and Organizational Reasons

Role-related expectations, and intra- and interrole dynamics also have a major influence on worker productivity and performance. As noted earlier in

214

this chapter, these role-related forces influence how workers think, feel, and behave when confronted with work-related demands and other stimuli. Early findings demonstrated that role-related conflicts are negatively associated with job performance, especially when role ambiguity is high (Fried et al., 1998; Fried & Tiegs, 1995). Additional research suggests that the effects of interrole interference or conflict may be mediated by negative effects on worker concentration (e.g., Demerouti et al., 2007). More recent research also shows that workers' perception of family-related support from their supervisors is positively linked with job performance (Odle-Dusseau et al., 2012; Odle-Dusseau et al., 2016).

Thus, understanding work and nonwork role dynamics is not just important to individual workers. Businesses and other organizations can also leverage the power of positive role dynamics to attract, develop, and retain workers. Consider that the process of recruiting applicants for open positions in organizations is really the selling of a role and its various facets, including its potential dynamics with other roles. Effective recruiting efforts help candidates develop a realistic sense of whether a role is likely to be a good fit for them or not (Uggerslev et al., 2012; Yu, 2014).

Organizations can appeal to candidates by offering and emphasizing policies and benefits that support positive work and nonwork interrole dynamics (Casper & Buffardi, 2004; Cunningham, 2008; Honeycutt & Rosen, 1997). Such marketing requires more than reframing existing policies and procedures; it requires real effort to build and support an organizational culture that facilitates healthy management of work and nonwork roles. This might involve highlighting an organization's commitment to flexible scheduling, telecommuting, and paid time off to care for dependants. These types of policies and practices really appeal to individuals who have complex nonwork lives to manage, such as parents and those who care for elderly parents or a disabled spouse or partner. After the period of recruitment and attraction, corresponding actual support from the organization can strengthen workers' attachment to the organization (Casper & Harris, 2008) and yield positive health benefits for workers and cost savings for organizations (e.g., Jennings et al., 2016).

Sometimes organizational leaders assume that single workers with no dependants and no nonwork commitments will be more likely to identify strongly with their work and therefore be the best performers. Instead, workers who are involved in multiple work and nonwork roles also tend to be more embedded in their communities and less likely to dramatically change course (and potentially turnover) than those who have no roots or likelihood of establishing roots (e.g., Fasbender et al., 2019; Porter et al., 2019). Given these logical extensions of role-related theory and research, organizations can benefit from facilitating healthy management of work and nonwork roles as a way of increasing more generalizable social capital (Direnzo et al., 2015; Hauser et al., 2015) and strengthening the connection between workers and their communities.

Broader Societal Reasons

There are other societal benefits in addition to the gains in social capital and general community involvement that are possible with healthier management of work and nonwork role dynamics. Workers' abilities to managing work and nonwork life roles are closely linked to decisions they make regarding where to live, work, and play. These decisions, in turn, affect the structure of geographic work areas, residential areas, schooling zones, etc. This has not been a major area of study within OHP yet, but the COVID-19 pandemic reminded many workers and organizations that as work roles change, so do nonwork roles and societal level functioning (and vice versa).

Another societal reason to understand work and nonwork role dynamics better pertains to issues of fairness, equity, and inclusion. Failing to address work and nonwork role dynamics can bias organizations and societal institutions against workers who simultaneously manage complex work and nonwork demands (e.g., child and adult dependants, service commitments, religious involvements, education, health and medical care). Such bias can prevent organizations from making real advancements toward diversity and inclusion.

Methodological Considerations and Practical Recommendations

There are many methodological techniques employed by OHP professionals who research and intervene to improve workers' abilities to manage work and nonwork role dynamics. As with all WHSWB challenges explored in this book, it is important to begin with good information or data, which can be gathered using a variety of measurement approaches. Working from this information, it is possible to take steps to improve work and nonwork role dynamics at the individual, work group, leader, and organizational level. We explore a variety of such strategies in this section.

Measuring and Monitoring Work and Nonwork Role Dynamics

Work and nonwork role dynamics are inherently personal, limiting our measurement and evaluation options to self-reported methods of quantifying and qualifying workers' and significant others' perceptions of extent to which role dynamics are positive, negative, facilitating/enhancing, conflicting, etc. For this reason, most research and evaluation efforts in this area have leveraged surveys and diary studies to gather data (e.g., Hewett et al., 2017; Kempen et al., 2019). There are also some situations in which interviews can be particularly valuable (e.g., Ford & Collinson, 2011). Finally, there may be some value in observational methods with the goal is high-level surveillance or monitoring of trends regarding, for instance, personal time and effort investment in specific role domains (e.g., Galizzi et al., 2010).

Intervening to Improve Work and Nonwork Role Dynamics

Improving workers' abilities to manage their own work and nonwork role dynamics is an important step toward addressing related WHSWB concerns. As highlighted throughout this chapter, however, pertinent theories and research suggest that such individual-level efforts are unlikely to be sufficient on their own. It is also, therefore, necessary to address role dynamics at the group, leader, and organization level.

Strategies for Individuals

Interventions can help workers develop and implement strategies for managing work and nonwork roles in a way that fits with personal and family needs, values, and available resources. Individual level interventions can also help workers to identify critical resources needed to meet role-related demands, and to develop strategies for addressing these needs. Because work and nonwork role-related decisions have ripple effects that extend beyond the worker, there is also value in translating strong OHP research about work and nonwork role-related decision making (e.g., Shockley & Allen, 2015) into interventions designed to help workers develop more realistic and accurate concepts of how the role-related choices they make are likely to create facilitation or create conflict. This type of intervention would also necessarily involve helping people understand who they are (in terms of role-related identity salience), what they value, and what the significant others in their lives also value and need. Interventions along these lines might focus on helping people identify and understand their values and priorities, and make authentic career, family, education, and service role-related decisions that are in alignment with these attributes.

It is important to teach workers that their own management of role dynamics is influenced by other people in their lives. As an example, Hahn and Dormann (2013) found that workers' detachment is linked to their partners' preferences for segmentation between work and nonwork roles, and ultimately to workers' own life satisfaction. All of these relationships were also affected by the presence or absence of children in the home. As Greenhaus and Callanan (2020) note, "Work-related decisions . . . that are informed by nonwork considerations . . . can help employees experience greater balance in life because a broader range of values or needs is taken into account than if the decision had been based exclusively on work-related factors" (p. 478). This is closely aligned with the notion of workers' developing characteristic boundary management styles, which can be seen as "a competency for personal and life effectiveness" (Kossek, 2016, p. 259). More insightful recommendations along these lines are presented in Greenhaus and Powell (2017).

Frequent interrole transitions increase the potential for blurring or lack of clarity to emerge in roles. Research suggests that maintaining some sense of role boundaries and segmentation between roles is associated with less

217

work-family conflict (Kossek et al., 2012; Powell & Greenhaus, 2010) and more work-family balance (e.g., Li et al., 2013). Workers may benefit from interventions that teach practical strategies for managing interruptions across and transitions between role domains, such as through more controlled use of communications-related technology (e.g., Berkowsky, 2013; Boswell & Olson-Buchanan, 2007).

Finally, workers struggling to manage competing work and nonwork role-related demands may also benefit from interventions that facilitate access to the necessary resources to meet those demands. We need look no further than the COVID-19 pandemic to understand that a major nonwork demand with direct work-related consequences is how to care for children while also being a working parent. Needed resources here might include flexibility in scheduling work meetings and delivering completed work, and resources to support child care. This simple example illustrates how a resource-based perspective can be leveraged to improve workers' ability to manage interrole dynamics.

Strategies for Groups, Leaders, and Organizations

Although work and nonwork role dynamics are inherently personal, such dynamics are not solely governed by person-level forces. There are several ways in which interventions designed for working groups, leaders, and broader organizations can be helpful in this area of OHP practice. Consider that for many workers, the work domain is less permeable than and therefore less likely to be impacted by demands and influences from other nonwork roles. This, combined with the consistent strength and intensity of demands most workers experience in their work roles, grants these roles major influence over WHSWB.

Simply put, it is easier to protect and promote WHSWB when work role demands and resources are well-aligned. In practice, this means that organizations must work toward building and sustaining a healthy work-life cultures (Foucreault et al., 2018). As outlined with a focus on work and family role domains by Thompson et al. (1999), such a culture is likely to be reflected in workers' perceptions of organizational time expectations, consequences linked to use of work-life programs, and managerial support for managing family/life responsibilities. In a general sense, interventions can be designed to address all three of these perceptions with initiatives operating at the group, leader, and organizational levels.

Indeed, most role dynamics intervention efforts that target groups, leaders, and organizations have focused on improving the supportiveness of organizational supervision (e.g., Lapierre & Allen, 2006), so that supervisors demonstrate, encourage, and facilitate use of an organization's policies and resources to support work and nonwork role management (e.g., paid leave for family care, flexible scheduling). The logic behind these types of policies and this kind of flexibility, is that such support allows workers to address nonwork demands without feeling like they are being penalized at work for doing so.

It is one thing for an organization to offer flexibility to workers, but if workers do not feel encouraged and supported to use that flexibility, it might as well not exist. Similarly, if workers do not perceive that they actually have control to exercise this flexibility, then additional demands may be perceived (as discussed in Chapters 6 and 7) and negative health and well-being related consequences may develop (e.g., Kossek et al., 2006).

A number of theoretically and empirically derived intervention efforts have been undertaken to address these points (for a helpful summary and discussion, see Kossek et al., 2014). Specifically, workers' control over work time and approach to work has been shown to improve perceptions of schedule control and reduce work-family conflict, among other benefits (Hill et al., 2013; Moen et al., 2011). It has also been shown that FSSB can be trained and as a result decrease workers' work-family conflict and improve a variety of health and well-being outcomes associated with positive work and nonwork role dynamics (Crain et al., 2014; Hammer et al., 2009). Building on these and other studies, a more comprehensive approach to improving workers' management of work and nonwork role demands by increasing work scheduling control and supervisor support for positive work and nonwork role dynamics has been developed and tested as the *Support-Transform-Achieve-Results (STAR)* intervention. This intervention has been strongly tested in a variety of settings and the results indicate many positive effects, including reductions in worker stress and burnout, and improvements in various indicators of worker well-being (e.g., Kossek et al., 2019; Moen et al., 2016). Notably, it has also been shown that this intervention can yield these positive WHSWB-related impacts without negatively impacting worker performance (Bray et al., 2018). Helpful resources associated with the preceding interventions are available for direct access through the Work, Family & Health Network (https://workfamilyhealthnetwork.org/toolkits-achieve-workplace-change).

There are a variety of ways the elements to the preceding intervention frameworks can be applied in organizational settings. One method might involve training and engaging supervisors to model healthy work and nonwork role management behaviors. Such training might involve helping supervisors, managers, and other leaders to recalibrate what they see as commitment, work ethic, and employee engagement so that they are more realistic when evaluating subordinates and colleagues. A good example of this is seen in the work of Koch and Binnewies (2015), which showed that supervisors who demonstrated work-home segmentation were perceived as supportive role models and their subordinate workers reported more work and nonwork role segmentation of their own and less exhaustion and disengagement. Similarly, supervisors who set clear boundaries to keep work separate from nonwork life signal organization-level consideration for workers' personal nonwork lives and needs (e.g., Friedman & Lobel, 2003).

Interventions related to the preceding concepts have empowered some workers and minimized the extent to which they experience conflict between work and nonwork roles. We want to emphasize, however that many of these types of group, leader, and organization-level interventions are still focused on changing perceptions, thoughts, and behaviors ultimately at the level of individual workers. We are hopeful that more organizations will seriously consider interventions that involve actual redesign to work roles and their associated demands to optimize workers' access to essential resources and minimize the likelihood of work-related demands spilling over into or interrupting workers' nonwork role engagements. This approach to intervention does not place the demand for managing work and nonwork role demands solely on workers, but rather involves the organization and its leadership in making changes to the way work is assigned, expectations are managed, and communications are controlled. An example of work along these lines is found in Stanko and Beckman (2015), who demonstrated the effectiveness of organization-level and technology-based efforts to improve workers' boundary control abilities and attentional focus.

Evaluating Interventions Addressing Work and Nonwork Role Dynamics

Methods and strategies for evaluating interventions in this domain do not differ much from those used in studying many other OHP-related phenomena. The material we presented in Chapter 2 of this book regarding evaluating OHP interventions can, therefore, be helpful when evaluating work and nonwork role dynamic interventions. It is important, however, to remember that interventions addressing work and nonwork role dynamics are ultimately aimed at changing qualities (perceived and actual) of workers' intra- and inter-role experiences. For this reason, most evaluation methods will necessarily involve data gathered through subjective self-reports. Inferences based on such data can be strengthened by designing evaluations that also gather corresponding information from spouses, significant others, and dependants in the nonwork domain, and coworkers and supervisors in the work domain. The most common methods for gathering such data efficiently include surveys, diary studies, and interviews. We have cited a number of strong examples of interventions that demonstrate these and other recommended evaluation techniques throughout this chapter, especially earlier in this section.

Concluding Thoughts and Reality Check

Work and nonwork role dynamics is an area of OHP research and practice for which the boundaries (ironic word choice, we know) are not entirely clear or strong. There is substantial overlap among the essential topics discussed in this chapter and in other chapters of this book. This should make sense, given that roles are the context in which all of these phenomena occur. Acknowledging these areas of overlap and considering the material presented in this chapter, it is easy to understand how WHSWB can be protected and promoted, and how workers will work better and be more committed and engaged in their work organizations, when they are encouraged and supported to also engage in healthy work and nonwork role engagements.

Achieving healthy work and nonwork role dynamics is increasingly difficult, as technologies and norms combine to make work roles continuously salient and accessible. Workers need strategies, opportunities, and organizational support to actively and effectively manage their work and nonwork role demands and resources. We are hopeful that increased understanding of these matters and our shared global experiences with the effects of the COVID-19 pandemic will help workers and organizations work together to reimagine what positive work and nonwork role dynamics are.

Media Resources

- Description of technology tools from Microsoft to monitor use of nonwork time:
  https://docs.microsoft.com/en-us/workplace-analytics/myanalytics/use/wellbeing
- Magazine-type article discussing how work-related stress can be bad for your children:
  https://www.fatherly.com/love-money/your-work-stress-could-be-affecting-your-kids-health/
- News article on why having a hobby outside of work is good for your career:
  https://www.cnbc.com/2017/08/02/3-science-backed-reasons-having-a-hobby-will-help-your-career.html

---

Discussion Questions

1) What is a role and what are different ways in which different roles can interact and/or relate to each other?
2) Is interrole conflict inevitable for all working adults? What about their spouses, children, neighbors, etc.?
3) Is interrole enhancement a feasible possibility for everyone? What might this look like?
4) When you think of work-family/work-nonwork balance, what comes to mind?
5) What is an organization's responsibility with respect to helping workers manage their work and nonwork role dynamics?

Professional Profile: Leslie Hammer, Ph.D.

Country/region: USA, Oregon
Current position title: Professor, Department of Psychology, Portland State University and Associate Director, OHP Training program. Professor, Oregon Institute of Occupational Health Sciences and the PSU-OHSU School of Public Health, Director, Oregon Healthy Workforce Center.
Background: I have a Ph.D. in Industrial-Organizational Psychology from Bowling Green State University where I also received my Masters Degree. I received my Bachelor's degree from James Madison University. I have been working to improve the health, safety, and/or well-being of workers since receiving my Ph.D. in 1991. I belong to the Society for Occupational Health Psychology (Founding President), the American Society of Safety Professionals, the Society for Industrial and Organizational Psychology (Fellow), the European Academy of Occupational Health Psychology (Fellow), the Academy of Management, and the American Psychological Association (Fellow). I accepted my academic position at Portland State University (PSU) in the Applied Psychology Ph.D. program before I had my dissertation data collected. I arrived with my dissertation data in hand in December 1990 as an Instructor and taught courses Winter and Spring quarters 1991. I defended my dissertation in the summer of 1991 and started as an Assistant Professor in the Department of Psychology at PSU in the Fall of 1991 and have been there ever since. In September 2015 I accepted a position as a Professor in the Oregon Institute of Occupational Health Sciences at OHSU and since then have split my time between OHSU and PSU.

I conduct research in the area of Occupational Health and I mentor students at both OHSU and PSU. My research focuses on ways in which organizations can help reduce work and family stress and improve positive spillover among employees by facilitating both formal and informal workplace supports, such as Family Supportive Supervisor Behavior (FSSB) training. I have worked with such employee populations as grocery workers, health care workers (specifically nursing aid workers), construction workers, information technology workers, and the military (active duty, national guard/ reserves, and veterans). My work focuses on workplace programs aimed at training supervisors how to better support workers and in turn reduce workers stress and improve health, safety, and well-being. I have been involved with numerous workplace intervention randomized controlled trials evaluating the effectiveness of supervisor support training. I am the Associate Director of the

Occupational Health Psychology graduate training program at Portland State University that is funded through a training program grant from NIOSH. I am also the Co-Director of the NIOSH-funded Oregon Healthy Workforce Center (OHWC), one of six centers of excellence in Total Worker Health®.

How my work impacts WHSWB: I develop supervisor support training and evaluate the effectiveness of that training health, safety and well-being outcomes of workers using randomized controlled trials in workplace settings. My research has been funded by the NIH, NIOSH, and the Department of Defense. I have developed and evaluated the Safety and Health Improvement Program (SHIP) as part of my work with the Oregon Healthy Workforce Center. The Study for Employment of Veterans (SERVe) was recently completed and focused on increasing supervisor support and enhancing employment outcomes for veterans reintegrating into the workforce, testing the effectiveness of a supervisor training workplace intervention. The Military Sleep and Health Study (MESH) is currently underway and focuses on increasing supervisor support for sleep, testing the effectiveness of a supervisor training sleep leadership intervention on sleep, health, and safety. And finally, the recently-funded behavioral health and resilience grant, Readiness Supportive Leadership Training (RESULT) is aimed at training Active Duty leaders about how to support their subordinates' efforts in the area of health and well-being. All of these studies grew out of my earlier work with the Work, Family and Health Network.

I was the director of one of six centers funded by the NIH and NIOSH- Work, Family, and Health Network (WFHN) that ran from 2005–2014 (see https://workfamilyhealthnetwork.org/). Dr. Ellen Kossek was my co-director and we worked very closely throughout our work with the Network. The Network completed a cluster randomized trial of workplace policies and programs that impact worker health, safety, and well-being. My research focuses on ways in which organizations can help reduce work and family stress and improve positive spillover among employees by facilitating both formal and informal workplace supports, such as the FSSB training. I have worked with such employee populations as grocery workers, health care workers (specifically nursing aid workers), construction workers, information technology workers, and am currently working with employment support and retention for our nation's military veterans. My work with the WFHN has involved a network of interdisciplinary research teams from the University of Minnesota, Penn State University, Harvard University, Portland State University, Michigan State University, Kaiser Permanente's Center for Health Research, and RTI International and the University of Southern California and has led to numerous published studies on the effectiveness of the workplace intervention that integrated the FSSB and face-to-face facilitated sessions on workers, the workplace, and family outcomes.

My motivation: I am passionate about improving the working conditions for workers and their families and especially among those more marginalized workers. I believe there are ways that organizations can be changed to better support workers and am passionate about facilitating those changes.

Chapter References

Allen, T. A., & Armstrong, J. (2006). Further examination of the link between work-family conflict and physical health: The role of health-related behaviors. *American Behavioral Scientist, 49*(9), 1204–1221. https://doi.org/10.1177/0002764206286386

Allen, T. D. (2001). Family-supportive work environments: The role of organizational perceptions. *Journal of Vocational Behavior, 58*(3), 414–435. https://doi.org/10.1006/jvbe.2000.1774

Allen, T. D., Cho, E., & Meier, L. L. (2014). Work–Family Boundary Dynamics. *Annual Review of Organizational Psychology and Organizational Behavior, 1*(1), 99–121. https://doi.org/10.1146/annurev-orgpsych-031413-091330

Allen, T. D., French, K. A., Dumani, S., & Shockley, K. M. (2020). A cross-national meta-analytic examination of predictors and outcomes associated with work-family conflict. *Journal of Applied Psychology, 105*(6), 539–576. https://doi.org/10.1037/apl0000442

Bacigalupe, G. (2002). Is balancing family and work a sustainable metaphor? *Journal of Feminist Family Therapy, 13*(2–3), 5–20. https://doi.org/10.1300/J086v13n02_02

Bakker, A. B., Westman, M., & van Emmerik, I. J. H. (2009). Advancements in crossover theory. *Journal of Managerial Psychology, 24*(3), 206–219. https://doi.org/10.1108/02683940910939304

Baltes, B. B., & Heydens-Gahir, H. A. (2003). Reduction of work-family conflict through the use of selection, optimization, and compensation behaviors. *Journal of Applied Psychology, 88*(6), 1005–1018. https://doi.org/10.1037/0021-9010.88.6.1005

Baltes, B. B., Briggs, T. E., Huff, J. W., Wright, J. A., & Neuman, G. A. (1999). Flexible and compressed workweek schedules: A meta-analysis of their effects on work-related criteria. *Journal of Applied Psychology, 84*(4), 496–513. https://doi.org/10.1037//0021-9010.84.4.496

Berkowsky, R. W. (2013). When you just cannot get away: Exploring the use of information and communication technologies in facilitating negative work/home spillover. *Information, Communication & Society, 16*(4), 519–541. https://doi.org/10.1080/1369118x.2013.772650

Biddle, B. J. (1986). Recent developments in role theory. *Annual Review of Sociology, 12*, 67–92. https://doi.org/10.1146/annurev.so.12.080186.000435

Bogaerts, Y., De Cooman, R., & De Gieter, S. (2018). Getting the work-nonwork interface you are looking for: The relevance of work-nonwork boundary management fit. *Frontiers in Psychology, 9*, 1158. https://doi.org/10.3389/fpsyg.2018.01158

Boswell, W. R., & Olson-Buchanan, J. B. (2007). The use of communication technologies after hours: The role of work attitudes and work-life conflict. *Journal of Management, 33*(4), 592–610. https://doi.org/10.1177/0149206307302552

Bray, J. W., Hinde, J. M., Kaiser, D. J., Mills, M. J., Karuntzos, G. T., Genadek, K. R., Kelly, E. L., Kossek, E. E., & Hurtado, D. A. (2018). Effects of a flexibility/support intervention on work performance: Evidence from the Work, Family, and Health Network. *American Journal of Health Promotion, 32*(4), 963–970. https://doi.org/10.1177/0890117117696244

Brenner, P. S., Serpe, R. T., & Stryker, S. (2014). The causal ordering of prominence and salience in identity theory: An empirical examination. *Social Psychology Quarterly, 77*(3), 231–252. https://doi.org/10.1177/0190272513518337

Brown, S. P., Jones, E., & Leigh, T. W. (2005). The attenuating effect of role over-load on relationships linking self-efficacy and goal level to work performance. *Journal of Applied Psychology, 90*(5), 972–979. https://doi.org/10.1037/0021-9010.90.5.972

Bulger, C. A., Matthews, R. A., & Hoffman, M. E. (2007). Work and personal life boundary management: Boundary strength, work/personal life balance, and the segmentation-integration continuum. *Journal of Occupational Health Psychology, 12*(4), 365–375. https://doi.org/10.1037/1076-8998.12.4.365

Burke, R. J. (2004). Work and personal life integration. *International Journal of Stress Management, 11*(4), 299–304. https://doi.org/10.1037/1072-5245.11.4.299

Byron, K. (2005). A meta-analytic review of work–family conflict and its anteced-ents. *Journal of Vocational Behavior, 67*(2), 169–198. https://doi.org/10.1016/j.jvb.2004.08.009

Capitano, J., & Greenhaus, J. H. (2018). When work enters the home: Antecedents of role boundary permeability behavior. *Journal of Vocational Behavior, 109*, 87–100. https://doi.org/10.1016/j.jvb.2018.10.002

Casper, W. J., & Buffardi, L. C. (2004). Work-life benefits and job pursuit intentions: The role of anticipated organizational support. *Journal of Vocational Behavior, 65*(3), 391–410. https://doi.org/10.1016/j.jvb.2003.09.003

Casper, W. J., & Harris, C. M. (2008). Work-life benefits and organizational attach-ment: Self-interest utility and signaling theory models. *Journal of Vocational Behav-ior, 72*(1), 95–109. https://doi.org/10.1016/j.jvb.2007.10.015

Casper, W. J., Vaziri, H., Wayne, J. H., DeHauw, S., & Greenhaus, J. (2018). The jingle-jangle of work-nonwork balance: A comprehensive and meta-analytic review of its meaning and measurement. *Journal of Applied Psychology, 103*(2), 182–214. https://doi.org/10.1037/apl0000259

Chen, Z., & Powell, G. N. (2012). No pain, no gain? A resource-based model of work-to-family enrichment and conflict. *Journal of Vocational Behavior, 81*(1), 89–98. https://doi.org/10.1016/j.jvb.2012.05.003

Clark, B., Chatterjee, K., Martin, A., & Davis, A. (2019). How commuting affects sub-jective wellbeing. *Transportation.* https://doi.org/10.1007/s11116-019-09983-9

Clark, S. C. (2000). Work/family border theory: A new theory of work/family balance. *Human Relations, 53*(6), 747–770. https://doi.org/10.1177/0018726700536001

Crain, T. L., Hammer, L. B., Bodner, T., Kossek, E. E., Moen, P., Lilienthal, R., & Buxton, O. M. (2014). Work-family conflict, family-supportive supervisor behav-iors (FSSB), and sleep outcomes. *Journal of Occupational Health Psychology, 19*(2), 155–167. https://doi.org/10.1037/a0036010

Cunningham, C. J. L. (2008). Keeping work in perspective: Work-nonwork consider-ations and applicant decision making. *Employee Responsibilities and Rights Journal, 21*(2), 89–113. https://doi.org/10.1007/s10672-008-9095-x

Demerouti, E., Bakker, A. B., Nachreiner, F., & Schaufeli, W. B. (2001). The job demands-resources model of burnout. *Journal of Applied Psychology, 86*(3), 499–512. https://doi.org/10.1037/0021-9010.86.3.499

Demerouti, E., Taris, T. W., & Bakker, A. B. (2007). Need for recovery, home–work interference and performance: Is lack of concentration the link? *Journal of Voca-tional Behavior, 71*(2), 204–220. https://doi.org/10.1016/j.jvb.2007.06.002

Dey, M., Frazis, H., Loewenstein, M. A., & Sun, H. (2020). Ability to work from home: evidence from two surveys and implications for the labor market in the COVID-19 pandemic, *Monthly Labor Review,* U.S. Bureau of Labor Statistics, June 2020, https://doi.org/10.21916/mlr.2020.14

Direnzo, M. S., Greenhaus, J. H., & Weer, C. H. (2015). Relationship between pro-tean career orientation and work-life balance: A resource perspective. *Journal of Organizational Behavior, 36*(4), 538–560. https://doi.org/10.1002/job.1996

Fasbender, U., Van der Heijden, B. I. J. M., & Grimshaw, S. (2019). Job satisfaction, job stress and nurses' turnover intentions: The moderating roles of on-the-job and off-the-job embeddedness. *Journal of Advanced Nursing, 75*(2), 327–337. https://doi.org/10.1111/jan.13842

Ford, J., & Collinson, D. (2011). In search of the perfect manager? Work-life balance and managerial work. *Work, Employment and Society, 25*(2), 257–273. https://doi.org/10.1177/0950017011398895

Foucreault, A., Ollier-Malaterre, A., & Ménard, J. (2018). Organizational culture and work–life integration: A barrier to employees' respite? *The International Journal of Human Resource Management, 29*(16), 2378–2398. https://doi.org/10.1080/09585192.2016.1262890

French, K. A., Allen, T. D., & Henderson, T. G. (2019). Challenge and hindrance stressors and metabolic risk factors. *Journal of Occupational Health Psychology, 24*(3), 307–321. https://doi.org/10.1037/ocp0000138

Fried, Y., & Tiegs, R. B. (1995). Supervisors' role conflict and role ambiguity dif-ferential relations with performance ratings of subordinates and the moderating effect of screening ability. *Journal of Applied Psychology, 80*(2), 282–291. https://doi.org/10.1037/0021-9010.80.2.282

Fried, Y., Ben-David, H. A., Tiegs, R. B., Avital, N., & Yeverechyahu, U. (1998). The interactive effect of role conflict and role ambiguity on job performance. *Journal of Occupational and Organizational Psychology, 71*, 19–27. https://doi.org/10.1111/j.2044-8325.1998.tb00659.x

Friedman, S. D., & Lobel, S. (2003). The happy workaholic: A role model for employ-ees. *Academy of Management Executive, 17*(3), 401–412. https://doi.org/10.5465/ame.2003.10954764

Frone, M. R. (2003). Work-family balance. In J. C. Quick & L. E. Tetrick (Eds.), *Handbook of occupational health psychology* (pp. 143–162). American Psychological Association. https://doi.org/10.1037/10474-007

Galizzi, M., Miesmaa, P., Punnett, L., & Slatin, C. (2010). Injured workers' underre-porting in the health care industry: An analysis using quantitative, qualitative, and observational data. *Industrial Relations: A Journal of Economy and Society, 49*(1), 22–43. https://doi.org/10.1111/j.1468-232X.2009.00585.x

Gerber, J. P., Wheeler, L., & Suls, J. (2018). A social comparison theory meta-analysis 60+ years on. *Psychological Bulletin, 144*(2), 177–197. https://doi.org/10.1037/bul0000127

Grandey, A. A., & Cropanzano, R. (1999). The Conservation of Resources model applied to work-family conflict and strain. *Journal of Vocational Behavior, 54*, 350–370. https://doi.org/10.1006/jvbe.1998.1666

Grawitch, M. J., Maloney, P. W., Barber, L. K., & Mooshegian, S. E. (2013). Exam-ining the nomological network of satisfaction with work-life balance. *Journal of Oc-cupational Health Psychology, 18*(3), 276–284. https://doi.org/10.1037/a0032754

Greenhaus, J. H., & Beutell, N. J. (1985). Sources of conflict between work and fam-ily roles. *The Academy of Management Review, 10*(1), 76–88. https://doi.org/10.5465/amr.1985.4277352

Greenhaus, J., & Callanan, G. (2020). Implications of the changing nature of work for the interface between work and nonwork roles. In B. Hoffman, M. Shoss, & L. Wegman (Eds.), *The Cambridge handbook of the changing nature of work*

(Cambridge Handbooks in Psychology, pp. 467–488). Cambridge: Cambridge University Press. https://doi.org/10.1017/9781108278034.022

Greenhaus, J. H., & Powell, G. N. (2003). When work and family collide: Deciding between competing role demands. *Organizational Behavior and Human Decision Processes*, *90*(2), 291–303. https://doi.org/10.1016/s0749-5978(02)00519-8

Greenhaus, J. H., & Powell, G. N. (2006). When work and family are allies: A theory of work-family enrichment. *The Academy of Management Review*, *31*(1), 72–92. https://doi.org/10.5465/amr.2006.19379625

Greenhaus, J. H., & Powell, G. N. (2017). *Making work and family work: From hard choices to smart choices.* New York, NY: Routledge.

Greenhaus, J. H., Allen, T. D., & Spector, P. E. (2006). Health consequences of work-family conflict: The dark side of the work-family interface. In P. L. Perrewé & D. C. Ganster (Eds.), *Employee health, coping and methodologies* (*Research in occupational stress and well being*, Vol. 5, pp. 61–98). Emerald Group Publishing Limited. https://doi.org/10.1016/s1479-3555(05)05002-x

Greenhaus, J. H., Collins, K. M., & Shaw, J. D. (2003). The relation between work–family balance and quality of life. *Journal of Vocational Behavior*, *63*(3), 510–531. https://doi.org/10.1016/s0001-8791(02)00042-8

Gutek, B. A., Searle, S., & Klepa, L. (1991). Rational versus gender role explanations for work-family conflict. *Journal of Applied Psychology*, *76*(4), 560–568. https://doi.org/10.1037/0021-9010.76.4.560

Hahn, V. C., & Dormann, C. (2013). The role of partners and children for employees' psychological detachment from work and well-being. *Journal of Applied Psychology*, *98*(1), 26–36. https://doi.org/10.1037/a0030650

Halbesleben, J. R. B., Harvey, J., & Bolino, M. C. (2009). Too engaged? A Conservation of Resources view of the relationship between work engagement and work interference with family. *Journal of Applied Psychology*, *94*(6), 1452–1465. https://doi.org/10.1037/a0017595

Hall, D. T., & Richter, J. (1988). Balancing work life and home life: What can organizations do to help? *The Academy of Management Executive*, *2*(3), 213–223. https://doi.org/10.5465/ame.1988.4277258

Hammer, L. B., Allen, E., & Grigsby, T. D. (1997). Work-family conflict in dual-earner couples: Within-individual and crossover effects of work and family. *Journal of Vocational Behavior*, *50*, 185–203. https://doi.org/10.1006/jvbe.1996.1557

Hammer, L. B., Cullen, J. C., Neal, M. B., Sinclair, R. R., & Shafiro, M. V. (2005). The longitudinal effects of work-family conflict and positive spillover on depressive symptoms among dual-earner couples. *Journal of Occupational Health Psychology*, *10*(2), 138–154. https://doi.org/10.1037/1076-8998.10.2.138

Hammer, L. B., Kossek, E. E., Yragui, N. L., Bodner, T. E., & Hanson, G. C. (2009). Development and validation of a multidimensional measure of Family Supportive Supervisor Behaviors (FSSB). *Journal of Management*, *35*(4), 837–856. https://doi.org/10.1177/0149206308328510

Hanson, G. C., Hammer, L. B., & Colton, C. L. (2006). Development and validation of a multidimensional scale of perceived work-family positive spillover. *Journal of Occupational Health Psychology*, *11*(3), 249–265. https://doi.org/10.1037/1076-8998.11.3.249

Hauser, C., Perkmann, U., Puntscher, S., Walde, J., & Tappeiner, G. (2015). Trust works! Sources and effects of social capital in the workplace. *Social Indicators Research*, *128*(2), 589–608. https://doi.org/10.1007/s11205-015-1045-z

Hewett, R., Haun, V. C., Demerouti, E., Rodríguez Sánchez, A. M., Skakon, J., & De Gieter, S. (2017). Compensating need satisfaction across life boundaries: A daily diary study. *Journal of Occupational and Organizational Psychology*, 90(2), 270–279. https://doi.org/10.1111/joop.12171

Hill, E. J., Allen, S., Jacob, J., Bair, A. F., Bikhazi, S. L., Van Langeveld, A., Martinengo, G., Parker, T. T., & Walker, E. (2007). Work family facilitation: Expanding theoretical understanding through qualitative exploration. *Advances in Developing Human Resources*, 9(4), 507–526. https://doi.org/10.1177/1523422307305490

Hill, R., Tranby, E., Kelly, E., & Moen, P. (2013). Relieving the time squeeze? Effects of a white-collar workplace change on parents. *Journal of Marriage and Family*, 75(4), 1014–1029. https://doi.org/10.1111/jomf.12047

Hobfoll, S. E. (1989). Conservation of resources: A new attempt at conceptualizing stress. *American Psychologist*, 44(3), 513–524. https://doi.org/10.1037/0003-066x.44.3.513

Honeycutt, T. L., & Rosen, B. (1997). Family friendly human resource policies, salary levels, and salient identity as predictors of organizational attraction. *Journal of Vocational Behavior*, 50(2), 271–290. https://doi.org/10.1006/jvbe.1996.1554

Janssen, P. P. M., Peeters, M. C. W., Jonge, J. d., Houkes, I., & Tummers, G. E. R. (2004). Specific relationships between job demands, job resources and psychological outcomes and the mediating role of negative work–home interference. *Journal of Vocational Behavior*, 65(3), 411–429. https://doi.org/10.1016/j.jvb.2003.09.004

Jennings, K. S., Sinclair, R. R., & Mohr, C. D. (2016). Who benefits from family support? Work schedule and family differences. *Journal of Occupational Health Psychology*, 21(1), 51–64. https://doi.org/10.1037/a0039651

Kahn, R. L., Wolfe, D. M., Quinn, R. P., Snoek, J. D., & Rosenthal, R. A. (1964). *Organizational stress: Studies in role conflict and ambiguity*. John Wiley.

Katz, D., & Kahn, R. L. (1966). *The social psychology of organizations*. Wiley.

Kempen, R., Roewekaemper, J., Hattrup, K., & Mueller, K. (2019). Daily affective events and mood as antecedents of life domain conflict and enrichment: A weekly diary study. *International Journal of Stress Management*, 26(2), 107–119. https://doi.org/10.1037/str0000104

King, L. A., & King, D. W. (1990). Role conflict and ambiguity: A critical assessment of construct validity. *Psychological Bulletin*, 107(1), 48–64. https://doi.org/10.1037/0033-2909.107.1.48

Koch, A. R., & Binnewies, C. (2015). Setting a good example: Supervisors as work-life-friendly role models within the context of boundary management. *Journal of Occupational Health Psychology*, 20(1), 82–92. https://doi.org/10.1037/a0037890

Kossek, E. E. (2016). Managing work-life boundaries in the digital age. *Organizational Dynamics*, 45(3), 258–270. https://doi.org/10.1016/j.orgdyn.2016.07.010

Kossek, E. E., Baltes, B. B., & Matthews, R. A. (2011). How work-family research can finally have an impact in organizations. *Industrial and Organizational Psychology*, 4, 352–369. https://doi.org/10.1111/j.1754-9434.2011.01353.x

Kossek, E. E., Hammer, L. B., Kelly, E. L., & Moen, P. (2014). Designing work, family & health organizational change initiatives. *Organizational Dynamics*, 43(1), 53–63. https://doi.org/10.1016/j.orgdyn.2013.10.007

Kossek, E. E., Lautsch, B. A., & Eaton, S. C. (2006). Telecommuting, control, and boundary management: Correlates of policy use and practice, job control, and work–family effectiveness. *Journal of Vocational Behavior*, 68(2), 347–367. https://doi.org/10.1016/j.jvb.2005.07.002

Kossek, E. E., Ruderman, M. N., Braddy, P. W., & Hannum, K. M. (2012). Work–nonwork boundary management profiles: A person-centered approach. *Journal of Vocational Behavior, 81*(1), 112–128. https://doi.org/10.1016/j.jvb.2012.04.003

Kossek, E. E., Thompson, R. J., Lawson, K. M., Bodner, T., Perrigino, M. B., Hammer, L. B., Buxton, O. M., Almeida, D. M., Moen, P., Hurtado, D. A., Wipfli, B., Berkman, L. F., & Bray, J. W. (2019). Caring for the elderly at work and home: Can a randomized organizational intervention improve psychological health? *Journal of Occupational Health Psychology, 24*(1), 36–54. https://doi.org/10.1037/ocp0000104

Kreiner, G. E. (2006). Consequences of work-home segmentation or integration: A person-environment fit perspective. *Journal of Organizational Behavior, 27*(4), 485–507. https://doi.org/10.1002/job.386

Kreiner, G. E., Hollensbe, E. C., & Sheep, M. L. (2009). Balancing borders and bridges: Negotiating the work-home interface via boundary work tactics. *Academy of Management Journal, 52*(4), 704–730. https://doi.org/10.5465/amj.2009.43669916

Lapierre, L. M., & Allen, T. A. (2006). Work-supportive family, family-supportive supervision, use of organizational benefits, and problem-focused coping: Implications for work-family conflict and employee well-being. *Journal of Occupational Health Psychology, 11*(2), 169–181. https://doi.org/10.1037/1076-8998.11.2.169

Li, Y., Miao, L., Zhao, X., & Lehto, X. (2013). When family rooms become guest lounges: Work–family balance of B&B innkeepers. *International Journal of Hospitality Management, 34,* 138–149. https://doi.org/10.1016/j.ijhm.2013.03.002

Lobel, S. A. (1991). Allocation of investment in work and family roles: Alternative theories and implications for research. *The Academy of Management Review, 16*(3), 507–521. https://doi.org/10.5465/amr.1991.4279467

Matthews, R. A., & Barnes-Farrell, J. L. (2010). Development and initial evaluation of an enhanced measure of boundary flexibility for the work and family domains. *Journal of Occupational Health Psychology, 15*(3), 330–346. https://doi.org/10.1037/a0019302

Matthews, R. A., Swody, C. A., & Barnes-Farrell, J. L. (2012). Work hours and work-family conflict: the double-edged sword of involvement in work and family. *Stress and Health, 28*(3), 234–247. https://doi.org/10.1002/smi.1431

Moen, P., Kelly, E. L., Fan, W., Lee, S., Almeida, D., Kossek, E. E., & Buxton, O. M. (2016). Does a flexibility/support organizational initiative improve high-tech employees' well-being? Evidence from the Work, Family, and Health Network. *American Sociological Review, 81*(1), 134–164. https://doi.org/10.1177/0003122415622391

Moen, P., Kelly, E. L., Tranby, E., & Huang, Q. (2011). Changing work, changing health: Can real work-time flexibility promote health behaviors and well-being? *Journal of Health and Social Behavior, 52*(4), 404–429. https://doi.org/10.1177/0022146511418979

Nicholson, N. (1984). A theory of work role transition. *Administrative Science Quarterly, 29*(2), 172–191. https://doi.org/10.2307/2393172

Nippert-Eng, C. (1996). Calendars and keys: The classification of "home" and "work". *Sociological Forum, 11*(3), 563–582. https://doi.org/10.1007/BF02408393

Odle-Dusseau, H. N., Britt, T. W., & Greene-Shortridge, T. M. (2012). Organizational work-family resources as predictors of job performance and attitudes: The process of work-family conflict and enrichment. *Journal of Occupational Health Psychology, 17*(1), 28–40. https://doi.org/10.1037/a0026428

Odle-Dusseau, H. N., Hammer, L. B., Crain, T. L., & Bodner, T. E. (2016). The influence of family-supportive supervisor training on employee job performance and attitudes: An organizational work-family intervention. *Journal of Occupational Health Psychology, 21*(3), 296–308. https://doi.org/10.1037/a0039961

Olson-Buchanan, J. B., & Boswell, W. R. (2006). Blurring boundaries: Correlates of integration and segmentation between work and nonwork. *Journal of Vocational Behavior, 68*(3), 432–445. https://doi.org/10.1016/j.jvb.2005.10.006

Perrone, K. M., Webb, L. K., & Blalock, R. H. (2005). The effects of role congruence and role conflict on work, marital, and life satisfaction. *Journal of Career Development, 31*(4), 225–238.

Porter, C. M., Posthuma, R. A., Maertz, C. P., Joplin, J. R. W., Rigby, J., Gordon, M., & Graves, K. (2019). On-the-job and off-the-job embeddedness differentially influence relationships between informal job search and turnover. *Journal of Applied Psychology, 104*(5), 678–689. https://doi.org/10.1037/apl0000375

Powell, G. N., & Greenhaus, J. H. (2010). Sex, gender, and the work-to-family interface: Exploring negative and positive interdependencies. *Academy of Management Journal, 53*(3), 513–534. https://doi.org/10.5465/amj.2010.51468647

Redmond, L. S., & Mokhtarian, P. L. (2001). The positive utility of the commute: modeling ideal commute time and relative desired commute amount. *Transportation, 28*, 179–205.

Reece, K. T., Davis, J. A., & Polatajko, H. J. (2009). The representations of work-life balance in Canadian newspapers. *Work, 32*(4), 431–442. https://doi.org/10.3233/WOR-2009-0854

Rizzo, J. R., House, R. J., & Lirtzman, S. I. (1970). Role conflict and ambiguity in complex organizations. *Administrative Science Quarterly, 15*(2), 150–163.

Salancik, G. R., & Pfeffer, J. (1978). A social information processing approach to job attitudes and task design. *Administrative Science Quarterly, 23*(2), 224–253. https://doi.org/10.2307/2392563

Shockley, K. M., & Allen, T. D. (2015). Deciding between work and family: an episodic approach. *Personnel Psychology, 68*(2), 283–318. https://doi.org/10.1111/peps.12077

Sieber, S. D. (1974). Toward a theory of role accumulation. *American Sociological Review, 39*(4), 567–578.

Spieler, I., Scheibe, S., & Stamov Roßnagel, C. (2018). Keeping work and private life apart: Age-related differences in managing the work-nonwork interface. *Journal of Organizational Behavior, 39*(10), 1233–1251. https://doi.org/10.1002/job.2283

Stanko, T. L., & Beckman, C. M. (2015). Watching you watching me: Boundary control and capturing attention in the context of ubiquitous technology use. *Academy of Management Journal, 58*(3), 712–738. https://doi.org/10.5465/amj.2012.0911

Stryker, S., & Serpe, R. T. (1994). Identity salience and psychological centrality: Equivalent, overlapping, or complementary concepts? *Social Psychology Quarterly, 57*(1), 16–35. https://doi.org/10.2307/2786972

Stuhlmacher, A. F., & Poitras, J. (2010). Gender and job role congruence: A field study of trust in labor mediators. *Sex Roles, 63*(7–8), 489–499. https://doi.org/10.1007/s11199-010-9844-9

Sturges, J. (2012). Crafting a balance between work and home. *Human Relations, 65*(12), 1539–1559. https://doi.org/10.1177/0018726712457435

Swanberg, J. E., McKechnie, S. P., Ojha, M. U., & James, J. B. (2011). Schedule control, supervisor support and work engagement: A winning combination for workers in hourly jobs? *Journal of Vocational Behavior, 79*(3), 613–624. https://doi.org/10.1016/j.jvb.2011.04.012

ten Brummelhuis, L. L., & Bakker, A. B. (2012). A resource perspective on the work-home interface: the work-home resources model. *American Psychologist, 67*(7), 545–556. https://doi.org/10.1037/a0027974

Thompson, C. A., Beauvais, L. L., & Lyness, K. S. (1999). When work-family benefits are not enough: The influence of work-family culture on benefit utilization, organizational attachment, and work-family conflict. *Journal of Vocational Behavior, 54*(3), 392–415. https://doi.org/10.1006/jvbe.1998.1681

Uggerslev, K. L., Fassina, N. E., & Kraichy, D. (2012). Recruiting through the stages: A meta-analytic test of predictors of applicant attraction at different stages of the recruiting process. *Personnel Psychology, 65*(3), 597–660. https://doi.org/10.1111/j.1744-6570.2012.01254.x

van Hooff, M. L. (2015). The daily commute from work to home: Examining employees' experiences in relation to their recovery status. *Stress and Health, 31*(2), 124–137. https://doi.org/10.1002/smi.2534

Voydanoff, P. (1988). Work role characteristics, family structure demands, and work/family conflict. *Journal of Marriage and Family, 50*(3), 749–761. https://doi.org/10.2307/352644

Wiese, B. S., Seiger, C. P., Schmid, C. M., & Freund, A. M. (2010). Beyond conflict: Functional facets of the work–family interplay. *Journal of Vocational Behavior, 77*(1), 104–117. https://doi.org/10.1016/j.jvb.2010.02.011

Yu, K. Y. T. (2014). Person–organization fit effects on organizational attraction: A test of an expectations-based model. *Organizational Behavior and Human Decision Processes, 124*(1), 75–94. https://doi.org/10.1016/j.obhdp.2013.12.005

10

PHYSICAL AND ENVIRONMENTAL DEMANDS AND RESOURCES

Kristen Jennings Black and Christopher J. L. Cunningham

Workers in all industries and organizations are exposed to physical and environmental demands and resources while working. Demands like heavy lifting, loud noise, chemical exposure, shiftwork or even sedentary work can lead to significant psychological and physical strain. These demands and their effects have to be understood in conjunction with psychological experiences and individual differences (e.g., self-efficacy, age, health conditions) that can both affect the impact of physical demands and one's ability to cope with them. Understanding work-related physical demands and resources helps occupational health psychology (OHP) professionals identify ways to structure and design work environments to provide and replenish essential resources, while minimizing unnecessary demands. This chapter discusses these topics, along with intervention approaches to help workers adapt to physical demands (e.g., trainings to build skills and efficacy for carrying out physical tasks properly) and improve the resource-richness of the physical work environment (e.g., positive workplace aesthetics).

When you are finished reading this chapter, you should be able to:

LO 10.1: Define and provide examples of physical and environmental work demands and resources present in most occupational contexts.

LO 10.2: Explain how and why not all workers are affected uniformly by similar physical and environmental work demands.

LO 10.3: Describe an optimal balancing of physical and environmental demands and resources for a specific occupation.

LO 10.4: Identify and evaluate the effectiveness of interventions to address physical and environmental work demands.

Workers and Physical and Environmental Work Demands

There are many extreme examples of how physical and environmental factors at work can impact worker health, safety, and well-being (WHSWB). These range from the brain trauma or torn ligaments due to physical contact among professional athletes, to the accumulated lung damage due to poor air quality among miners. Some work, like mining and farming, has even acquired the label of *"3D work"* – dirty, dangerous, demanding, and sometimes degrading (Moyce & Schenker, 2018). Even work not characterized by these features exposes workers to a combination of physical and environmental characteristics that may function as demands or resources (a distinction we discussed more fully in Chapter 6).

We can begin to understand the effects of the physical work environment on workers through the lens of environmental psychology, which like many OHP perspectives, emphasizes fit between person and environment as key to optimal functioning (Canter & Craik, 1981; Leather et al., 2010). Specific to the present chapter, Wohlwill (1974) also emphasized the importance of ensuring optimal person-environment fit with respect to sensory stimulation, social stimulation, and movement. Over-stimulation (e.g., too much noise, too many distractions, too much movement) or under-stimulation (e.g., silent offices, isolation, sedentariness) at work is bad for worker productivity and well-being.

This concept of fitting workers to their work environments relates to the challenge of balancing demands and resources we outlined in Chapter 6. Physical and environmental demands place a sometimes heavy and often very consistent load on workers to which they must adapt. These demands affect workers if they simply do not have the resources to respond (e.g., their physical strength or self-efficacy is not sufficient for a task) or if they have exhausted a particular resource (e.g., physical endurance for completing a strenuous task). Serious strain can occur over time and from repeated exposure to these types of demands (e.g., years of nearly continuous keyboard typing or shiftwork-related disturbance of natural sleep-wake cycles). Thankfully, physical and environmental resources that facilitate effective management of demands can offset some of the negative impacts (e.g., adequate technology and resources, time, natural lighting, cleanliness).

Our goal in this chapter is to help you see how organizations can maximize fit between workers and their physical work environments. This involves ensuring that workers have sufficient resources to meet the physical and environmental demands associated with their work. Many essential characteristics of work environments have the potential to function as demands or resources; the challenge is optimizing workers' exposures to these forces to protect WHSWB (note that we explore the topic of safety more fully in Chapter 11).

Common Physical and Environmental Demands and Resources at Work

Understanding the design of work and its implications for WHSWB is essential to creating healthy work organization (Wilson et al., 2004). Thus, we begin our discussion here by building familiarity with work requirements and environmental characteristics that function as demands and/or resources. As we noted in Chapter 6, the absence of resources can act as a demand and the absence of a demand can be a resource. This is particularly true when considering work-related physical and environmental factors, which can easily be both a demand (e.g., poor lighting) or a resource (e.g., sufficient, natural lighting). This chapter provides a high-level overview of these topics. To learn more, explore the vast evidence base pertaining to these topics in associated research and practice literature from industrial medicine, ergonomics and human factors, and industrial hygiene.

Heavy Lifting, Exertion, and Repetitive Motions

A common set of physical demands are those that directly place a burden on the body's musculoskeletal system. *Physical demands* can include: lifting, pushing, or pulling heavy objects; performing the same motion over and over again; applying force with your fingers; twisting, bending, squatting, or kneeling; standing in one position for a long time; and holding your arms in one position for a long time (e.g., Sinclair et al., 2010). These demands are especially common among workers involved in physical labor (e.g., construction and manufacturing, agriculture, oil extraction, mining). Physically strenuous work often coincides with environmental exposure risks (e.g., heat, noise, exposure to pathogens), discussed a bit later.

Physical demands are present in many other occupations that may require shorter periods of physical work (e.g., delivery drivers who bend and lift heavy packages, nursing professionals who assist with moving a patient). "Hidden" physical demands can affect workers in other professions, often due to repetitive and unnatural body movement. For instance, dental professionals experience physical demands associated with the unnatural, static positions of their arms when performing cleanings or exams (Morse et al., 2010). Similarly, operators of heavy equipment and power tools experience physical demands in the form of vibrations, which have been linked to back pain, among other consequences (Burstrom et al., 2015).

Opportunities for Movement

Exerting oneself physically at work depletes certain types of resources. Counterintuitively, being too still can also be quite demanding. Prolonged standing can be a requirement for jobs that also involve much physical exertion, like construction and manufacturing, but it is also common in fields like healthcare, retail, and food service (Shockey et al., 2018), where cashiers or surgeons may remain standing in the same spot for hours. Excessive standing poses a risk for musculoskeletal complaints in particular, but resources like supportive shoes or floor mats can lessen the impact (Coenen et al., 2018; Waters & Dick, 2015). Similarly, sitting for long periods, although easier to maintain than standing, can also come with effects on the body, like lower back pain (Chen et al., 2009). There is evidence that sedentary behaviors in and outside of work are risk factors for many physical health outcomes, like diabetes and heart disease (van Uffelen et al., 2010), as well as metabolic syndrome (i.e., a clustering of several metabolic risk factors, like abdominal obesity, high blood pressure, high cholesterol; Edwardson et al., 2012).

Workers can benefit when their work allows for a variety of different positions and movements (i.e., not always sitting or standing). Although responding to the physical demands of work can drain some resources, when work provides opportunities for movement, this can facilitate development of other resources. For example, even if tangible physical health benefits are minimal, movement variety while working (i.e., facilitated by flexible or active workstations) can help generate positive psychological states and improvements in concentration (Carr et al., 2016; Kilpatrick et al., 2013). Generating these types of resources, at essentially zero cost to the organization, can be as simple as standing for an office meeting or discussing meeting topics over a walk (Clayton et al., 2015; Danquah & Tolstrup, 2020).

Environmental Demands and Exposures

Physical job demands often coexist with potentially harmful environmental characteristics and conditions, which are often subject to regulation from national or international agencies (e.g., International Labor Organization [ILO], Occupational Safety and Health Administration [OSHA], American National Standards Institute [ANSI]). Here we review several common environmental characteristics that often function as physical demands for workers.

Noise

The effects of noise in a work environment can range from simple annoyances that impede work performance (Lamb & Kwok, 2016) to long-term physical health consequences, such as actual hearing loss, which is an unfortunately common condition for workers in many industries (National Institute for Occupational Safety and Health [NIOSH], 2019). Some bursts of loud noise (i.e., over 120 decibels) can cause immediate damage to our auditory systems; regular exposure to noise over 70 or 80 decibels (i.e., just below the level of raising your voice to talk with someone an arm's length away) can cause damage over time (Centers for Disease Control and Prevention [CDC], 2019). As we mentioned in Chapter 5, noise is not the only concern for hearing loss; it can also result from exposure to chemicals labeled as ototoxicants (NIOSH, 2018b).

Most studies suggest that quieter work environments are a resource for workers. However, it is important to keep in mind Wohlwill's (1974) emphasis on balance that we mentioned earlier. Workers often differ in terms of preference and tolerance for noise; the absence of noise may be a resource to some workers and a demand to others who desire at least some background auditory stimulation.

Air Quality and General Respiratory Demands

Workers in any work environment may be exposed to poor air quality and more serious forms of exposure or demands on their respiratory systems. These might include exposure to mold, chemicals, and general lack of fresh air (NIOSH, 2013). Work environments associated with obvious respiratory demands are subject to targeted regulations. Some examples would be OSHA's (1990, 2017) standards for exposure to beryllium and asbestos, chemicals that have been linked with respiratory illness. Air quality is also necessarily monitored in the mining industry; in America, this is done primarily through the Mining Safety and Health Administration (https://msha.gov). More broadly, laboratory and field studies confirm that workers in buildings with better environmental air conditions (i.e., proper ventilation, temperature, humidity, carbon dioxide concentrations) had higher cognitive functioning and reported fewer "sick building" symptoms, like headaches or respiratory concerns (MacNaughton et al., 2015; MacNaughton et al., 2017). Attending to the relatively invisible impact of air quality can also have financial implications for organizations (DeAngelis, 2017).

Temperature

Many occupations involve working in environments (indoors or outdoors) that expose workers to extremely cold or hot temperatures. The effects of either extreme can be serious, including dehydration, heat exhaustion/hyperthermia, or hypothermia; in more mild cases, heat and cold can be uncomfortable and performance-limiting (Redden & Larkin, 2015). There is actually a very narrow range of temperatures shown to be comfortable and safe for indoor workers (between 68 and 76 degrees Fahrenheit; OSHA, 1999). Beyond being uncomfortable, environments that expose workers to temperatures outside this range can impose real practical limitations on worker functioning (e.g., impairing manual dexterity and judgment). There are many excellent resources available to help address risks of exposure to extreme temperatures (e.g., Jacklitsch et al., 2016; NIOSH, 2018a).

Lighting

Inadequate or excessively intense lighting conditions at work impose significant physical demands on workers (Redden & Larkin, 2015). When followed, safety regulations identify the minimum number of lighting sources and the visibility that should be present in work environments (OSHA, 2001). Although the importance of adequate lighting in manufacturing and other physical labor settings is obvious, performance in office work environments is also significantly and negatively impacted by poor lighting (Lamb & Kwok, 2016). In contrast, the presence of sufficient and/or natural lighting, may function as a powerful resource that improves and sustains workers' positive attitudes (Galasiu & Veitch, 2006; Leather et al., 1998).

Workspace Aesthetics and Design

Attention to *workplace aesthetics* has increased in recent years, as organizations consider how the look and feel of the workplace can be a resource to workers. Workers themselves are more likely to perceive needs for improvements in workplace aesthetics than ergonomics (Schell et al., 2011). Appropriate lighting and color schemes are specific workspace elements that appear to influence mood across cultural contexts and occupational samples (e.g., Kuller et al., 2006). From a worker's perspective, a window that provides natural light is one of the most consistently desired aesthetic features (Galasiu & Veitch, 2006). If natural light is not an option, incorporating more color or nature elements (e.g., plants, pictures of nature) in a workspace can have some positive effects on worker attitudes and psychological states (An et al., 2016; Dravigne et al., 2008; Kuller et al., 2006).

Organizations have to be thoughtful when implementing workspace changes, as work arrangements may not always have their intended effects. As one example, Robinson (2017) noted that one company's transition to a sleek, stylish, open-office space resulted in *less* collaboration among employees, as they were all wearing headphones to control excess noise. Workspaces should be designed to support (as a resource) workers' efforts to meet the demands inherent in their work (e.g., Wohlers et al., 2017). For example, most office workers need comfortable spaces to concentrate and spaces that facilitate collaboration when needed. Keep in mind, though, that workers also differ in their need for aesthetics, which appears to strengthen the relationship between such features and worker attitudes and well-being (Johnson & Cunningham, 2019) – it is important to understand workers' needs before spending thousands of dollars on new paint, office plants, or other aesthetic improvements.

Work Schedules

Although perhaps not an obvious physical feature of a work environment, work schedules play a major role in work experiences and directly trigger a number of other physical and environmental demands for workers. This is especially true when long working hours, insufficient breaks, and nonstandard schedules are the norm. Long work hours and the fatigue they cause can affect worker well-being (Angrave & Charlwood, 2015; Ganster et al., 2016). Shiftwork in particular has received a good deal of research attention, given its necessity in continuously functioning organizations (e.g., hospitals, manufacturing settings with continuous productions, 24-hour stores and restaurants). *Shiftwork* can be considered any schedule outside of "typical" 9AM–5PM weekday work hours (Smith et al., 2011). There is strong evidence that shiftwork negatively affects worker physical health, primarily through disruptions to normal sleep-wake cycles (i.e., circadian rhythm; Smith et al., 2011) and an impact on cognition that may take years to reverse (Marquie et al., 2015). Shiftwork is also linked to higher risks of safety incidents (Folkard & Tucker, 2003). Although there have been some efforts to minimize risks of this form of environmental demand through regulation (e.g., European Union Law [EUR-Lex], 2003), ultimate control over work schedules belongs to organizational leaders.

Influential Individual Differences, Psychological States, and Social Norms

Up to this point, we have highlighted a number of physical and environmental work demands that can affect WHSWB. It is also important to understand that workers differ in terms of their exposure and response to these demands. The effects of such demands are also influenced by a variety of person-specific psychological and social factors.

Some groups of individuals (e.g., males; racial and ethnic minorities) disproportionately occupy jobs that expose workers to high levels of physical and environmental demands, such as in the manufacturing, construction, and agricultural industries (U.S. Bureau of Labor Statistics [BLS], 2019). Underlying differences in personal traits and motivations also affect how we respond to demands in our environment. For instance, differences in motivation influence sedentary behavior at work (Gaston et al., 2016) and tendencies to experience positive emotion can weaken the effects of some environmental demands on worker health (Kożusznik et al., 2017). Other individual differences (e.g., mindfulness) affect the extent to which positive resources in the work environment matter to workers (Johnson & Cunningham, 2019) and the extent to which workers use resources to combat physical demands (e.g., self-efficacy affects use of proper lifting techniques; Asante et al., 2007).

The physical environment can also directly affect psychological health (e.g., lighting and noise annoyances correlate with mood, anger, and impulsive behavior; Lamb & Kwok, 2016; Simister & Cooper, 2005; Stansfeld & Matheson, 2003). Psychological processes also help explain how factors in the work environment affect workers (linking once again to the expanded stressor-response model outlined in Chapter 2; cf., Kożusznik et al., 2017; Rashid & Zimring, 2008). As examples, studies find that psychological variables (e.g., appraisals of one's work environment, motivation, distraction) explain why environmental demands affect both physical health and one's productivity (Kożusznik et al., 2017; Lamb & Kwok, 2016). Summarizing, "physical conditions matter together and in interaction with phenomena that occur at the cognitive and affective levels" (Kożusznik et al., 2017, p. 107). In addition, the social context at work influences how workers interpret and respond to physical and environmental demands, with workplace norms often dictating what is typical, expected, or normal when it comes to physical and environmental exposures (Bandura & Walters, 1977). The point to emphasize here is that truly understanding the effects of physical and environmental demands at work requires us to also attend to how workers perceive, appraise, and respond to such demands.

Why Physical and Environmental Demands and Resources Matter

Managing physical and environmental demands and resources at work is an essential component to a broad strategy to reduce work-related illnesses and injuries. We discuss physical health and safety costs more directly in Chapters 5 and 11, so our focus here is on the more general argument for intentional design and control of physical and environmental demands.

Benefits for Workers

Workers benefit from aesthetically pleasing and physically comfortable work environments, and when they have access to the physical and environmental resources needed to meet work demands (e.g., collaborative workspaces, quiet areas, necessary technology and tools). Some more obvious benefits are avoiding pain, discomfort, and injury associated with demands, while more subtle benefits include positive psychological states when workers can function well, and even enjoy, their work environment. These potential benefits are an incentive for workers to make changes that are under their control, such as incorporating movement variety into the workday or adding some personal touches to their workspace. The research we have cited throughout this chapter tells us that these little changes are worth it.

Benefits to the Organization

Reducing burdens on workers from physical and environmental demands can minimize risks of negative WHSWB issues and improve workers' productivity and *quality* of work. One study of automotive assembly-line workers showed that workers who consistently stayed at high-workload workstations made more mistakes that affected product quality than those at low-workload workstations (Ivarsson & Eek, 2016). Simply rotating workstations (and thereby modulating workload) helped maintain quality. Organizations that more broadly consider physical and environmental demands and provide protective equipment, rest breaks, and schedule flexibility are signaling clearly that they care for their workers (cf., Eisenberger et al., 1986). Social exchange perspectives tell us that workers tend to reciprocate good treatment, so workers who receive resources to meet their physical and environmental demands are likely to be more committed to their organization and less likely to leave (An et al., 2016; Leather et al., 1998). This same reciprocity principle is apparent in findings that workers find counterproductive workplace behaviors less acceptable in clean versus unclean work environments (Huangfu et al., 2017).

Improving less visible physical aspects of the work environment also may be financially smart for organizations. MacNaughton and colleagues (2015) illustrate this point in their calculations that improving ventilation rates in an office building would cost less than $40 per person, but yield an estimated $6,500 increase in employee productivity each year. This estimate does not sound so extreme when you consider how air quality affects comfort, cognition, and sickness. For readers who want to learn more, the Healthy Buildings Program within the Harvard School of Public Health (https://forhealth.org/) provides a number of useful resources regarding how work environments affect workers and organizations.

Work environments that meet workers' needs help to retain healthier and safer workers, which ultimately support healthier and stronger organizations. We want to emphasize that physical and environmental improvements at work do not have to cost a fortune (e.g., allowing workers to personalize their workspaces or offer suggestions for physical or environmental resource needs). Even when efforts targeting physical and environmental demands are expensive (e.g., major technology or infrastructure upgrades), few interventions at work are likely to have a greater or longer-lasting effect on WHSWB.

Methodological Considerations and Practical Recommendations

Physical and environmental demands and resources are evaluated using various methods. Observation and objective metrics may be particularly valuable in this area of OHP research and practice, but other methods are also needed to dig into the closely related psychological and social factors that influence our experience of these physical features of work environments. In the following subsections, we discuss several essential methodological considerations and intervention recommendations for managing physical and environmental demands and resources.

Measuring and Monitoring Physical and Environmental Demands

More objective metrics (e.g., pertaining to air quality, noise levels, temperature), are necessary for understanding many environmental demands. Trained evaluators (typically occupational health professionals with various educational backgrounds) can also identify less obvious physical and environmental demands that may be present in a work environment or embedded in a work process (e.g., observing tasks that involve a great deal of physical effort, identifying machinery that excessively dangerous to operate). Trained evaluators can also identify physical and environmental demands that workers themselves might not even notice (e.g., chemical exposures, repetitive motions). One observational method that may be particularly useful when studying these types of demands is *process mapping*, which involves identifying and documenting how work is done. The resulting step-by-step understanding or "map" helps identify points in the process that could be improved, such as by changing the process to minimize demands or offset demands with additional resources (e.g., additional workers, training, new technology).

Observational methods and objective metrics do not tell the whole story when it comes to physical and environmental demands. Comparing objective data (e.g., noise measured in decibels) with employee perceptions and reactions to such demands provides a more comprehensive picture of workers' actual experiences (e.g., Kożusznik et al., 2017). There are helpful measures available for this type of evaluation, including NIOSH's generic job stress questionnaire (Hurrell & McLaney, 1988) and components of the Work Design Questionnaire (Morgeson & Humphrey, 2006) that include questions about perceptions of noise, temperature, humidity, exposure to dangerous substances, cleanliness, and other factors. In some situations, more contextualized assessments of demands are pertinent, such as for retail (e.g., Sinclair et al., 2010) or manufacturing settings (e.g., Ivarsson & Eek, 2016). Leiter and Robichaud (1997) provide another example of multifaceted assessment of physical and environmental demands when they asked Canadian Air Force personnel to rate a list of hazards (e.g., electrical shock, heavy lifting, interaction with aircraft engines) in terms of the prevalence, risk for personal harm, and perceived level of personal preparation for encountering these demands. These sorts of approaches can provide rich and meaningful data for understanding how workers are affected by their environment and how equipped they feel to meet their demands.

A multifaceted measurement and monitoring approach also can identify differing perceptions of environmental and physical demands among workers. As examples, experienced workers may feel quite comfortable in a risky environment or organizational leaders may view a work environment as relatively safe and free of physical hazards; these perceptions may not match what other workers perceive. Because objective and subjective perceptions of these demands and resources can differ among workers and other organizational stakeholders, focus groups, interviews, and diary studies can provide a higher level of detail, which can be especially valuable when planning and evaluating interventions in this domain (e.g., Pehkonen et al., 2009; Quandt et al., 2013).

Interventions to Address Physical and Environmental Demands

The general goals of interventions to address physical and environmental demands and resources are to maximize comfort and safety in the work environment and increase workers' access to resources to meet demands. We focus more attention on safety-related concerns in Chapter 11, but in the remainder of this section, we present several practical strategies for generally improving the balance of work-related physical and environmental demands and resources.

Strategies for Individuals

In most occupations, it is impossible (and honestly, counterproductive) to entirely remove physical and environmental demands. The challenge is balancing such demands with the resources workers need to do their jobs and stay healthy and safe. One often overlooked, but very important person-centered strategy here is to modify and constrain work demands so that workers, given their actual resources and abilities, can meet them. A context-specific functional assessment can be an effective approach for establishing whether adjustments or accommodations could help to reduce or prevent avoidable strain from physical or environmental demands. Frings-Dresen and Sluiter (2003) developed an example of this type of approach to assess nurses' functional capacity to carry out realistic work tasks. This approach provides a practical method for helping to ensure a proper fit between workers and their work environments.

Another important and generalizable individual-level intervention strategy is to ensure individual workers have the appropriate knowledge and skills (i.e., through proper selection and training) to do their work in the safest and most efficient way possible. Training should make workers aware of risks associated with the physical and environmental demands they will face while doing their work (Smith & Carayon, 2011). This is important for any job, regardless of the severity of risks. Training should also help workers to better manage their demands. Examples of trainings that focus on safely managing physical and environmental hazards are vast.

For some issues, simple knowledge-based training can be quite effective. For instance, ergonomic education regarding proper posture and workstation arrangements has been found to affect worker behaviors and musculoskeletal risks, even if the tangible features of the workspace are not changed (Robertson et al., 2009). To address more complex physical and environmental demands, a combination of knowledge-based training (e.g., safe lifting techniques) and assistive technologies (e.g., lift assist devices) may improve the likelihood of achieving an intervention's goals (e.g., reducing injuries and associated costs; Aslam et al., 2015; Li et al., 2004). An important point here is that person-level interventions are more likely to be effective and lasting when supported by changes to the broader work context. Related to this, person-level interventions can empower workers to be more aware of physical and environmental demands at work and to work cooperatively with their leaders to make positive changes when they notice areas for improvement (e.g., Pehkonen et al., 2009).

Many training-related interventions pertaining to physical and environmental demands ultimately focus on changing worker behavior. As an example, Johnson and Hall (2005) leveraged the theory of planned behavior (Ajzen, 1991) to reduce lifting-related injuries among manufacturing workers. Their evaluation study of this intervention highlighted the importance of behavioral control as the primary antecedent predicting safe lifting, followed by supportive social norms and then individual attitudes. These findings suggest that training workers in relevant knowledge and skills will only partially address this challenge. It is also essential to ensure workers have the resources they need (e.g., time, assistive devices, and social supports) to adhere to their training and ultimately work safely.

There are increasingly clever ways to enhance workers' sense of control over physical and environmental demands. One illustration relevant to safe lifting is an application which enables workers to enter real-world data and determine in real-time the risks involved in a manual lifting task (NIOSH, 2017). We are very encouraged at what advancements in assistive and augmented reality technologies will bring to workers in environments characterized by significant physical and environmental demands.

Strategies for Groups, Leaders, and Organizations

As noted earlier, efforts to address physical and environmental demands are more likely to be effective when they are not only directed at individual workers. Several general intervention strategies can help by targeting groups and leaders within an organization. For starters, changing work group norms can be an effective way to build a resource that reinforces "good", healthy, and safe worker responses to physical and environmental work demands. Unfortunately, group norms can also function more like a demand than a resource if they encourage workers to push through pain, take on workloads that exceed personal capacity, or refrain from wearing protective equipment. The typical behaviors and expectations of work group members influence the behavior of the entire group. This has been shown with respect to multiple behaviors associated with physical and environmental demands at work, including safe lifting in manufacturing settings (Johnson & Hall, 2005) and proper use of respirator masks among smelters (Robertsen et al., 2018).

It is important to ensure that group norms provide more of a resource than a demand for workers. More specifically, interventions may need to focus on adjusting group norms that downplay or obscure the real risks associated with physical and environmental work demands. Strong interventions to improve workers' ability to manage physical and environmental demands at work may develop from a deep understanding of why such norms exist. For example, maybe workers in a particular unit do not use the provided protective equipment because it is uncomfortable or cumbersome. The intervention need is to reevaluate the equipment and the work practices that require it to see if there is another way to get the work done or a different form of protection that might be more effective in that situation.

Leaders also play an important role in setting and managing group norms and the overall physical and environmental conditions of the workplace (e.g., Zohar & Luria, 2003). Educating and empowering leaders about how to monitor and assess the demands and resources present in their work areas can have tremendous positive ripple effects. The seriousness with which leaders take on these tasks can also then reinforce positive group norms for handling physical and environmental demands. For example, if leaders value limiting exposure and hazards and do not sacrifice these values for the sake of other goals, it signals to employees that their health and safety matters. Leaders can also incorporate time for feedback and discussion regarding physical and environmental demands and resource needs during team meetings and regular performance evaluations.

One of the most direct ways leaders and organizations can control workers' exposures to physical and environmental demands is through modifications to work assignments and work schedules. Individual leaders make some of these adjustments, but some cases may require a more comprehensive change to the way multiple areas in an organization operate. In terms of work assignments, job rotations can be helpful for cross-training workers by alternating their tasks and thereby mixing up the physical and environmental demands to which they are exposed. Such rotations can also help to build broader functional knowledge, skill sets, and social networks, which are all forms of work-related resources that may benefit workers and the organization over time. These methods have been shown to benefit workers and improve the quality of their work (Ivarsson & Eek, 2016).

With respect to work schedules, simply allowing regular breaks to replenish resources can go a long way toward reducing fatigue, strain, and injuries (e.g., Fischer et al., 2017). Going a bit deeper, Folkard and Lombardi (2004) developed a practical list of factors that each amplify risks linked to work scheduling (e.g., night shifts and longer work periods are associated with significant increased incident risk; incident risk increases with number of days of consecutive work). Knowledge of these risks may help organizations to adjust working arrangements and schedules for entire groups of workers.

When shiftwork is necessary, there are some practical considerations that can lessen the burden on workers. For example, some individuals tolerate shiftwork better than others (i.e., some workers can more easily adapt their circadian rhythm and remain on fixed, rather than rotating shift schedules; Smith et al., 2011) and some shift rotation patterns are more favorable than others (e.g., Bushnell et al., 2010). Regardless of scheduling form, one of the most direct ways to improve workers' ability to manage physical and environmental demands at work is to provide workers some control over their work schedule (Smith et al., 2011). Scheduling that considers worker preferences and limitations can help keep workers engaged and able to complete tasks in their best physical and mental state.

Many physical and environmental demands and resources can only be managed at the level of an organization because that is the level at which control over these aspects of work exists. Certain types of physical and environmental demand exposures can be eliminated or prevented with the help of physical barriers (e.g., enclosing dangerous or loud machinery to reduce noise levels and minimize direct worker contact) or advanced technology (e.g., using robotics to remove workers from hazardous work environments). Many physical and environmental demands, however, are inseparable from workers' daily reality. In these instances, it may be necessary to consider the actual design of the work environment; helpful interactive tools pertaining to comprehensive healthy building design are available from the Harvard School of Public Health's Healthy Buildings program (https://9foundations.forhealth.org/) and facilitate consideration of environmental factors like noise, lighting, air quality, temperature, and ventilation.

With respect to the physical layout of the work space, organizations are also encouraged to do whatever is possible to ensure the work environment matches the tasks being carried out and that employees are involved in any work (re) design efforts (e.g., Wohlers et al., 2017). Work design decisions that minimize physical and environmental demands, and maximize access to resources needed to meet these demands can benefit workers and organizations in many ways. Along these lines, organizations can also make adjustments to address less obvious physical and environmental demands, such as repetitive motions (e.g., typing) or sedentary work assignments. In a typical office setting, ergonomic interventions have resulted in favorable outcomes when workers are educated on and provided the right workspace for good posture, or when given flexible workstations for sitting or standing (Agarwal et al., 2018; Robertson et al., 2009). More guidance regarding ergonomic changes is readily available through organizations like NIOSH (e.g., https://www.cdc.gov/niosh/topics/ergonomics/).

When it is not possible to prevent or reduce workers' exposure to physical and environmental work demands, other strategies can limit exposure and ultimate risks for negative effects on WHSWB. A common example of this is emphasizing consistent and proper use of *personal protective equipment* (PPE). Most PPE (e.g., hard hats, goggles, earplugs, face masks, gloves, kneepads, lifting support belts, specialized footwear) serves a specific purpose and therefore matches a particular physical or environmental demand very closely, essentially functioning as a resource that can protect a worker and facilitate their successful functioning in the presence of that demand. Here again, however, is a situation where worker psychology plays an essential role. Even when PPE is available, workers may choose not to use it because workgroup norms do not support it, they are personally unmotivated to use it, or the gear is uncomfortable (e.g., too hot, not properly fit) or performance-limiting (e.g., not being able to hear a co-worker with ear plugs; not being able to perform manual tasks with bulky gloves). These non-physical barriers to PPE use are documented in research across a variety of occupations ranging from healthcare workers to farmers to manufacturing workers (Damalas & Abdollahzadeh, 2016; Kang et al., 2017; Robertsen et al., 2018). In sum, providing the right resources may not be enough to ensure workers use them consistently and properly.

Evaluating Physical and Environmental Demand Interventions

Extending from the intervention design and evaluation principles we discussed in Chapter 2, it is especially important to remember that the influence of interventions to address physical and environmental demands and resources cannot be fully understood without also considering associated psychological and social factors. The best evaluation efforts for this area of OHP include some evaluation of person and environment interaction that also includes consideration of underlying individual differences, existing social forces and norms, and the actual physical and environmental demands and resources you may be most directly trying to address. Even if changes in the environment are purely physical (e.g., new equipment; reduced noise), you want to know how and why workers are or are not responding to those changes. Because many of the intervention strategies outlined in this chapter also require some level of financial investment, evaluation strategies need to be designed to generate at least rough estimates of return on these investments.

A good example here is a participatory ergonomic intervention among kitchen workers; interestingly, no significant pre-post changes were observed in musculoskeletal symptoms and physical workload for their intervention group compared to a control group (Haukka et al., 2008). The intervention group even reported more psychosocial stress following the intervention, which was unexpected (Haukka et al., 2010). The good news is these researchers gathered quite a bit of additional data that helped shed light on actual benefits of the intervention (Pehkonen et al., 2009). For example, focus groups with participants revealed that workers really were motivated by the approach and they perceived that they had fewer symptoms and their work had been improved (even though the objective reports did not necessarily support this). Workers also reported a possible barrier to the intervention's effectiveness in that they did not feel supported by management during the intervention. Data like these make it possible to improve upon and strengthen an intervention for greater impact in the future.

Concluding Thoughts and Reality Check

All workers respond to a variety of physical and environmental demands while working. Many of these demands are inherent and obvious in some occupations (e.g., noise, poor air quality), but quite subtle in others (e.g., sitting still, repetitive motion). There are strategies at the person, group, leader, and organization level to reduce worker exposure to these demands and improve workers' access to the resources necessary to respond to these demands. Interventions to address these forms of demand-resource relationships will be most effective if they also consider workers' psychological and social realities at work.

All organizations can design workspaces and work assignments that support and facilitate WHSWB, even while responding to legitimate work-related demands. When there is good alignment between physical and environment demands and resources, workers can be healthy, engaged, and committed to the work they are doing. This alignment can begin with simple opportunities for workers to control and personalize their own workspaces, to take short breaks throughout work periods, and to connect with nature or at least real sunlight while at work. These and larger-scale efforts can transform work from being only a source of physical and environmental demands, to also being a source for essential physical and environmental resources.

Media Resources

- Poor working conditions for migrant workers receive additional attention during the COVID-19 Pandemic: https://www.washingtonpost.com/outlook/2020/05/22/immigrant-workers-have-born-brunt-covid-19-outbreaks-meatpacking-plants/
- Robotics taking away some of the dangerous aspects of work: https://www.safetyandhealthmagazine.com/articles/16789-robots-in-the-workplace
- Heart disease risk associated with exposure to metals and pesticides: https://www.sciencedaily.com/releases/2018/12/181211190008.htm

---

Discussion Questions

1) What aspects of the physical work environment constitute "demands"? What aspects can be resources instead?

2) Are all workers affected similarly by physical characteristics of work environments? If not, what are some individual differences or other factors that also play a role?

3) What would characteristics of an "optimal physical work environment" be? How might these differ by occupation, industry, etc.?

4) What are pros/cons and best-practice design recommendations to keep in mind for different types of work environments (e.g., traditional office, open-office, manufacturing)?

Professional Profile: Stacy Croghan, M.S.

Country/region: USA, Alabama
Current position title: Principal at Labyrinth Solutions, LLC
Background: I have been working to improve the health, safety, and well-being of workers for 20+ years. My education includes a B.S. in Biochemistry from Auburn University and a M.S. of Public Health in Occupational and Environmental Hygiene from Johns Hopkins University. I am a Certified Industrial Hygienist (CIH), credentialed by the American Board of Industrial Hygiene, and a Certified Safety Professional (CSP), credentialed by the American Society of Safety Professionals. I am a member of the American Industrial Hygiene Association, American Conference of Governmental Industrial Hygienists, American Board of Industrial Hygiene, and the Board of Certified Safety Professionals.

As the former wife of a military officer, frequent moves to different states exposed me to a wide variety of opportunities to grow and expand my career. I began as an analytical chemist conducting water and soil quality research. After a move from the east coast, my undergraduate experience in microbiology helped me land my first consulting job focusing on indoor air quality. From there, I picked up other industrial hygiene skills as well as environmental skills related to petroleum-related water and soil quality. Eventually, I decided to focus on industrial hygiene and safety and sought out employment under CIH's and CSP's to grow my knowledge base and experience. After 20 years, I opened my own consulting firm which has the capability of addressing most industrial hygiene, safety and environmental issues experienced by my clients. I currently provide a full range of services in the industrial hygiene and safety field including risk assessments, inspections/audits, testing/sampling, gap analyses, program management and policy/program development.

I provide technical guidance in the areas of industrial hygiene, occupational safety and health, and environmental impacts to my clients. Throughout my career, I have worked with a broad range of Federal, state and commercial clients, across most industries, at varying levels within the client organizations. I have extensive executive-level, field and training experience covering a variety of EHS areas including regulatory compliance and reporting, management systems, industrial hygiene exposure assessments, program audits, chemical hygiene, and hazardous materials handling and disposal. My expertise spans from development of national level, agency-wide guidance down to site-specific data gathering and solution development. I specialize in policy and program development.

How my work impacts WHSWB: The services I provide focus on getting to the root cause of the client's issues and modifying the systems or developing controls to mitigate hazards and risks. This ultimately provides a safer work environment.

My motivation: I believe individuals should be able to go to work and know that their employer has done and continues to do everything possible to protect their workers so that they may be able to return home to their families each day, unharmed. I am driven to help employers achieve a safer work environment so that workers are not taken from their families by the very job they do to provide for their families.

Chapter References

Agarwal, S., Steinmaus, C., & Harris-Adamson, C. (2018). Sit-stand workstations and impact on low back discomfort: A systematic review and meta-analysis. *Ergonomics*, *61*(4), 538–552. https://doi.org/10.1080/00140139.2017.1402960

Ajzen, I. (1991). The theory of planned behavior. *Organizational Behavior and Human Decision Processes*, *50*, 179–211. https://doi.org/10.1016/0749-5978(91)90020-T

An, M., Colarelli, S. M., O'Brien, K., & Boyajian, M. E. (2016). Why we need more nature at work: Effects of natural elements and sunlight on employee mental health and work attitudes. *PLoS One*, *11*(5), e0155614. https://doi.org/10.1371/journal.pone.0155614

Angrave, D., & Charlwood, A. (2015). What is the relationship between long working hours, over-employment, under-employment and the subjective well-being of workers? Longitudinal evidence from the UK. *Human Relations*, *68*(9), 1491–1515. https://doi.org/10.1177/0018726714559752

Asante, A. K., Brintnell, E. S., & Gross, D. P. (2007). Functional self-efficacy beliefs influence functional capacity evaluation. *Journal of Occupational Rehabilitation*, *17*(1), 73–82. https://doi.org/10.1007/s10926-007-9068-1

Aslam, I., Davis, S. A., Feldman, S. R., & Martin, W. E. (2015). A review of patient lifting interventions to reduce health care worker injuries. *Workplace Health and Safety*, *63*(6), 267–275. https://doi.org/10.1177/2165079915580038

Bandura, A., & Walters, R. H. (1977). *Social learning theory* (Vol. 1). Englewood Cliffs, NJ: Prentice-Hall.

Burstrom, L., Nilsson, T., & Wahlstrom, J. (2015). Whole-body vibration and the risk of low back pain and sciatica: A systematic review and meta-analysis. *International Archives of Occupational and Environmental Health*, *88*(4), 403–418. https://doi.org/10.1007/s00420-014-0971-4

Bushnell, T., Colombi, A., Caruso, C. C., & Tak, S. (2010). Work schedules and health behavior outcomes at a large manufacturer. *Industrial Health*, *48*, 395–405. https://doi.org/10.2486/indhealth.MSSW-03

Canter, D. V., & Craik, K. H. (1981). Environmental psychology. *Journal of Environmental Psychology*, *1*, 1–11. https://doi.org/10.1016/S0272-4944(81)80013-8

Carr, L. J., Leonhard, C., Tucker, S., Fethke, N., Benzo, R., & Gerr, F. (2016). Total Worker Health intervention increases activity of sedentary workers. *American Journal of Preventive Medicine*, *50*(1), 9–17. https://doi.org/10.1016/j.amepre.2015.06.022

CDC. (2019). *What noises cause hearing loss?* https://www.cdc.gov/nceh/hearing_loss/what_noises_cause_hearing_loss.html

Chen, S. M., Liu, M. F., Cook, J., Bass, S., & Lo, S. K. (2009). Sedentary lifestyle as a risk factor for low back pain: A systematic review. *International Archives of Occupational and Environmental Health*, *82*(7), 797–806. https://doi.org/10.1007/s00420-009-0410-0

Clayton, R., Thomas, C., & Smothers, J. (2015). How to do walking meetings right. *Harvard Business Review*. https://hbr.org/2015/08/how-to-do-walking-meetings-right

Coenen, P., Willenberg, L., Parry, S., Shi, J. W., Romero, L., Blackwood, D. M., Maher, C. G., Healy, G. N., Dunstan, D. W., & Straker, L. M. (2018). Associations of occupational standing with musculoskeletal symptoms: A systematic review with meta-analysis. *British Journal of Sports Medicine*, *52*(3), 176–183. https://doi.org/10.1136/bjsports-2016-096795

Damalas, C. A., & Abdollahzadeh, G. (2016). Farmers' use of personal protective equipment during handling of plant protection products: Determinants of implementation. *Science of the Total Environment*, *571*, 730–736. https://doi.org/10.1016/j.scitotenv.2016.07.042

Danquah, I. H., & Tolstrup, J. S. (2020). Standing meetings are feasible and effective in reducing sitting time among office workers-walking meetings are not: Mixed-methods results on the feasibility and effectiveness of active meetings based on data from the "Take a stand!" study. *International Journal of Environmental Research and Public Health*, *17*(5). https://doi.org/10.3390/ijerph17051713

DeAngelis, D. (2017). Health buildlings, productive people. *Monitor on Psychology*, *48*(5), 40. https://www.apa.org/monitor/2017/05/cover-healthy-buildings

Dravigne, A., Waliczek, T. M., Lineberger, R. D., & Zajicek, J. M. (2008). The effect of live plants and window views of green spaces on employee perceptions of job satisfaction. *Horticulture Science*, *43*(1), 183–187. https://doi.org/10.21273/HORTSCI.43.1.183

Edwardson, C. L., Gorely, T., Davies, M. J., Gray, L. J., Khunti, K., Wilmot, E. G., Yates, T., & Biddle, S. J. (2012). Association of sedentary behaviour with metabolic syndrome: A meta-analysis. *PLoS One*, *7*(4), e34916. https://doi.org/10.1371/journal.pone.0034916

Eisenberger, R., Huntington, R., Hutchison, S., & Sowa, D. (1986). Perceived organizational support. *Journal of Applied Psychology*, *71*(3), 500–507. https://doi.org/10.1037/0021-9010.71.3.500

EUR-Lex. (2003). *Directive 2003/88/EC of the European Parliament and of the Council of 4 November 2003 concerning certain aspects of the organisation of working time.* https://eur-lex.europa.eu/legal-content/en/TXT/?uri=CELEX%3A32003L0088

Fischer, D., Lombardi, D. A., Folkard, S., Willetts, J., & Christiani, D. C. (2017). Updating the "Risk Index": A systematic review and meta-analysis of occupational injuries and work schedule characteristics. *Chronobiology International*, *34*(10), 1423–1438. https://doi.org/10.1080/07420528.2017.1367305

Folkard, S., & Lombardi, D. A. (2004). Toward a "Risk Index" to assess work schedules. *Chronobiology International*, *21*(6), 1063–1072. https://doi.org/10.1081/cbi-200036919

Folkard, S., & Tucker, P. (2003). Shift work, safety and productivity. *Occupational Medicine*, *53*(2), 95–101. https://doi.org/10.1093/occmed/kqg047

Frings-Dresen, M. H. W., & Sluiter, J. K. (2003). Development of a job-specific FCE protocol: The work demands of hospital nurses as an example. *Journal of Occupational Rehabilitation*, *13*(4). https://doi.org/10.1023/A:1026268620904

Galasiu, A. D., & Veitch, J. A. (2006). Occupant preferences and satisfaction with the luminous environment and control systems in daylit offices: A literature review. *Energy and Buildings*, *38*(7), 728–742. https://doi.org/10.1016/j.enbuild.2006.03.001

Ganster, D. C., Rosen, C. C., & Fisher, G. G. (2016). Long working hours and well-being: What we know, what we do not know, and what we need to know. *Journal of Business and Psychology*, *33*(1), 25–39. https://doi.org/10.1007/s10869-016-9478-1

Gaston, A., De Jesus, S., Markland, D., & Prapavessis, H. (2016). I sit because I have fun when I do so! Using self-determination theory to understand sedentary behavior motivation among university students and staff. *Health Psychology and Behavioral Medicine*, *4*(1), 138–154. https://doi.org/10.1080/21642850.2016.1170605

Haukka, E., Leino-Arjas, P., Viikari-Juntura, E., Takala, E. P., Malmivaara, A., Hopsu, L., Mutanen, P., Ketola, R., Virtanen, T., Pehkonen, I., Holtari-Leino, M., Nykanen, J., Stenholm, S., Nykyri, E., & Riihimaki, H. (2008). A randomised controlled trial on whether a participatory ergonomics intervention could prevent musculoskeletal disorders. *Occupational and Environmental Medicine*, 65(12), 849–856. https://doi.org/10.1136/oem.2007.034579

Haukka, E., Pehkonen, I., Leino-Arjas, P., Viikari-Juntura, E., Takala, E. P., Malmivaara, A., Hopsu, L., Mutanen, P., Ketola, R., Virtanen, T., Holtari-Leino, M., Nykanen, J., Stenholm, S., Ojajarvi, A., & Riihimaki, H. (2010). Effect of a participatory ergonomics intervention on psychosocial factors at work in a randomised controlled trial. *Occupational and Environmental Medicine*, 67(3), 170–177. https://doi.org/10.1136/oem.2008.043786

Huangfu, G., Lv, F., Sheng, C., & Shi, X. (2017). Effect of workplace environment cleanliness on judgment of counterproductive work behavior. *Social Behavior and Personality: An International Journal*, 45(4), 599–604. https://doi.org/10.2224/sbp.6083

Hurrell, J. J., & McLaney, M. A. (1988). Exposure to job stress: A new psychometric instrument. *Scandinavian Journal of Work, Environment & Health*, 14(1), 27–28.

Ivarsson, A., & Eek, F. (2016). The relationship between physical workload and quality within line-based assembly. *Ergonomics*, 59(7), 913–923. https://doi.org/10.1080/00140139.2015.1105303

Jacklitsch, B., Williams, W. J., Musolin, K., Coca, A., Kim, J. H., & Turner, N. (2016). *Occupational exposure to heat and hot environments: Revised criteria 2016* (2016-106). https://www.cdc.gov/niosh/docs/2016-106/pdfs/2016-106.pdf?id=10.26616/NIOSHPUB2016106

Johnson, L. F., & Cunningham, C. J. L. (2019). *Do workplace aesthetics matter? Testing the moderating effects of need for aesthetics and general mindfulness*. 13th International Conference on Work, Stress, & Health, Philadelphia, PA.

Johnson, S. E., & Hall, A. (2005). The prediction of safe lifting behavior: An application of the theory of planned behavior. *Journal of Safety Research*, 36(1), 63–73. https://doi.org/10.1016/j.jsr.2004.12.004

Kang, J., O'Donnell, J. M., Colaianne, B., Bircher, N., Ren, D., & Smith, K. J. (2017). Use of personal protective equipment among health care personnel: Results of clinical observations and simulations. *American Journal of Infection Control*, 45(1), 17–23. https://doi.org/10.1016/j.ajic.2016.08.011

Kilpatrick, M., Sanderson, K., Blizzard, L., Teale, B., & Venn, A. (2013). Cross-sectional associations between sitting at work and psychological distress: Reducing sitting time may benefit mental health. *Mental Health and Physical Activity*, 6(2), 103–109. https://doi.org/10.1016/j.mhpa.2013.06.004

Kożusznik, M. W., Peiró, J. M., Soriano, A., & Navarro Escudero, M. (2017). "Out of sight, out of mind?": The role of physical stressors, cognitive appraisal, and positive emotions in employees' health. *Environment and Behavior*, 50(1), 86–115. https://doi.org/10.1177/0013916517691323

Kuller, R., Ballal, S., Laike, T., Mikellides, B., & Tonello, G. (2006). The impact of light and colour on psychological mood: A cross-cultural study of indoor work environments. *Ergonomics*, 49(14), 1496–1507. https://doi.org/10.1080/00140130600858142

Lamb, S., & Kwok, K. C. (2016). A longitudinal investigation of work environment stressors on the performance and wellbeing of office workers. *Applied Ergonomics*, 52, 104–111. https://doi.org/10.1016/j.apergo.2015.07.010

Leather, P., Pyrgas, M., Beale, D., & Lawrence, C. (1998). Windows in the workplace: Sunlight, views, and occupational stress. *Environment and Behavior*, 30(6), 739–762. https://doi.org/10.1177/001391659803000601

Leather, P., Zarola, T., & Santos, A. (2010). The physical workplace: An OHP perspective. In S. Leka & J. Houdmont (Eds.), *Occupational health psychology* (pp. 225–249). Blackwell Publishing, Ltd.

Leiter, M. P., & Robichaud, L. (1997). Relationships of occupational hazards with burnout: An assessment of measures and models. *Journal of Occupational Health Psychology, 2*(1), 35–44. https://doi.org/10.1037//1076-8998.2.1.35

Li, J., Wolf, L., & Evanoff, B. (2004). Use of mechanical patient lifts decreased musculoskeletal symptoms and injuries among health care workers. *Injury Prevention, 10*(4), 212–216. https://doi.org/10.1136/ip.2003.004978

MacNaughton, P., Pegues, J., Satish, U., Santanam, S., Spengler, J., & Allen, J. (2015). Economic, environmental and health implications of enhanced ventilation in office buildings. *International Journal of Environmental Research and Public Health, 12*(11), 14709–14722. https://doi.org/10.3390/ijerph121114709

MacNaughton, P., Satish, U., Laurent, J. G. C., Flanigan, S., Vallarino, J., Coull, B., Spengler, J. D., & Allen, J. G. (2017). The impact of working in a green certified building on cognitive function and health. *Building and Environment, 114*, 178–186. https://doi.org/10.1016/j.buildenv.2016.11.041

Marquie, J. C., Tucker, P., Folkard, S., Gentil, C., & Ansiau, D. (2015). Chronic effects of shift work on cognition: Findings from the VISAT longitudinal study. *Occupational and Environmental Medicine, 72*(4), 258–264. https://doi.org/10.1136/oemed-2013-101993

Morgeson, F. P., & Humphrey, S. E. (2006). The Work Design Questionnaire (WDQ): Developing and validating a comprehensive measure for assessing job design and the nature of work. *Journal of Applied Psychology, 91*(6), 1321–1339. https://doi.org/10.1037/0021-9010.91.6.1321

Morse, T., Bruneau, H., & Dussetschleger, J. (2010). Musculoskeletal disorders of the neck and shoulder in the dental professions. *Work, 35*(4), 419–429. https://doi.org/10.3233/WOR-2010-0979

Moyce, S. C., & Schenker, M. (2018). Migrant workers and their occupational health and safety. *Annual Review of Public Health, 39*, 351–365. https://doi.org/10.1146/annurev-publhealth-040617-013714

NIOSH. (2013). *Indoor environmental quality.* https://www.cdc.gov/niosh/topics/indoorenv/default.html

NIOSH. (2017). *NIOSH Lifting equation app: NLE Calc.* https://www.cdc.gov/niosh/topics/ergonomics/nlecalc.html

NIOSH. (2018a). *Cold stress.* https://www.cdc.gov/niosh/topics/coldstress/default.html

NIOSH. (2018b). *Preventing hearing loss caused by chemical (ototoxicity) and noise exposure.* https://www.cdc.gov/niosh/docs/2018-124/default.html

NIOSH. (2019). *Occupational Hearing Loss (OHL) Surveillance.* https://www.cdc.gov/niosh/topics/ohl/default.html

OSHA. (1990). *Asbestos* (Toxic and Hazardous Substances, Issue). https://www.osha.gov/laws-regs/regulations/standardnumber/1910/1910.1001

OSHA. (1999). *OSHA technical manual.* https://www.osha.gov/dts/osta/otm/otm_iii/otm_iii_2.html#5

OSHA. (2001). *Illumination.* https://www.osha.gov/laws-regs/regulations/standardnumber/1926/1926.56

OSHA. (2017). *Protecting workers from exposure to Beryllium and Beryllium compounds: Final rule overview* (3821). https://www.osha.gov/Publications/OSHA3821.pdf

Pehkonen, I., Takala, E. P., Ketola, R., Viikari-Juntura, E., Leino-Arjas, P., Hopsu, L., Virtanen, T., Haukka, E., Holtari-Leino, M., Nykyri, E., & Riihimaki, H. (2009). Evaluation of a participatory ergonomic intervention process in kitchen work. *Applied Ergonomics, 40*(1), 115–123. https://doi.org/10.1016/j.apergo.2008.01.006

Quandt, S. A., Arcury-Quandt, A. E., Lawlor, E. J., Carrillo, L., Marin, A. J., Grzywacz, J. G., & Arcury, T. A. (2013). 3-D jobs and health disparities: The health implications of Latino chicken catchers' working conditions. *American Journal of Industrial Medicine, 56*(2), 206–215. https://doi.org/10.1002/ajim.22072

Rashid, M., & Zimring, C. (2008). A review of the empirical literature on the relationships between indoor environment and stress in health care and office settings: Problems and prospects on sharing evidence. *Environment and Behavior, 40*(2), 151–190. https://doi.org/10.1177/0013916507311550

Redden, E. S., & Larkin, G. B. (2015). Environmental conditions and physical stressors. In *APA handbook of human systems integration.* (pp. 193–209). https://doi.org/10.1037/14528-013

Robertsen, O., Siebler, F., Eisemann, M., Hegseth, M. N., Foreland, S., & Vangberg, H. B. (2018). Predictors of respiratory protective equipment use in the Norwegian smelter industry: The role of the Theory of Planned Behavior, safety climate, and work experience in understanding protective behavior. *Frontiers in Psychology, 9,* 1366. https://doi.org/10.3389/fpsyg.2018.01366

Robertson, M., Amick, B. C., 3rd, DeRango, K., Rooney, T., Bazzani, L., Harrist, R., & Moore, A. (2009). The effects of an office ergonomics training and chair intervention on worker knowledge, behavior and musculoskeletal risk. *Applied Ergonomics, 40*(1), 124–135. https://doi.org/10.1016/j.apergo.2007.12.009

Robinson, J. (2017). A not-so-quiet little problem: Noise! *Human Resource Executive, July/August,* 15–18.

Schell, E., Theorell, T., & Saraste, H. (2011). Workplace aesthetics: Impact of environments upon employee health? *Work, 39*(3), 203–213. https://doi.org/10.3233/WOR-2011-1182

Shockey, T. M., Luckhaupt, S. E., Groenewold, M. R., & Lu, M. L. (2018). Frequent exertion and frequent standing at work, by industry and occupation group – United States, 2015. *CDC: Morbidity and Mortality Weekly Report, 67*(1), 1–6. https://www.ncbi.nlm.nih.gov/pmc/articles/PMC5769795/pdf/mm6701a1.pdf

Simister, J., & Cooper, C. (2005). Thermal stress in the U.S.A.: Effects on violence and on employee behaviour. *Stress and Health, 21*(1), 3–15. https://doi.org/10.1002/smi.1029

Sinclair, R. R., Martin, J. E., & Sears, L. E. (2010). Labor unions and safety climate: Perceived union safety values and retail employee safety outcomes. *Accident Analysis and Prevention, 42*(5), 1477–1487. https://doi.org/10.1016/j.aap.2009.11.003

Smith, C. S., Folkard, S., Tucker, P., & Evans, M. S. (2011). Work schedules, health, and safety. In J. C. Quick & L. E. Tetrick (Eds.), *Handbook of occupational health psychology* (pp. 185–204). American Psychological Association.

Smith, M. J., & Carayon, P. (2011). Controlling occupational safety and health hazards. In J. C. Quick & L. E. Tetrick (Eds.), *Handbook of occupational health psychology* (Vol. 2, pp. 75–93). American Psychological Association.

Stansfeld, S. A., & Matheson, M. P. (2003). Noise pollution: Non-auditory effects on health. *British Medical Bulletin, 68,* 243–257. https://doi.org/10.1093/bmb/ldg033

U.S. BLS. (2019). *Household data annual averages: Employed persons detailed by occupation, sex, race, and hispanic or latino ethnicity.* https://www.bls.gov/cps/cpsaat11.htm

van Uffelen, J. G., Wong, J., Chau, J. Y., van der Ploeg, H. P., Riphagen, I., Gilson, N. D., Burton, N. W., Healy, G. N., Thorp, A. A., Clark, B. K., Gardiner, P. A., Dunstan, D. W., Bauman, A., Owen, N., & Brown, W. J. (2010). Occupational sitting and health risks: A systematic review. *American Journal of Preventative Medicine*, 39(4), 379–388. https://doi.org/10.1016/j.amepre.2010.05.024

Waters, T. R., & Dick, R. B. (2015). Evidence of health risks associated with prolonged standing at work and intervention effectiveness. *Rehabilitation Nursing*, 40(3), 148–165. https://doi.org/10.1002/rnj.166

Wilson, M. G., DeJoy, D. M., Vandenberg, R. J., Richardson, H. A., & McGrath, A. L. (2004). Work characteristics and employee health and well-being: Test of a model of healthy work organization. *Journal of Occupational and Organizational Psychology*, 77, 565–588. https://doi.org/10.1348/0963179042596522

Wohlers, C., Hartner-Tiefenthaler, M., & Hertel, G. (2017). The relation between activity-based work environments and office workers' job attitudes and vitality. *Environment and Behavior*, 51(2), 167–198. https://doi.org/10.1177/0013916517738078

Wohlwill, J. F. (1974). Human adaptation to levels of environmental stimulation. *Human Ecology*, 2(2), 127–147. https://doi.org/10.1007/BF01558117

Zohar, D., & Luria, G. (2003). The use of supervisory practices as leverage to improve safety behavior: A cross-level intervention model. *Journal of Safety Research*, 34(5), 567–577. https://doi.org/10.1016/j.jsr.2003.05.006

11

SAFETY AT WORK

Kristen Jennings Black and Christopher J. L. Cunningham

Workplace safety behaviors and norms around safety within organizations are important topics for occupational health psychology (OHP) researchers and practitioners. Some work environments have inherent risks that create an obvious need for close monitoring and intervention, but safety challenges can also include devastating incidents that are difficult to anticipate and can happen anywhere, such as workplace violence. In this chapter, we discuss factors that influence workplace safety behaviors, including individual worker characteristics, work environment characteristics, and organizational practices and climate. Intervention strategies that target these different factors to improve worker safety are presented. Our goal is to consider these issues in a broad sense, making sure to consider how safety-related concerns affect *all* workers (regardless of industry/occupation, status within the organization, and even legal status within a particular country).

When you are finished reading this chapter, you should be able to:

LO 11.1: Describe key environmental, organizational, and individual characteristics that contribute to safety in the workplace.
LO 11.2: Define safety climate and describe the primary contributors to and outcomes of a strong safety climate.
LO 11.3: Describe metrics that can be used to monitor safety and evaluate changes in safety.
LO 11.4: Evaluate interventions strategies that can be used to improve worker safety.

Overview of Work-Related Safety

If asked, most organizational leaders and workers will say that worker safety is important. Who wants anyone to get hurt, seriously injured, or killed while doing their job? Of course, the reasons different stakeholders care about safety can vary substantially, from simply complying with industry or governmental regulations, to avoiding financial costs of accidents, to proactively protecting workers to demonstrate they are valued more than as a means to production. This last reason is unfortunately not as common as it should be. Safety at work has to be about more than accident and cost avoidance. Real safety at work comes from building and sustaining a climate in which workers can be confident that they are valued and as protected as possible.

Acting with this perspective requires monitoring and addressing psychological and physical forms of safety. The principles, methods, and evidence presented in this chapter are relevant to targeted and broad-based efforts to build and sustain safety at work. This includes understanding how safety is defined and monitored, and the antecedents and consequences of safety-related events at work. We discuss these elements, while also reviewing workplace safety research and practical strategies to foster safety at work.

What Does it Mean to Work Safely?

Safety at work can be defined and explained in various ways. One approach is to define safety performance in terms of safety compliance and safety participation behaviors (Griffin & Neal, 2000). *Safety compliance* refers to the required aspects of the job (similar to "in-role behavior" or task performance in the general performance literature), while *safety participation* refers to behaviors that may not be required, but that support the safety context (similar to extra role behaviors or contextual performance). These types of safety behaviors are influenced by slightly different factors (e.g., Griffin & Neal, 2000): Safety compliance is influenced more by *safety knowledge* (i.e., of safe practices and procedures), while safety participation is influenced more by *safety motivation* (i.e., desire to follow safe practices). Both safety knowledge and motivation are influenced by the *safety climate* of an organization, defined as the shared perceptions that workers have about their safety at work and the extent to which safety is valued by the organization (Griffin & Neal, 2000; Zohar, 1980). We discuss safety climate more later in this chapter, when we explore organizational influences on safety.

In practice, safety is often defined and measured following guidance and standards from formal governing bodies. At perhaps the broadest level, there are international guidelines that provide recommended and typically evidence-based safety standards, such as the International Labour Organization's (ILO) health and safety convention protocols (ILO, 2002). Many countries or regions have government-run organizations tasked with developing and/or enforcing safety-related policy and practices. A few examples are the Standards Council of Canada (SCC; https://www.scc.ca/), Safe Work Australia (https://www.safeworkaustralia.gov.au/), the Occupational Safety and Health Administration or OSHA in the United States (US; https://www.osha.gov/), and the European Agency for Safety and Health at Work (EU-OSHA; https://osha.europa.eu/en) and its national focal points for member countries.

These types of governing bodies take varying approaches and can have different levels of success in protecting worker safety. For instance, America is not globally recognized as a leader in workplace safety, but the American OSHA does provide general safety standards to drive at least a base level of organizational compliance. The Canadian SCC (2020) exemplifies a more holistic approach, encouraging organizations to integrate values of safety and wellness into the organizational climate and regular practices (e.g., safety-related standards in job descriptions; formal mechanisms for gathering and addressing employee feedback about safety). A similarly proactive perspective is evident in guidance from EU-OSHA, which emphasizes the need to continually update safety standards to apply to the changing nature of work, changing health risks, and unique demands for certain employers (e.g., small businesses; EU-OSHA, 2017).

Holistic approaches to safe work encourage organizations to set safety-related goals that extend beyond simple compliance. This is important, because compliance is not a strong motivator for change – organizations may not be inspected regularly enough or fined heavily enough to actually achieve large-scale, industry-wide safe work practices. As Patton (2017) remarks and news stories illustrate, some organizations would rather pay a very hefty fine than implement changes that could slow down production (e.g., BP issued additional fines after not correcting problems related to an explosion; Greenhouse, 2010).

What Predicts Safety Behavior?

Antecedents or risk factors that may lead to accidents and injuries generally arise from the work environment itself, influences of the broader organization (e.g., policies, procedures, norms), and characteristics of individual workers. In this section, we dig into each of these sets of predictors of safety at work. As we discuss in Chapter 10, there are a number of physical and environmental demands associated with work that can impact worker health, safety, and well-being (WHSWB). Our focus here is on safety-related antecedents specifically.

Organizational and Cultural Factors

Worker safety is strongly influenced by an organization's *safety climate*, which provides norms about the appropriateness of behaviors in the work environment (Clarke, 2006; Griffin & Neal, 2000). Among researchers, there is an effort to distinguish between climate and culture (e.g., Petitta et al., 2017), but most definitions share common features and in practice these terms are often used almost interchangeably (Guldenmund, 2000). Given this, and to keep things simple, we use the safety climate label here to refer to the shared perceptions that workers have about safety in their work environments, and the extent to which safety is valued by their organizations (Griffin & Neal,

2000; Zohar, 1980) and reflected in organizational policies and practices (Zohar & Luria, 2004). When these elements are present and consistently reflected in worker perceptions, organizations have a strong and positive safety climate; organizations missing these elements and/or consensus in worker perceptions have a negative safety climate.

Positive safety climates encourage safe work behaviors and signal to workers that they are valued and supported by the organization. This is why positive safety climates are linked to general improvements in WHSWB (including job- and organization-related attitudes) and reductions in workers' intentions to leave the organization (e.g., Clarke, 2010; Huang et al., 2016). A positive safety climate may even function as a supportive work-related resource for workers in risky environments (e.g., nurses dealing with aggressive patients; Horan et al., 2019).

Related to safety climate, but perhaps more tangible is the presence and enforcement of safety policies and procedures within organizations. Enforcement of safety procedures is, unsurprisingly, a consistent predictor of workers' safety behaviors (Petitta et al., 2017; Vinodkumar & Bhasi, 2010). Safety-related policies and procedures are often linked to rewards or incentives for safe behavior, or penalties for being involved in a safety incident. Such incentive systems may positively influence worker safety behavior (Mattson et al., 2014; Saracino et al., 2015), but ensuring the success of these systems can be tricky. Kerr's (1975) classic article, *On the folly of rewarding A while hoping for B* exemplifies how these types of reward systems can lead to unintended effects. In a safety context specifically, incentivizing safety may encourage workers to underreport actual safety incidents, especially if part of the incentive program is some sort of penalty when safety incidents occur (Atkinson, 2000; Geller, 1996). Even in organizations with a positive safety climate, safety incidents are often underreported (Probst et al., 2008). A variety of motives influence this type of *safety silence* (e.g., workers perceiving an incident as insufficiently severe, concern about retaliation for reporting; Manapragada & Bruk-Lee, 2016). When it comes to using reward systems, it is generally recommended that these systems are clear in rewarding safe practices (i.e., as demonstrated through actual behaviors, reporting incidents, reporting near misses), rather than focusing on critical incidents as the outcome to be avoided (Geller, 1996). This is psychology in action – just think about the worker who breaks a company's "325 days without an accident" streak. Would they be motivated to report the incident if it triggers a team reward or penalty?

Work Task and Environment Factors

Worker safety is also contingent on a variety of work task- and environment-related factors. Organizations are responsible for factors such as the design and cleanliness of the work environment, maintenance of machinery, and design of work tasks, as well as ensuring workers at least have access to the equipment and proper safety training needed to work safely (e.g., Smith &

Carayon, 2011; Vinodkumar & Bhasi, 2010). Work scheduling is another organizational practice that impacts safety; long work hours and fatigue have been associated with impairments to attention and risks for accidents (Spurgeon et al., 1997; Williamson et al., 2011). Rest breaks during and between shifts, can reduce accident and injury risks (Fischer et al., 2017). Managing work demands more generally is also a critical organizational responsibility, as chronic work-related stressor exposures (especially to role ambiguity and time pressures) are associated with increased injuries, unsafe behaviors, risky decisions, and near misses (Clarke, 2012; Nahrgang et al., 2011). Leader behaviors can also have a direct impact on workers' safety behaviors (Clarke, 2013; Griffin & Hu, 2013). Specifically, workers learn how seriously an organization values safety based on leaders' behaviors. Supervisors who consistently emphasize safety over speed, profit, or other motivations send a clear message that safety is valued (Zohar & Luria, 2004).

Many occupations involve work that is physically demanding and inherently full of safety risks. This is especially true in construction, agriculture, manufacturing, and transportation services, which are industries that consistently report high numbers of work-related injuries and fatalities (Eurostat, 2019; Safe Work Australia, 2020; U.S. Bureau of Labor Statistics [BLS], 2018, 2019a). In these types of industries, the risks come from environmental hazards that limit workers' ability to work safely (e.g., excessive noise or poor lighting affecting sensory capabilities, dangerous nature of work tasks, exposure to toxic materials; Smith & Carayon, 2011). Emergency responders, for example, regularly confront safety risks in the midst of disasters or acts of violence to protect others; workplace fatality statistics are unfortunate evidence of this risk for first responders (e.g., U.S. BLS, 2019b). Fortunately, many helpful resources have been developed by human factors psychologists, industrial hygienists, and ergonomists to help improve the safe functioning of workers in these environments. Even in characteristically unsafe work environments, these resources and proactive strategies (e.g., crime prevention through environmental design; Casteel & Peek-Asa, 2000) can help reduce the occurrence of safety incidents (Hoel et al., 2001). Such efforts include critical incident preparations and trainings for rare, but especially devastating safety events (e.g., active-shooter trainings, disaster response drills; McGraw, 2016).

Worker Characteristics

All the preceding factors contribute to safety at work, but none matter more than workers' own attention and commitment to safe behaviors. Thankfully, workers generally care about safety and often want to be involved in safety-related decisions, to the point that they may consider leaving an organization if risks are not well-managed (Cree & Kelloway, 1997; Rogers & Kelloway, 1997). For this reason, it is beneficial to facilitate workers' active engagement in managing safety risks and discourage safety silence. This can be done by enabling and supporting workers' *safety voice* (i.e., the extent to which workers feel they can speak up about safety at work; Tucker et al., 2008). Tucker and Turner (2015) found that safety voice was related to fewer injuries among teenage workers, but only when supervisors were open to hearing workers' concerns. Without supervisor openness, safety voice was actually related to *more* injuries over time. These findings underscore how important it is to engage with workers regarding safety matters and to provide supportive organizational climates that encourage workers to be aware and to share.

Several personal characteristics and attributes of workers may also influence safety behaviors, including intelligence, physical or motor capabilities, health status, and personality (Smith & Carayon, 2011). Obviously, workers must be able to understand their work responsibilities to safely execute tasks, so it is unsurprising that safety knowledge is one of the strongest predictors of safety performance (Christian et al., 2009). Some knowledge may depend on workers' underlying cognitive abilities and experiences (reason for proper candidate selection practices), but there is also an important place for targeted safety training that provides workers with the necessary information in an appropriate format that they can translate into their work environments (Smith & Carayon, 2011). Having the right knowledge, skills, and abilities to do the job will only take workers part of the way toward working safely; workers also must actually be motivated to behave safely (i.e., safety motivation), potentially in ways that go above and beyond basic job requirements (i.e., *safety participation*; Griffin & Neal, 2000; Neal & Griffin, 2006). Together, good training and a positive safety climate can enhance both safety knowledge and safety motivation (Griffin & Neal, 2000; Vinodkumar & Bhasi, 2010).

Workers' own physical or psychological health states or conditions can further affect safe functioning at work. For example, *fatigue* (i.e., physical and mental energy depletion) and burnout can both contribute to unsafe behaviors and accidents (Nahrgang et al., 2011; Smith & Carayon, 2011). More specifically, construction workers experience difficulty completing physical and cognitive tasks when fatigued (Zhang et al., 2015), healthcare workers may make more errors in patient care when they are sleep deprived (Caruso, 2014; Landrigan et al., 2004), and truck drivers are more likely to be involved in an accident when fatigued and unable to stop for a break (Bunn et al., 2019).

Workers' health conditions or behaviors that may not be directly related to work, but could impact safety, should also be monitored. For instance, worker insomnia has been positively linked to work injuries; thankfully, supervisor attention to safety concerns can help alleviate some of this type of risk (Kao et al., 2016). Substance use can also present safety concerns, given the effects controlled substances can have on cognitive, motor, and emotion regulation abilities, affecting performance and risks for workplace aggression (Barling et al., 2009; Frone, 2004, 2011). Reviews in this realm conclude that substance use certainly plays a role in safety, but can rarely be pinpointed as the sole cause of an accident or injury – rather, substance use can be one behavioral manifestation of a pattern of *deviance proneness* (e.g., Ramchand et al., 2009). Ultimately, it makes sense for organizations to have clear policies on substance use, both to reduce direct risks and to eliminate permissive norms toward such behaviors that can make other workers feel less safe and more stressed (Frone, 2009).

Finally, there are several additional demographic and personality-related worker characteristics that may influence safety. Men tend to be involved in more accidents, injuries, and occupational fatalities than women (U.S. BLS, 2018), which is at least partially explained by the disproportionate number of men that work in physically demanding and/or high-risk professions. Young workers, often employed in entry-level work, also tend to have above average injury rates (Salminen, 2004; Tucker & Turner, 2015); however, older workers have a higher risk for fatal incidents compared to younger workers (Peng & Chan, 2019; Salminen, 2004). Education can affect safety, depending on whether the workers' educational background, reading level, and language skills are adequate for carrying out the tasks or for comprehending training material in the format provided (Smith & Carayon, 2011). In terms of underlying personality traits, agreeableness, conscientiousness, risk taking, impulsivity, and sensation seeking have demonstrated fairly consistent relationships with safety behavior and safety attitudes (Beus et al., 2015; Clarke & Robertson, 2005; Henning et al., 2009).

Why Workplace Safety Matters

An obvious reason to address safety matters at work is to reduce workers' likelihood of serious injury or mortality. These risks are unfortunately not trivial, as evidenced by the data: 552 fatal accidents in the European Union in 2017 (Eurostat, 2019); 5,250 fatal work injuries in 2015 in the US (U.S. BLS, 2019a); and untold millions of non-fatal accidents and near-misses every year in general. Workplace safety incidents impact workers, and their families, communities, organizations, and society.

Impact on Workers

Workers pay the ultimate price for safety incidents of any severity, which limit their ability to work and live. Although it is often assumed that the cost of work-related safety incidents is mostly absorbed by the employing organization, workers and their families and communities often end up carrying a heavy financial burden when such events occur. Safe Work Australia (2015) estimates that work-related injury and disease cost the Australian economy $61.8 billion in 2012 and 2013 and that 77% of that cost (i.e., from absences, incapacity) was endured by workers. You can imagine the magnitude of these costs in other, more populated regions and countries, such as in the US where workplace accidents and injuries are estimated to cost between $128 and $155 billion per year (Schulte, 2005).

In terms of worker well-being, it is not uncommon for workers to experience psychological symptoms following an occupational injury, like PTSD or depression. Such symptoms can further impair workers' ability to return to work (Lin et al., 2016) and functioning outside of work. The financial insecurity that may arise from workplace injuries can also result in work-family conflict (Lawrence et al., 2013). Alternatively, when workplace safety is managed well, positive safety climates at work can actually facilitate positive psychological states and positive work attitudes, like engagement and job satisfaction (Huang et al., 2016). These types of positive states are good for the worker and for the organization, given the connections between engagement and work performance (reviewed more in Chapter 4). All of this together demonstrates the broad impacts of workplace safety on workers and their families.

Business and Financial Outcomes

Direct organizational costs associated with lapses in safety are typically linked to worker absences and lost productivity at work, increased workers compensation and healthcare costs, and potentially costs associated with lawsuits or fines linked to organizational non-compliance with safety regulations. There are also indirect costs, like increased training time and costs to ready temporary workers, as well as lingering costs to accommodate and support workers who may be disabled for the long-term. Ultimately, an organization's investment in active safety management is a smart investment from a risk and cost avoidance perspective. This type of investment can also send a strong signal to workers that the organization cares about their well-being.

Some organizations may consider paying non-compliance fines (if required) as the "cheaper" option compared to making changes to enhance safety. However, there is good evidence that investing in worker safety can create positive returns and a competitive advantage (Fabius et al., 2013). Zou and colleagues (2010) provide an example of how return on safety investment estimates can be calculated and applied in the construction industry, with one case study showing that a one million dollar investment in safety led to an estimated 1.5 million dollars in safety-related cost avoidance. Another example from the ship building industry provides a detailed account of the costs of a fairly comprehensive health and safety intervention and the quantifiable (e.g., dramatic reduction in injury rates, missed time at work, and treatment costs) and intangible benefits that resulted (Thiede & Thiede, 2015). Efforts to actively manage and enhance worker safety will cost organizations money, but even low-cost interventions have potential to be effective (e.g., training supervisors to monitor and support safety), and any investment builds important organizational resources that can sustain safe worker behaviors over time.

Societal Costs

The examples of national costs related to worker illnesses and injuries we mentioned earlier are likely underestimates of the true costs of unsafe work, given difficulties inherent in understanding the full *social welfare costs* for workers who are impaired in their abilities to participate in the labor force (Weil, 2001). No matter how these costs are estimated, unsafe work creates a society-level burden. Complicating this matter further is the fact that work-related safety risks are not evenly shared throughout the working populations of most countries. Instead, they tend to be highest among marginalized and often ignored groups of workers, such as migrant workers (Moyce & Schenker, 2018) and illegal immigrants. These same groups often have the least access to and financial support for adequate healthcare insurance and workers compensation programs that can support them and their families if they experience a safety incident. This is true globally and constitutes a serious threat to the protection of human rights and dignity. Our hope is that OHP professionals will work to ensure the safety of *all* workers, regardless of their economic or citizenship status, or field of employment.

Methodological Considerations and Practical Recommendations

Workplace safety is an OHP-related area for which multi-faceted measurement and intervention strategies are essential, given the complex ways worker characteristics, work environment features, and the organizational context influence safety behaviors. In this section we review the deep knowledge base regarding how to monitor and improve safety at work.

Measuring and Monitoring Workplace Safety

The SCC (2020) identifies a healthy organizational enterprise as one that is aware of its state of health and safety. This is only possible when organizations regularly gather and review data that reflect safety-related incidents and behaviors among workers, and work toward addressing areas in need of improvement. Safety-related data can be gathered various ways, with many objective metric options (e.g., accident, injury, and fatality rates; days missed due to an incident). Objective metrics, however, are not infallible and best-practice guidance is to view safety-related metrics as lower-end estimates, given the prevalence of underreporting (Probst et al., 2008). Accidents are often only reported when they involve days missed from work, but *microaccidents*, or those that only involve basic first aid, are also meaningful data points to monitor (Zohar, 2000). Similarly, organizations can benefit from documenting *near misses*, where serious incidents are narrowly avoided, often through worker attention and fast action.

Organizations may also want to know about workers' daily safety behaviors and experiences, rather than relying on data that reflects when things go wrong. Such data can come from regular observations or inspections and can be particularly helpful for identifying training or maintenance needs. This is especially true when such observations are normative and workers do not feel pressure to be on their "best behavior". Surveys, focus groups, interviews, or even experience sampling approaches can also be informative for learning about the workers' perceptions of the state of safety in terms of what they are doing, thinking, and perceiving (Sonnentag et al., 2013). It may also be valuable for organizations to administer self-report assessments, which might include tests of safety knowledge, or attitudinal measures of safety motivation, perceptions of the safety climate, and safety voice. Many validated safety-related assessments are available, though the most helpful will likely be contextualized to the specific environment and industry (Zohar, 2010). As one example, Probst and colleagues (2019) developed a rubric-based safety climate assessment for the construction industry (https://safetyclimateassessment.com/; described more in Dr. Probst's professional profile at the end of this chapter).

Zohar (2010) provides some other practical suggestions for assessing safety, like measuring patterns of organizational priority for safety versus other goals and capturing any discrepancies between espoused and enacted values. Zohar (2010) also suggests measuring agreement in safety perceptions within and between workgroups. This can be done using focus groups and surveys to see whether workers in the same unit are saying similar or different things regarding safety perceptions. More technically, *multilevel modeling* can statistically examine group-level safety perceptions and their impacts (see Bliese & Jex, 2002 for a review of this method).

Intervening to Support Workplace Safety

To wrap-up this chapter, we provide some practical ways organizations can promote worker safety. We have highlighted that safety is affected by numerous factors operating at various levels. Safety is possible at work when these factors are aligned and there is a consistent organization-wide focus on these matters for the benefit of workers and the organization.

Strategies for Individuals

Because safety knowledge and safety motivation are important predictors of safety behaviors (Griffin & Neal, 2000; Neal & Griffin, 2006), selecting and training workers properly are essential intervention strategies. This may involve selecting workers with specific knowledge and skill sets required for safe work, as well as individual differences linked to safer work behaviors (e.g., conscientiousness or low impulsivity; Clarke & Robertson, 2005; DeJoy et al., 2000). Effective safety training should be realistic when compared to the actual work environment (Salas et al., 2012), based on a *training needs analysis* to identify what content needs to be presented, and aligned with key trainee characteristics (e.g., baseline knowledge and skills; learning styles). Such training does not have to be complicated or expensive. As an example, Scharf and colleagues (2011) worked to improve safety among ironworkers using stereoscopic (i.e., realistic and three-dimensional) images of real work sites as a mechanism for training these workers to more effectively identify and plan to address common safety hazards. Broadly speaking, safety trainings should address how to identify and respond appropriately to safety risks and incidents, and how improving awareness and reporting can reduce safety silence. With respect to reporting, workers need to understand that they may still be in trouble if they report an incident and it is due to their carelessness or disregard for safety protocol; however, it is important to ensure workers feel secure enough to report incidents without fear of excessive repercussions or immediate termination.

Safety-related training should not be a one-time, immediately post-hire event for workers. Actual safety incidents provide meaningful contexts and opportunities for informal learning and development or re-training. From a developmental perspective, discussing near misses may be just as valuable as reviewing actual accidents (Patton, 2017). *Critical incident debriefing* can be implemented as an ongoing practice to build shared understanding of factors that contribute to a safety-related incident or near miss, and to identify opportunities and action steps to take toward preventing more such events in the future. Post-event debriefing is a powerful tool for learning and development, and for coping with psychologically distressing incidents, and thus has been applied in a variety of settings, such as among surgical teams, first responders, or military units (e.g., Adler et al., 2008; Ahmed et al., 2012; Tuckey & Scott, 2014).

We want to reiterate that the effectiveness of knowledge-based safety training will be limited if safety motivation is lacking (i.e., workers know what to do, but do not see value or reason to do it). Safety training must be properly framed and contextualized to clearly communicate that safety is an organizational imperative. It is unreasonable to expect such training to be effective if delivered in an outdated, impersonal format and/or facilitated by a disinterested trainer who leads in with a comment like, "Let's get this over with." More effective safety training appeals to basic elements of worker motivation, like helping workers feel empowered to act safely within an organization that values their input. Emphasizing safety voice can be motivating by instilling a sense that workers are collectively responsible for safety and that their observations and perspectives matter. Safety voice is influenced by a number of structural and cultural features in the workplace (e.g., clear incident reporting systems, empowerment and motivation to speak up and protect others; Noort et al., 2019).

Strategies for Groups and Leaders

Arguably, the most impactful work-related safety interventions are multi-faceted and multi-level efforts to change workgroup norms and safety climate (Christian et al., 2009). Strategies targeting safety climate include creating dialogue among workgroups and training leaders to be more involved in safety (Lee et al., 2019). Trainings that encourage supervisors to be more involved with monitoring and rewarding safety, or to incorporate messages about safety into daily conversations can improve safety climate and increase safety behaviors (Zohar & Luria, 2003; Zohar & Polachek, 2014). As an interesting example of this being done in a more holistic WHSWB intervention, the Safety and Health Involvement Program (SHIP; Hammer et al.,

2019; Hammer et al., 2015) involved training supervisors on safety behavior monitoring, as well as supportive behaviors for the work-family and safety domain. This included implementing work-group interventions where construction teams discussed safety-related concerns and resolution strategies. The intervention benefited worker health and improved worker safety in workgroups that had previously had poor relationships with their supervisors.

Workgroup interventions are powerful because they can harness the influence of peer accountability. Practices emphasizing workgroup accountability have demonstrated promising results. A relevant example is the 20-20-20 technique in the construction industry to stop for 20 seconds, every 20 minutes and look for and address any hazards within 20 feet (Patton, 2017). Qualitative evaluations of such safety interventions show that workers can initially feel awkward calling one another out for safety incidents, but that over time it becomes more normal and even motivating to hold each other accountable with workgroup incentives for safe behavior (Sparer et al., 2016). When you think about it, safety incidents are themselves stressful, so a culture that is sensitive to safety concerns may make it all the more stressful to actually be involved in an incident, feeling concern that your career or social relationships could be threatened because of the event (e.g., Black et al., 2019). Getting comfortable with identifying and addressing safety hazards is important to preventing incidents and reducing fears associated with reporting.

Strategies for Organizations

Much of the responsibility for ensuring worker safety rests with organizational leadership and requires organization-wide efforts and resources. In general, organizations have the power (even in inherently risky work settings) to design and structure work so as to reduce, minimize, and sometimes eliminate safety hazards. As research shows, work-related stress, time pressures, and fatigue can all relate to safety behaviors or accidents (Clarke, 2012; Williamson et al., 2011), so careful management of work demands and environmental exposures is a strategy for organizations to directly improve worker safety. Providing sufficient resources to meet work demands and limit unnecessary risks and demands (e.g., avoiding tight deadlines that do not have to be; scheduling that allows for recovery between work periods) are other concrete strategies for improving worker safety. Key resources to work safely often include access to necessary *personal protective equipment* (e.g., eye protection, proper footwear, gloves, respirators). The use-case for such gear is often quite obvious; less obvious is the need for this equipment to be high-quality and well-fitting. Gear that is not well-designed and makes work more difficult (e.g., bulky gloves limiting manual dexterity, ear protection that impedes communication) may not be properly used and offer minimal protection (Redden & Larkin, 2015).

Two other essential organization-wide strategies are *hazard identification and incident monitoring* (Smith & Carayon, 2011). As noted earlier, organizations must keep accurate records of injuries, incidents, and fatalities, as well as (ideally) microaccidents and near misses. These data can reveal safety risk "hot spots" and hazards that need to be addressed (e.g., dangerous equipment that needs maintenance or replacement). This type of recording can be done with basic paper-based report forms and checklists, which can be developed from government or industry regulations, and adapted to include organization-specific nuances and details gathered from the input of front-line workers regarding what should be checked, by whom, and how often (Smith & Carayon, 2011). Some organizations may benefit from contracting with evaluation and assessment professionals to develop a strong and comprehensive hazard identification plan. Overall, observation and monitoring plans seem to work best when done frequently (e.g., daily, weekly), when peer accountability is emphasized, and when feedback is regular (Cooper, 2009). Monitoring practices become ineffective if they are rushed or not seen as valued; "pencil whipping" can occur when checklist or inspection items are mindlessly completed or ignored altogether (Ludwig, 2014).

Despite monitoring, some safety risks are difficult to minimize, so organizations also have a responsibility to prepare workers to carry out their work in the safest manner possible. One fairly obvious, but sometimes overlooked, intervention along these lines is to ensure workers truly understand the risks involved in their work and are able to essentially provide informed consent to work even in higher risk environments (cf., Smith & Carayon, 2011). In addition to leveraging selection and training, organizations can continue to motivate safe work behavior through well-designed reward systems (i.e., that do not encourage underreporting and that do not encourage unsafe work practices), integrating safety into performance expectations and goal-setting, and creating a positive safety culture (e.g., Geller, 2005; Mattson et al., 2014). As noted early in this chapter, a positive safety climate also requires clear and consistent enforcement of safety-related expectations; good safety management means holding to policies and expectations if safety is disregarded (Vinodkumar & Bhasi, 2010).

Finally, we must acknowledge that there are times in all work settings in which safety risks arise due to disasters or violence. Many of these, unfortunately highly publicized incidents, often appear to develop with no warning. In the case of violence, although attempting to prepare an organization may seem like a fool's errand, all workplaces can invoke some risk management strategies to reduce the potential for such acts by limiting organizational access to authorized personnel and providing training on organizational protocols for dealing with potential critical events (e.g., active-shooter training; McGraw, 2016). In work settings with a higher than average risk for violence, there are several environmental designs and protocols that can reduce the potential for harm (e.g., Casteel & Peek-Asa, 2000; OSHA, 2016). Several more specific interventions for workplace violence are reviewed in Chapter 8. Regardless of the circumstances, incidents of violence have immediate and lasting effects on workers' psychological health, such as fear of future or recurring events (Rogers & Kelloway, 1997). Managing the aftermath of violence can be accomplished by providing accommodations and resources to support affected workers and their psychological or physical health needs (e.g., Employee Assistance Programs, access to local mental health professionals).

Evaluating Workplace Safety Interventions

Evaluating safety-related interventions at work is often more straightforward than evaluating interventions targeting many other WHSWB challenges. The material from Chapter 2 can help, but there are some unique challenges that come along with evaluating this type of intervention. First, it may be difficult to observe changes in safety-related data because of the low base-rate nature of many of these incidents. Dramatic changes in uncommon events (e.g., fatalities, severe injuries) are especially unlikely to emerge quickly, so evaluators may need to look for reductions in metrics reflecting more frequent safety lapses, like microaccidents or near misses. This is yet another reason for organizations to regularly track these safety-related events. Near miss data is particularly interesting for evaluation purposes, in that increased reporting of such events can indicate that workers are speaking up more about safety concerns. Of course, the ultimate goal is to see rates of near misses decrease along with improvements in other safety-related metrics.

There are a number of safety-related interventions that have been referenced in this chapter that provide sound examples of evaluating a range of behavioral and psychological outcomes (e.g., Hess et al., 2020; Zohar & Luria, 2003, 2004), as well as more complex evaluations of cost-benefit analyses (e.g., Thiede & Thiede, 2015). A range of data targeting safety climate and worker knowledge, motivation, and behaviors can help explain bizarre or complex effects following a safety intervention (e.g., increased injury reports due not to more injuries, but to increased frequency of reporting). Similarly, workers accustomed to unsafe procedures, like poor lifting practices, may work more slowly and experience more initial discomfort when trained in appropriate and safer techniques. These realities need to be considered and incorporated into evaluation plans for safety-related interventions so that the ultimate effects can be identified. More advanced evaluation designs may also be needed to uncover an intervention's true effects (e.g., Hammer et al., 2019).

Concluding Thoughts and Reality Check

Workplace safety is an incredibly important topic for OHP professionals given the devastating impacts of work-related injuries, accidents, and deaths on workers, and their families, organizations, communities, and society. Safety at work is supported by organizational policies and practices, workers' abilities and characteristics, and workgroup norms that are shaped by organizational leaders and daily interactions in the workplace. You might assume that the strategies we highlighted in this chapter are commonplace in all high-risk organizations, but this is unfortunately not the case. Especially in smaller or less financially secure organizations, old and cheaply purchased equipment, retrofitted workstations, and mismatched tools and technologies are present and the risks they present are compounded by the absence of safety monitoring efforts and often critical safety-related training. Over time, these factors create a generally unsafe work environment that is reinforced by social norms that do not place a high value on working safely.

Empowering workers, workgroups, and organizational leaders to identify and report safety-related risks and incidents, and to hold each other accountable is an essential strategy for improving safety at work. Investing in and emphasizing safety signals care and compassion for workers. Taking the time to identify and correct hazards, select and train workers to work safely, and consistently adhere to safety-related policies and procedures shows workers that their lives and their safety are valued. Organizations that take these steps can avoid disruptions and costs associated with safety incidents, and create a work environment in which workers feel safe, secure, and motivated to do good work.

Media Resources

- High consumer demands conflict with worker safety: https://www.theatlantic.com/technology/archive/2019/11/amazon-warehouse-reports-show-worker-injuries/602530/
- Using AI to identify injury risks: https://www.forbes.com/sites/suziedundas/2018/10/25/kinetic-uses-a-i-to-monitor-workplace-movement-but-it-isnt-aiming-to-be-big-brother/#5d6f04fb21a7
- Campbell Soup's investment in safety: https://hbr.org/2018/09/how-we-reduced-our-injury-rate-by-90-at-campbell-soup-company?registration=success

---

Discussion Questions

1) Do safety incidents "just happen" or do they develop over time? How does this affect our ability to manage or control safety-related risks?

2) What are common safety-related risk factors and challenges present in work environments (in general and within particularly high-risk industry settings) and what can be done to manage these risks?

3) What are examples of safety-related risks that affect workplaces, but may not initiate at work? What can be done to manage these types of risks?

4) What are challenges associated with evaluating safety-related interventions in work environments? What are some strategies for overcoming these challenges?

Professional Profile: Tahira Probst, Ph.D.

Country/region: USA
Current position title: Professor
Background: I have been working to improve the health, safety, and well-being of
workers for 25 years. I completed my B.A.
in Psychology and German at the University of Notre Dame, and completed a
Fulbright Teaching Assistantship, in Aus-
tria after graduating. I earned a Ph.D. in
Industrial and Organizational (I-O) Psychology from the University of Illinois at
Urbana-Champaign (UIUC) with minors
in Social Psychology and Quantitative Psy-
chology. I currently belong to the Society for I-O Psychology and the Society
for Occupational Health Psychology.

When entering college, I had no idea what I wanted to do with my life. But, as
soon as I stepped foot in my first class, I knew I wanted to be a professor. Fortunately, my university had several undergraduate I-O psychology courses and the
instructor for those classes, Dr. Chris Anderson, eventually became my honors
thesis mentor. Using research funds, he flew me to San Francisco by myself to
collect data in an organization on goal setting and feedback. I was a 21-year-old
newbie being hosted by the company's CEO, gathering data, and having the
chance to change and hopefully improve workers' working lives. I was immediately hooked on research and knew that I-O psychology was the path for me.

At UIUC, I worked with giants in their fields (e.g., Fritz Drasgow, Chuck
Hulin, Louise Fitzgerald, Harry Triandis, Peter Carnevale, and Joe Martoc-
chio) and learned different perspectives and research paradigms from each.
Upon graduation, I took a tenure-track position at Washington State University (WSU) and have never looked back! I direct the Coalition for Healthy and
Equitable Workplaces research lab, where my research focuses on the health,
safety, and performance-related outcomes of economic stressors (including
job insecurity, financial strain, economic deprivation, and underemployment),
as well as multilevel contextual variables that influence these relationships. I
also have strong secondary research interests in the areas of workplace safety,
specifically accident under-reporting and safety climate. For my professional
service, I am currently Editor in Chief of *Stress & Health* and sit on the Editorial Boards of the *Journal of Occupational Health Psychology, Occupational
Health Science, Military Psychology, International Journal of Environmental
Research and Public Health,* and the *Journal of Business and Psychology.*

How my work impacts WHSWB: Most of my research is extremely applied, meaning that I often partner with local, regional, and national organizations that reach out to me in order to improve the health, safety,
and well-being of their workers. Typically, after collecting data within an

organization, my research assistants and I will compile a feedback report for the company that provides a snapshot of how their workers are doing, relative to others in our database that we have compiled over the past 2 decades. In addition, we point out potential areas for improvement and provide them with recommendations for achieving that improvement. Academicians typically publish their research in peer-reviewed journals, where it is often consumed by other researchers rather than practitioners who may be able to make changes based upon those findings. Therefore, I am also gratified when my research is highlighted in trade or industry publications, where it can potentially benefit people in real organizations. To date, my research has received national attention from major news outlets (e.g., *NY Times*, National Public Radio, Reuters) and has been disseminated in trade magazines such as *Aerosafety World*, *Industrial Engineer*, *Occupational Hazards*, *HR News*, *OH&S Canada*, *Occupational Health Management*, and *WebMD*.

My recent collaboration with CPWR – The Center for Construction Research and Training –involved developing and validating a rubric-based assessment of jobsite safety climate in the construction industry. The Safety Climate Assessment Tool (S-CAT) is a free online tool (https://safetyclimate assessment.com/) available to any construction contractor or safety and health professional who wants tailored and actionable information to improve the safety of every worker at every job-site. With just a few clicks, workers can answer questions about each safety climate dimension and then receive a personalized feedback report with benchmarking and comparative information indicating their current areas of success and ideas for making improvements. Companies can also have their workers take the S-CAT periodically to track their progress at improving their jobsite safety climate. To date, over 5000 workers in the construction industry have completed the S-CAT.

We wanted to use a rubric approach so that the feedback companies would receive would be actionable [i.e., they can see that what it would look like to move from "reactive" (management only comes to the jobsite after an incident has occurred) to "exemplary" (management frequently visits the jobsite and seeks out interactions with workers)]. We also wanted our data to be benchmarked, so that companies can see how they compare to all of the other companies in our database. We have found that companies who score higher on the S-CAT have lower recordable rates of workplace injuries and illnesses (Probst et al., 2019). Eventually, we hope to demonstrate that their continued use of the S-CAT and the ancillary provided intervention materials will improve their safety climate and decrease their recordable rates over time.

My motivation: People often criticize academia as an "ivory tower." However, the applied research that I do not only advances scientific understanding of the antecedents of workplace health and safety, but it also directly benefits that organizations that reach out to me for assistance. In addition, the training that I provide my graduate and undergraduate students (who often go directly into industry positions) directly influences the future human resources and environmental health and safety leaders and consultants.

Chapter References

Adler, A. B., Litz, B. T., Castro, C. A., Suvak, M., Thomas, J. L., Burrell, L., McGurk, D., Wright, K. M., & Bliese, P. D. (2008). A group randomized trial of critical incident stress debriefing provided to U.S. peacekeepers. *Journal of Traumatic Stress*, *21*(3), 253–263. https://doi.org/10.1002/jts.20342

Ahmed, M., Sevdalis, N., Paige, J., Paragi-Gururaja, R., Nestel, D., & Arora, S. (2012). Identifying best practice guidelines for debriefing in surgery: A tricontinental study. *The American Journal of Surgery*, *203*(4), 523–529. https://doi.org/10.1016/j.amjsurg.2011.09.024

Atkinson, W. (2000). The dangers of safety incentive programs. *Risk Management*, *47*(8), 32–38.

Barling, J., Dupre, K. E., & Kelloway, E. K. (2009). Predicting workplace aggression and violence. *Annual Review of Psychology*, *60*, 671–692. https://doi.org/10.1146/annurev.psych.60.110707.163629

Beus, J. M., Dhanani, L. Y., & McCord, M. A. (2015). A meta-analysis of personality and workplace safety: Addressing unanswered questions. *Journal of Applied Psychology*, *100*(2), 481–498. https://doi.org/10.1037/a0037916

Black, K. J., Munc, A., Sinclair, R. R., & Cheung, J. H. (2019). Stigma at work: The psychological costs and benefits of the pressure to work safely. *Journal of Safety Research*, *70*, 181–191. https://doi.org/10.1016/j.jsr.2019.07.007

Bliese, P. D., & Jex, S. M. (2002). Incorporating a mulitilevel perspective into occupational stress research: Theoretical, methodological, and practical implications. *Journal of Occupational Health Psychology*, *7*(3), 265–276. https://doi.org/10.1037/1076-8998.7.3.265

Bunn, T. L., Slavova, S., & Rock, P. J. (2019). Association between commercial vehicle driver at-fault crashes involving sleepiness/fatigue and proximity to rest areas and truck stops. *Accident Analysis & Prevention*, *126*, 3–9. https://doi.org/10.1016/j.aap.2017.11.022

Caruso, C. C. (2014). Negative impacts of shiftwork and long work hours. *Rehabilitation Nursing*, *39*(1), 16–25. https://doi.org/10.1002/rnj.107

Casteel, C., & Peek-Asa, C. (2000). Effectiveness of crime prevention through environmental design (CPTED) in reducing robberies. *American Journal of Preventive Medicine*, *18*(4S), 99–114. https://doi.org/10.1016/S0749-3797(00)00146-X

Christian, M. S., Bradley, J. C., Wallace, J. C., & Burke, M. J. (2009). Workplace safety: A meta-analysis of the roles of person and situation factors. *Journal of Applied Psychology*, *94*(5), 1103–1127. https://doi.org/10.1037/a0016172

Clarke, S. (2006). The relationship between safety climate and safety performance: A meta-analytic review. *Journal of Occupational Health Psychology*, *11*(4), 315–327. https://doi.org/10.1037/1076-8998.11.4.315

Clarke, S. (2010). An integrative model of safety climate: Linking psychological climate and work attitudes to individual safety outcomes using meta-analysis. *Journal of Occupational and Organizational Psychology*, *83*(3), 553–578. https://doi.org/10.1348/096317909x452122

Clarke, S. (2012). The effect of challenge and hindrance stressors on safety behavior and safety outcomes: A meta-analysis. *Journal of Occupational Health Psychology*, *17*(4), 387–397. https://doi.org/10.1037/a0029817

Clarke, S. (2013). Safety leadership: A meta-analytic review of transformational and transactional leadership styles as antecedents of safety behaviours.

Journal of Occupational and Organizational Psychology, 86(1), 22–49. https://doi. org/10.1111/j.2044-8325.2012.02064.x

Clarke, S., & Robertson, I. T. (2005). A meta-analytic review of the Big Five personality factors and accident involvement in occupational and non-occupational settings. *Journal of Occupational and Organizational Psychology, 78,* 355–376. https://doi. org/10.1348/096317905X26183

Cooper, M. D. (2009). Behavioral safety interventions: A review of process design factors. *Professional Safety, 54*(2), 36–45, Article ASSE-09-02-36.

Cree, T., & Kelloway, E. K. (1997). Responses to occupational hazards: Exit and participation. *Journal of Occupational Health Psychology, 2*(4), 304–311. https:// doi.org/10.1037/1076-8998.2.4.304

DeJoy, D. M., Searcy, C. A., Murphy, L. R., & Gerson, R. R. M. (2000). Behavioral-diagnostic analysis of compliance with universal precautions among nurses. *Journal of Occupational Health Psychology, 5*(1), 127–141. https://doi. org/10.1037/1076-8998.5.1.127

EU-OSHA. (2017). *Safer and healthier work for all: Modernisation of the EU occupational safety and health legislation and policy.* https://osha.europa.eu/en/ safety-and-health-legislation/european-directives

Eurostat. (2019, November 2019). *Accidents at work statistics.* https://ec.europa. eu/eurostat/statistics-explained/index.php/Accidents_at_work_statistics# Analysis_by_activity

Fabius, R., Thayer, R. D., Konicki, D. L., Yarborough, C. M., Peterson, K. W., Isaac, F., Loeppke, R. R., Eisenberg, B. S., & Dreger, M. (2013). The link between workforce health and safety and the health of the bottom line: Tracking market performance of companies that nurture a "culture of health". *Journal of Occupational & Environmental Medicine, 55*(9), 993–1000. https://doi.org/10.1097/ JOM.0b013e3182a6bb75

Fischer, D., Lombardi, D. A., Folkard, S., Willetts, J., & Christiani, D. C. (2017). Updating the "Risk Index": A systematic review and meta-analysis of occupational injuries and work schedule characteristics. *Chronobiology International, 34*(10), 1423–1438. https://doi.org/10.1080/07420528.2017.1367305

Frone, M. R. (2004). Alcohol, drugs, and workplace safety outcomes: A view from a general model of employee substance abuse and productivity. In J. Barling & M. R. Frone (Eds.), *The psychology of workplace safety* (pp. 127–156). American Psychological Association.

Frone, M. R. (2009). Does a permissive workplace substance use climate affect employees who do not use alcohol and drugs at work? A U.S. national study. *Psychology of Addictive Behaviors, 23*(2), 386–390. https://doi.org/10.1037/a0015965

Frone, M. R. (2011). Alcohol and illicit drug use in the workforce and workplace. In J. C. Quick & L. E. Tetrick (Eds.), *Handbook of occupational health psychology* (2nd ed., pp. 277–296). American Psychological Association.

Geller, E. S. (1996). The truth about safety incentives. *Professional Safety, 41*(10), 34–38.

Geller, E. S. (2005). Behavior-based safety and occupational risk management. *Behavior Modification, 29*(3), 539–561. https://doi.org/10.1177/0145445504273287

Greenhouse, S. (2010, Aug. 12). BP to pay record fine for refinery. *New York Times.* https://www.nytimes.com/2010/08/13/business/13bp.html

Griffin, M. A., & Hu, X. (2013). How leaders differentially motivate safety compliance and safety participation: The role of monitoring, inspiring, and learning. *Safety Science, 60,* 196–202. https://doi.org/10.1016/j.ssci.2013.07.019

Griffin, M. A., & Neal, A. (2000). Perceptions of safety at work: A framework for linking safety climate to safety performance, knowledge, and motivation. *Journal of Occupational Health Psychology*, 5(3), 347–358. replace with: https://doi.org/10.1037/1076-8998.5.3.347

Guldenmund, F. W. (2000). The nature of safety culture: A review of theory and research. *Safety Science*, 34, 215–257. https://doi.org/10.1016/S0925-7535(00)00014-X

Hammer, L. B., Truxillo, D. M., Bodner, T., Pytlovany, A. C., & Richman, A. (2019). Exploration of the impact of organisational context on a workplace safety and health intervention. *Work & Stress*, 33(2), 192–210. https://doi.org/10.1080/02678373.2018.1496159

Hammer, L. B., Truxillo, D. M., Bodner, T., Rineer, J., Pytlovany, A. C., & Richman, A. (2015). Effects of a workplace intervention targeting psychosocial risk factors on safety and health outcomes. *Biomed Research International*, 2015, 836967. https://doi.org/10.1155/2015/836967

Henning, J. B., Stufft, C. J., Payne, S. C., Bergman, M. E., Mannan, M. S., & Keren, N. (2009). The influence of individual differences on organizational safety attitudes. *Safety Science*, 47(3), 337–345. https://doi.org/10.1016/j.ssci.2008.05.003

Hess, J. A., Kincl, L., Weeks, D. L., Vaughan, A., & Anton, D. (2020). Safety voice for ergonomics (SAVE): Evaluation of a masonry apprenticeship training program. *Applied Ergonomics*, 86, 103083. https://doi.org/10.1016/j.apergo.2020.103083

Hoel, H., Sparks, K., & Cooper, C. L. (2001). *The cost of violence/stress at work and the benefits of a violence/stress-free working environment.* https://www.ilo.org/safework/info/publications/WCMS_108532/lang--en/index.htm

Horan, K. A., Singh, R. S., Moeller, M. T., Matthews, R. A., Barratt, C. L., Jex, S. M., & O'Brien, W. H. (2019). The relationship between physical work hazards and employee withdrawal: The moderating role of safety compliance. *Stress & Health*, 35(1), 81–88. https://doi.org/10.1002/smi.2844

Huang, Y. H., Lee, J., McFadden, A. C., Murphy, L. A., Robertson, M. M., Cheung, J. H., & Zohar, D. (2016). Beyond safety outcomes: An investigation of the impact of safety climate on job satisfaction, employee engagement and turnover using social exchange theory as the theoretical framework. *Applied Ergonomics*, 55, 248–257. https://doi.org/10.1016/j.apergo.2015.10.007

ILO. (2002). *P155 – Protocol of 2002 to the Occupational Safety and Health Convention, 1981.* https://www.ilo.org/dyn/normlex/en/f?p=NORMLEXPUB:12100:0::NO::P12100_ILO_CODE:P155

Kao, K. Y., Spitzmueller, C., Cigularov, K., & Wu, H. (2016). Linking insomnia to workplace injuries: A moderated mediation model of supervisor safety priority and safety behavior. *Journal of Occupational Health Psychology*, 21(1), 91–104. https://doi.org/10.1037/a0039144

Kerr, S. (1975). On the folly of rewarding A while hoping for B. *The Academy of Management Journal*, 18(4), 769–783. https://doi.org/10.5465/255378

Landrigan, C. P., Rothschild, J. M., Cronin, J. W., Kaushal, R., Burdick, E., Katz, J. T., Lilly, C. M., Stone, P. H., Lockley, S. W., Bates, D. W., & Czeisler, C. A. (2004). Effect of reducing interns' work hours on serious medical errors in intensive care units. *New England Journal of Medicine*, 351(18), 1838–1848. https://doi.org/10.1056/NEJMoa041406

Lawrence, E. R., Halbesleben, J. R. B., & Paustian-Underdahl, S. C. (2013). The influence of workplace injuries on work-family conflict: Job and financial insecurity as mechanisms. *Journal of Occupational Health Psychology*, *18*(4), 371–383. https://doi.org/10.1037/a0033991

Lee, J., Huang, Y. H., Cheung, J. H., Chen, Z., & Shaw, W. S. (2019). A systematic review of the safety climate intervention literature: Past trends and future directions. *Journal of Occupational Health Psychology*, *24*(1), 66–91. https://doi.org/10.1037/ocp0000113

Lin, K. H., Lin, K. Y., & Siu, K. C. (2016). Systematic review: Effect of psychiatric symptoms on return to work after occupational injury. *Occupational Medicine*, *66*(7), 514–521. https://doi.org/10.1093/occmed/kqw036

Ludwig, T. D. (2014). The anatomy of pencil whipping. *Professional Safety*, 47–50.

Manapragada, A., & Bruk-Lee, V. (2016). Staying silent about safety issues: Conceptualizing and measuring safety silence motives. *Accident Analysis & Prevention*, *91*, 144–156. https://doi.org/10.1016/j.aap.2016.02.014

Mattson, M., Torbiörn, I., & Hellgren, J. (2014). Effects of staff bonus systems on safety behaviors. *Human Resource Management Review*, *24*(1), 17–30. https://doi.org/10.1016/j.hrmr.2013.08.012

McGraw, M. (2016). Equipped for the unthinkable. *Human Resource Executive*, May, 40–41.

Moyce, S. C., & Schenker, M. (2018). Migrant workers and their occupational health and safety. *Annual Review of Public Health*, *39*, 351–365. https://doi.org/10.1146/annurev-publhealth-040617-013714

Nahrgang, J. D., Morgeson, F. P., & Hofmann, D. A. (2011). Safety at work: A meta-analytic investigation of the link between job demands, job resources, burnout, engagement, and safety outcomes. *Journal of Applied Psychology*, *96*(1), 71–94. https://doi.org/10.1037/a0021484

Neal, A., & Griffin, M. A. (2006). A study of the lagged relationships among safety climate, safety motivation, safety behavior, and accidents at the individual and group levels. *Journal of Applied Psychology*, *91*(4), 946–953. https://doi.org/10.1037/0021-9010.91.4.946

Noort, M. C., Reader, T. W., & Gillespie, A. (2019). Speaking up to prevent harm: A systematic review of the safety voice literature. *Safety Science*, *117*, 375–387. https://doi.org/10.1016/j.ssci.2019.04.039

OSHA. (2016). *Guidelines for preventing workplace violence for healthcare and social service workers*. https://www.osha.gov/Publications/osha3148.pdf

Patton, C. (2017). The safety imperative. *Human Resource Executive*, July/August, 19–21.

Peng, L., & Chan, A. H. S. (2019). A meta-analysis of the relationship between ageing and occupational safety and health. *Safety Science*, *112*, 162–172. https://doi.org/10.1016/j.ssci.2018.10.030

Petitta, L., Probst, T. M., Barbaranelli, C., & Ghezzi, V. (2017). Disentangling the roles of safety climate and safety culture: Multi-level effects on the relationship between supervisor enforcement and safety compliance. *Accident Analysis & Prevention*, *99*(A), 77–89. https://doi.org/10.1016/j.aap.2016.11.012

Probst, T. M., Brubaker, T. L., & Barsotti, A. (2008). Organizational injury rate underreporting: The moderating effect of organizational safety climate. *Journal of Applied Psychology*, *93*(5), 1147–1154. https://doi.org/10.1037/0021-9010.93.5.1147

Probst, T. M., Goldenhar, L. M., Byrd, J. L., & Betit, E. (2019). The safety climate assessment tool (S-CAT): A rubric-based approach to measuring construction safety climate. *Journal of Safety Research*, *69*, 43–51. https://doi.org/10.1016/j. jsr.2019.02.004

Ramchand, R., Pomeroy, A., & Arkes, J. (2009). *The effects of substance use on workplace injuries* (OP-247-ADHS). https://www.rand.org/pubs/occasional_papers/OP247.html

Redden, E. S., & Larkin, G. B. (2015). Environmental conditions and physical stressors. In *APA Handbook of human systems integration*. (pp. 193–209). https://doi.org/10.1037/14528-013

Rogers, K. A., & Kelloway, E. K. (1997). Violence at work: personal and organizational outcomes. *Journal of Occupational Health Psychology*, *2*(1), 63–71. https://doi.org/10.1037/1076-8998.2.1.63

Safe Work Australia. (2015). *The cost of work-related injury and illness for Australian employers, workers, and the community: 2012–13*. https://www.safeworkaustralia.gov.au/doc/cost-work-related-injury-and-illness-australian-employers-workers-and-community-2012-13

Safe Work Australia. (2020). *Fatal Statistics by Industry*. https://www.safeworkaustralia.gov.au/statistics-and-research/statistics/fatalities/fatality-statistics-industry

Salas, E., Tannenbaum, S. I., Kraiger, K., & Smith-Jentsch, K. A. (2012). The science of training and development in organizations: What matters in practice. *Psychological Science in the Public Interest*, *13*(2), 74–101. https://doi.org/10.1177/1529100612436661

Salminen, S. (2004). Have young workers more injuries than older ones? An international literature review. *Journal of Safety Research*, *35*(5), 513–521. https://doi.org/10.1016/j.jsr.2004.08.005

Saracino, A., Curcuruto, M., Antonioni, G., Mariani, M. G., Guglielmi, D., & Spadoni, G. (2015). Proactivity-and-consequence-based safety incentive (PCBSI) developed with a fuzzy approach to reduce occupational accidents. *Safety Science*, *79*, 175–183. https://doi.org/10.1016/j.ssci.2015.06.011

Scharf, T., Hunt, J., McCann, M., Pierson, K., Repmann, R., Miglicaccio, F., Limanowski, J., Creegan, J., Bowers, D., Happe, J., & Jones, A. (2011). *Hazard recognition for ironworkers: Preventing falls and close calls – updated findings*. 2010 International Conference on Fall Prevention and Protection.

Schulte, P. A. (2005). Characterizing the burden of occupational injury and disease. *Journal of Occupational & Environmental Medicine*, *47*(6), 607–622. https://doi.org/10.1097/01.jom.0000165086.25595.9d

Smith, M. J., & Carayon, P. (2011). Controlling occupational safety and health hazards. In J. C. Quick & L. E. Tetrick (Eds.), *Handbook of occupational health psychology* (Vol. 2, pp. 75–93). American Psychological Association.

Sonnentag, S., Binnewies, C., & Ohly, S. (2013). Event-sampling methods in occupational health psychology. In R. R. Sinclair, M. Wang, & L. E. Tetrick (Eds.), *Research methods in occupational health psychology: Measurement, design, and data analysis* (pp. 208–228). Routledge/Taylor & Francis Group.

Sparer, E. H., Catalano, P. J., Herrick, R. F., & Dennerlein, J. T. (2016). Improving safety climate through a communication and recognition program for construction: A mixed methods study. *Scandanavian Journal of Work, Environment, and Health*, *42*(4), 329–337. https://doi.org/10.5271/sjweh.3569

Spurgeon, A., Harrington, J. M., & Cooper, C. L. (1997). Health and safety problems associated with long working hours: A review of the current position.

Occupational and Environmental Medicine, 54, 367–375. https://doi.org/10.1136/oem.54.6.367

SCC. (2020). *Health enterprise – Prevention, promotion, and organizational practices contributing to health and wellness in the workplace* (CAN/BNQ 9700–800/2020).

Thiede, I., & Thiede, M. (2015). Quantifying the costs and benefits of occupational health and safety interventions at a Bangladesh shipbuilding company. *International Journal of Occupational and Environmental Health, 21*(2), 127–136. https://doi.org/10.1179/2049396714Y.0000000100

Tucker, S., & Turner, N. (2015). Sometimes it hurts when supervisors don't listen: The antecedents and consequences of safety voice among young workers. *Journal of Occupational Health Psychology, 20*(1), 72–81. https://doi.org/http://dx.doi.org/10.1037/a0037756

Tucker, S., Chmiel, N., Turner, N., Hershcovis, M. S., & Stride, C. B. (2008). Perceived organizational support for safety and employee safety voice: The mediating role of coworker support for safety. *Journal of Occupational Health Psychology, 13*(4), 319–330. https://doi.org/10.1037/1076-8998.13.4.319

Tuckey, M. R., & Scott, J. E. (2014). Group critical incident stress debriefing with emergency services personnel: A randomized controlled trial. *Anxiety, Stress and Coping, 27*(1), 38–54. https://doi.org/10.1080/10615806.2013.809421

U.S. BLS. (2019a). *Census of Fatal Occupational Injuries Summary, 2018* https://www.bls.gov/news.release/cfoi.nr0.htm

U.S. BLS. (2019b). *Fatal occupational injuries to emergency responders.* Retrieved April 22, 2020 from https://www.bls.gov/iif/oshwc/cfoi/er_fact_sheet.htm

Vinodkumar, M. N., & Bhasi, M. (2010). Safety management practices and safety behaviour: Assessing the mediating role of safety knowledge and motivation. *Accident Analysis & Prevention, 42*(6), 2082–2093. https://doi.org/10.1016/j.aap.2010.06.021

Weil, D. (2001). Valuing the Economic Consequences of work injury and illness: A comparison of methods and findings. *American Journal of Industrial Medicine, 40,* 418–437. https://doi.org/10.1002/ajim.1114

Williamson, A., Lombardi, D. A., Folkard, S., Stutts, J., Courtney, T. K., & Connor, J. L. (2011). The link between fatigue and safety. *Acciddent Analysis & Prevention, 43*(2), 498–515. https://doi.org/10.1016/j.aap.2009.11.011

Zhang, M., Murphy, L. A., Fang, D., & Caban-Martinez, A. J. (2015). Influence of fatigue on construction workers' physical and cognitive function. *Occupational Medicine, 65*(3), 245–250. https://doi.org/10.1093/occmed/kqu215

Zohar, D. (1980). Safety climate in industrial organizations: Theoretical and applied implications. *Journal of Applied Psychology, 65*(1), 96–102. https://doi.org/10.1037/0021-9010.65.1.96

Zohar, D. (2000). A group-level model of safety climate: testing the effect of group climate on microaccidents in manufacturing jobs. *Journal of Applied Psychology, 85*(4), 587–596. https://doi.org/10.1037/0021-9010.85.4.587

Zohar, D. (2010). Thirty years of safety climate research: Reflections and future directions. *Acciddent Analysis & Prevention, 42*(5), 1517–1522. https://doi.org/10.1016/j.aap.2009.12.019

Zohar, D., & Luria, G. (2003). The use of supervisory practices as leverage to improve safety behavior: a cross-level intervention model. *Journal of Safety Research, 34*(5), 567–577. https://doi.org/10.1016/j.jsr.2003.05.006

Zohar, D., & Luria, G. (2004). Climate as a social-cognitive construction of super-visory safety practices: Scripts as proxy of behavior patterns. *Journal of Applied Psychology, 89*(2), 322–333. https://doi.org/10.1037/0021-9010.89.2.322

Zohar, D., & Polachek, T. (2014). Discourse-based intervention for modifying super-visory communication as leverage for safety climate and performance improvement: A randomized field study. *Journal of Applied Psychology, 99*(1), 113–124. https://doi.org/10.1037/a0034096

Zou, P. X. W., Sun, A. C. S., Long, B., & Marix-Evans, P. (2010). Return on investment of safety risk management system in construction. 18th CIB World Building Con-gress: CIB W099- Safety and Health in Construction. Salford, United Kingdom.

12

BROADENING OHP IMPACT
BEYOND THE WORKPLACE

Christopher J. L. Cunningham and Kristen Jennings Black

In this final chapter, we consider several ways in which Occupational Health Psychology (OHP) professionals can more broadly apply their knowledge and methods to positively impact society, not just workers and organizations. Specifically, we examine ways in which OHP can contribute to research and practice that: (a) protects human dignity and rights; (b) addresses social determinants of health, safety, and well-being; and (c) supports peace and social justice. These areas of broader impact are now and will continue to be critically important even as working arrangements and societal pressures change in the future.

When you are finished reading this chapter, you should be able to:

LO 12.1: Describe several ways in which OHP professionals can positively impact society-level health, safety, and well-being concerns.

LO 12.2: Explain the importance of protecting human dignity and rights in all OHP research and practice efforts.

LO 12.3: Explain why reducing poverty and providing a living wage are essential to the health, safety, and well-being of all members of society.

Opportunities for Broader Impact

Throughout this book we have focused on ways in which OHP professionals can use their knowledge and methods to protect and promote worker health, safety, and well-being (WHSWB). Up to this point, we have examined essential OHP theories, research findings, and intervention methods with an emphasis on workers in specific work settings. Many of the essential OHP topics, theories, research evidence, and methods discussed in this book, however, can be more broadly studied and applied to address a variety of major societal challenges.

This broader utility of OHP knowledge and methods is true now and will continue to be true even as changes occur to the nature of work, organizations, and workers. The anticipated future of work, as outlined by futurists and respected institutions (e.g., International Labour Organization [ILO], 2019a), will likely involve new technologies and ways of working, as well as new forms of work altogether. What will not be so different in the future, is that these advancements will result in new demands to which workers still must adapt. Also, some workers will continue to have more opportunities and advantages than others. OHP professionals can help organizations and workers adapt to future work-related demands while improving the overall WHSWB situation for all workers.

While the future unfolds, we must not ignore current opportunities we have to develop and shape workers, organizations, and society in ways that support sustainable work, which protects workers' rights and dignity, and ensures equitable and safe working experiences for all (ILO, 2019a). With this chapter, we hope to encourage you to broaden your thinking about how OHP knowledge and methods can be extended to several important societal-level issues, including: (a) protecting human dignity and rights; (b) addressing social determinants of health, safety, and well-being; and (c) contributing to peace and social justice.

Protecting Human Dignity and Rights

As emphasized by governments, policy institutions, and even religious traditions and teachings, human dignity and rights should be protected in *and* out of work environments. Providing such protections requires coordinated efforts between organizations and governments, as well as supportive forces operating within society more generally. There is much work to be done within our global society to fully protect human dignity and rights. Even just focusing on work contexts, many workers worldwide do not receive sufficient and equivalent protections for the work-related risks they regularly face. Entire subpopulations of workers are basically exploited for the value they can create for their employers (e.g., migrant farm workers in unsafe conditions, sex workers, child laborers). Sometimes worker exploitation results from organizational cost management efforts, thinly veiled as limited, part-time, or even "flexible" work opportunities. This is often the situation when organizations leverage *contingent labor* (e.g., migrant workers, temporary or part-time help) or *gig workers* (e.g., independent contractors, online crowd-sourcing workers) as a strategy for controlling administrative and operational costs. These employment arrangements can become exploitative if workers are prevented from obtaining adequate pay, benefits, other rewards, safety protections, and job security that would otherwise be available to full-time workers. Indeed, gig workers often earn less than minimum wage to engage in work that offers little security or formalized benefits (O'Connor et al., 2020), and in many cases involves more work-related risks for interpersonal mistreatment, sexual harassment, and even assault (e.g., Ravenelle, 2019). Sometimes environmental circumstances create situations in which exploitative working scenarios unfold. As an example, consider the experiences of essential workers during national lockdowns due to the COVID-19 pandemic. Many individuals willingly and heroically continued to support essential societal functions (e.g., healthcare, education). Many others, however, had no real choice but to keep working so they could put food on the table. Regardless of how they are justified, these types of labor relationships flout human dignity and rights, and damage societal functioning.

Some workers are drawn (sometimes out of interest, sometimes out of necessity) to socially undesirable occupations (i.e., *dirty work*). Ashforth and Kreiner (1999) defined such work as involving physical contact with undesirable things (e.g., janitor, mortician), questionable morality (e.g., exotic dancer, pawn broker), or social taints (e.g., tabloid reporter, telemarketer). Morally dirty work tends to be associated with more purely negative views (i.e., "dirtier" work) than socially or physically dirty work, which may be undesirable, but seen as necessary to societal functioning (Ashforth & Kreiner, 2015). Workers in numerous occupations face demands that are not only undesirable, but disproportionately risky in physical, psychological, and social ways. These characteristics are present in so-called *3-D* (i.e., dirty, dangerous, and demanding) work roles and commonly experienced by migrant and undocumented workers. It is important to note, however, that these characteristics are also often experienced in more mainstream roles, including in healthcare or military occupations (e.g., de Boer et al., 2011; Porter et al., 2018).

Dirty and 3D work is demanding in and of itself; a main objective for OHP professionals supporting workers in these occupations is to help them reduce the amount of additional personal effort they need to put into legitimizing their work for themselves and others. Existing research shows us that within these occupations, workers and their managers use a number of strategies, like reframing or distancing from critical others, confronting negative stereotypes, and finding meaning in the work as ways of coping with work-related stigma (Ashforth & Kreiner, 1999; Ashforth et al., 2007; Bosmans et al., 2016). OHP professionals can help organizations and society to prepare, support, and respect the rights of such workers who willingly sacrifice their physical, social, and psychological health.

Maintaining a positive identification with one's work is important to all workers, even those who do not struggle with negative stigma due to the nature of their work. With this in mind, one other area in which OHP professionals can help in protecting human dignity and rights involves assisting workers as they transition into retirement, often coping with the lasting consequences of prolonged strain, and a loss of identity and purpose that comes along with stopping work (Alimujiang et al., 2019; Froidevaux et al., 2016). A number of OHP professionals and organizations are already demonstrating commitment to addressing the issues outlined in this section, but there is still much work to be done. Collaborating with organizations and policy-making institutions, we can help to ensure that protecting WHSWB is a priority for *all* workers. Keep in mind that by protecting workers' dignity and rights, we are also reinforcing broader community and societal norms supporting protections for human dignity and rights that extend far beyond the boundaries of a single organization.

Addressing Social Determinants of Health, Safety, and Well-Being

As noted in an excellent review chapter by Donkin et al. (2014), there are documented and substantial disparities in the health status and quality of life for different groups of people within society; these "health inequities arise from the conditions in which people are born, grow, live, work, and age and inequities in power, money, and resources that give rise to these conditions of daily life" (p. 1). In line with this, the Centers for Disease Control and Prevention (CDC) proposed five key social determinants of health as part of its Healthy People 2030 campaign: economic stability, education access and quality, healthcare access and quality, neighborhood environment, and community context (CDC, 2020). A major source of social influence with the power to address these types of health determinants and inequities are the organizations in which people work. Healthy, safe, meaningful, and fairly compensated work support a number of essential elements highlighted by this CDC campaign. There are at least three ways in which OHP research and practice can directly address social determinants to health, safety, and well-being.

Comprehensive Management of WHSWB

Protecting and promoting WHSWB requires targeted efforts to address the topics discussed throughout this book. Also valuable are efforts to more holistically consider and address the factors that influence WHSWB. An increasingly visible example of this latter type of comprehensive approach is the *Total Worker Health*® initiative (National Institute for Occupational Safety and Health [NIOSH], 2020; Schill & Chosewood, 2013). This program acknowledges the need to holistically care for workers using methods that transcend the boundaries of work and nonwork role domains. Also along these lines is the World Health Organization's (WHO) Healthy Workplace model, which emphasizes WHSWB as a priority for organizations and policy makers that not only benefits workers and organizations, but also families and communities (Burton, 2010; WHO, n.d.).

The European Working Conditions survey, through the European Foundation for the Improvement of Living and Working Conditions (Eurofound) is another tangible example of changing society-level attention to worker well-being (Eurofound, 2020). This recurring survey began in 1990 and gathers data on factors affecting well-being, including traditional measures of safety, hazards, and work hours, as well as factors that transcend the workplace, such as work-life balance and feelings of financial security. Collectively, these examples highlight a broader way of thinking about and responding to WHSWB, its antecedents, and its consequences. This perspective also helps us understand the importance of our next couple of areas in which OHP can help to address social determinants of health, safety, and well-being.

Occupational Health Inequity

Ensuring WHSWB for all workers includes addressing *occupational health inequities* or, "avoidable differences in work-related disease incidence, mental illness, or morbidity and mortality that are closely linked with social, economic, and/or environmental disadvantage such as work arrangements (e.g., contingent work), socio-demographic characteristics (e.g., age, sex, race, and class), and organizational factors (e.g., business size)" (NIOSH, 2019; para. 1). Numerous studies document such health inequities at work, highlighting instances of disproportionate exposure to job hazards and insecurity, as well as disproportionate access to valuable resources like health insurance and paid leave (Landsbergis et al., 2014; Lipscomb et al., 2006). An implication of such inequity is that the burden of suffering, morbidity, and mortality is disproportionately experienced by some members of society more than others (Macik-Frey et al., 2016).

Some efforts to advocate for workers and reduce disparities require the action of local, national, and international governing bodies. Formal policies and initiatives (e.g., availability of workers unions, wage-related legislation) have already made an impact for many communities (Siqueira et al., 2014). However, attempts to address occupational health inequity and disparities within specific organizational settings do not have to be all that complicated. An elegantly simple example comes from a "chair campaign" program introduced in Korea to provide service workers with the opportunity for rest during work shifts (Lee et al., 2011). This program successfully improved seating options for many workers and more generally increased societal awareness of this basic need. As discussed throughout this book, there are many demands inherent in just about every occupation, but there are also ways to match these demands with necessary resource and limit exposure to these demands.

Economic and Basic Needs Insecurity

Another major social determinant of health, safety, and well-being for workers and their families is consistent and sufficient resources to meet income, healthcare, food, shelter, and other basic needs. Unfortunately, many people are not able to meet at least some of these needs. It is estimated that approximately 9–10% of the world's population lives in extreme poverty, with two-thirds living on less than $10 per day (Our World in Data, 2019; World Bank, 2020). Although managing food, shelter, and other needs may be a bit beyond typical OHP areas of practice, our knowledge and expertise can help address income and benefits security, two resources that enable workers to access other resources necessary for survival and maintaining a decent quality of life. Organizations can serve as platforms through which needs identification and screening can occur. Organizations can also facilitate distribution of care, support, and other resources where needed. OHP professionals can help with these efforts, given that they require organizational decision makers to stop believing that who and how we are at work can somehow be segmented or kept separate from who we are outside of work. As we discussed in Chapter 9, work and nonwork roles are typically much more integrated than we often like to believe.

Digging into a particularly essential area of need, the effects of *income (in)security* and *economic stress* on workers, and their dependent families and communities are not commonly studied in the OHP or even in most applied behavioral or social sciences. This is particularly true when it comes to economically vulnerable populations, which historically have not been the focus of most OHP attention (for an excellent discussion of this history from a closely related Industrial-Organizational psychology perspective, see Gloss et al., 2017 and Saxena, 2017). The paradox here is that often ignored low-income workers regularly confront some of the most complex work and nonwork demands.

In our research and practice, OHP professionals can broaden our attentional scope to consider the experiences of low-income workers, particularly those who would be considered *working poor* (Klein & Rones, 1989). As defined by the U.S. Bureau of Labor Statistics (BLS; 2020), these are individuals who are working or actively seeking work for at least 27 weeks of the year, but whose income still falls below national poverty rates. In 2018, this applied to 4.5% of the American workforce, with higher rates among part-time workers, workers in the service industry, women (particularly single women maintaining families), racial minorities (Black/African American or Hispanic/Latino), and less educated workers (BLS, 2020). As noted by the ILO (2019b), the numbers of individuals in poverty who are working versus not working are surprisingly similar, suggesting the opportunity to work is not enough to prevent poverty. This conclusion is supported by other studies that have shown that the working poor are often more in need of supports like long-term food assistance than those who do not work (e.g., Berner et al., 2008). Workers (and their families) in these situations can feel quite hopeless. Consider working parents that work full weeks, yet their earned income does not even amount to the costs of the childcare needed to enable them to work. This can feel like a losing battle, where the "solution" (to get a job) creates new demands on already limited financial resources. OHP professionals can help to inform and support business leaders and policy makers who are attempting to generate real solutions to these types of challenges.

There really are numerous opportunities for OHP professionals to be involved in fighting against economic insecurity. This includes advocating for formal policies to ensure workers have a living wage (as noted in Chapter 3) and are not denied wages they rightly earn in more ambiguous or contingent work situations (Siqueira et al., 2014). Even when such public policies do not exist, we can encourage organizational leaders to provide workers with fair compensation and access to benefits as strategies for protecting WHSWB and strengthening communities. Organizations can also provide non-monetary resources that can improve financial security within communities, like financial education and literacy workshops and career development opportunities (e.g., skills training, continuing education for career growth). OHP professionals and organizations can even intentionally help those who are not in the workforce through connections with hiring agencies and non-profit organizations that support those often overlooked in our society, such as those who are homeless or previously incarcerated. Finally, OHP professionals can advocate for smart investments in communities that can improve individuals' ways of responding to income- and work-related stress (e.g., Probst et al., 2018). An essential element to such efforts is creating support resources, in organizations and communities, where individuals experiencing financial hardship can obtain resources they need without experiencing shame.

While we have focused mainly on the financially vulnerable here, we want to acknowledge that all workers can experience the poor health effects of stressors like job insecurity, unemployment, and underemployment (Creed & Klisch, 2005; Friedland & Price, 2003; Park & Baek, 2019; Shoss, 2017). There is also growing evidence of strong relationships between income insecurity and psychological and physical health (Kopasker et al., 2018; Vandoros et al., 2019). From the perspective of the conservation of resources (COR) theory (Hobfoll, 1989), which we have referred to many times in this text, income is a resource in and of itself, but also a means to acquisition of many other more directly useful resources. Consider the stability and access to resources a decent income affords (e.g., ability to pay bills; self-esteem; peace of mind). In contrast, think about the loss of personal comfort, efficacy, and tangible resources that comes along with insufficient income. When examining these issues, there are other theoretical perspectives that are also relevant and deserve more careful consideration by OHP researchers and practitioners, including the health capital (i.e., "Healing") and "Breaking Point" models of mental health that are well-described by Watson and Osberg (2017).

OHP professionals who decide to dig more into these topics should note that simple measures of employment and objective income fall short of capturing workers' often complex actual financial experiences. Perceived income adequacy measures have been developed to understand the perceptual nature of income-related stress (Sears, 2008). Other complexities include accounting for stress associated with personal debt, which has received relatively little attention in the organizational psychology literature (Sinclair & Cheung, 2016), despite increased attention to the weight of debt covered in the general media. Debt may be a contextual factor that alters the personal value and meaning of income as a resource, and can make threats to employment and income much more salient and debilitating. Keeping these points in mind can help to propel future OHP research and practice efforts in this area.

Contributing to Social Justice and Peace

Protecting workers' dignity and rights, ensuring basic needs are met, and comprehensively tackling issues of WHSWB are all ways in which organizations and OHP professionals can support social justice. OHP professionals have much to offer this area of research and practice, though this is another area that is not currently mainstream within this discipline (for a fascinating history of these issues in a global context since the early 1900s, we encourage you to read the special centenary issue of the *World at Work* magazine; ILO 2019c). Work organizations are often microcosms of their surrounding communities. If and when an organization figures out how to demonstrate and act in a socially responsible and just manner internally, this can have many positive and stabilizing ripple effects externally. Unfortunately, when an organization and its members do not demonstrate or model such values and behaviors, the effects can be further division and instability in the external community. In this way, there are opportunities for organizations to be mechanisms for increasing social justice and even peace in their broader communities.

One such opportunity is linked to efforts to address issues of inclusion and diversity within organizations. Often this involves developing and implementing policies and practices that support tolerance and fair pathways to development and advancement within these organizations. These efforts can have effects that transcend the boundaries of any particular organization. Inclusion is a critical social justice issue for organizations, and a critical WHSWB issue for OHP professionals, as experiences of bias, discrimination, or even subtle mistreatment toward minority group members can do more than create frustration or discomfort. These sorts of experiences can relate to real, substantial psychological and physical health outcomes for workers, as well as harm to the organization that can accrue from turnover costs, productivity loss, and withdrawal (e.g., Cortina et al., 2017). Studies even suggest that working conditions can affect the expression of bias and discrimination toward members of the community that workers serve. For instance, one study found that non-black physicians exhibited more implicit and explicit bias when they experienced higher levels of burnout, which has further implications for the proper treatment of colleagues and patients (Dyrbye et al., 2019).

Organizations with a culture that demands inclusion and mutual respect can go a long way in the treatment of workers and those affected by their work. Perhaps, this sort of commitment can even lead to a broader sense of peace in our communities. Consider initiatives to modify policing work role boundaries by embedding law enforcement officers into communities as a way of having a greater impact by building community partnerships (Crowl, 2017). Other examples of this type of deep community engagement also abound, from internship and apprenticeship arrangements between manufacturing companies and local schools, to medical and legal practices donating expertise and support to free health clinics, to grocery stores helping to sustain local food banks. There are a lot of possibilities to do good in the world when organizations think about the true reach they have into the communities and societies in which they exist.

Research, Intervention, and Evaluation Considerations for Broader Impact

OHP professionals have a rich knowledge base and methodological skillset that can help with identifying and explaining ways in which working and work organizations can be leveraged as mechanisms for broader societal benefit and change. In this section, we highlight special considerations for OHP professionals who venture out of traditional areas of research and practice to understand and advocate for the topics outlined in this chapter.

First, regardless of where this type of work may take you, we encourage all OHP professionals to operate within the guidelines of the American Psychological Association's Ethical Principles of Psychologists and Code of Conduct (www.apa.org/ethics) or a similar set of guidelines that may be more specifically suited to your region of the world. Even if you are not licensed as a psychologist yourself, adhering to such standards helps to protect the broader profession of psychology and the dignity and rights of others with whom we work. At a high level, this means doing our best to operate within our own boundaries of professional competence, and to operate with the intention of truly bettering the lives of workers and their communities. This means demonstrating respect and reserving personal judgment when we work with vulnerable and marginalized groups (e.g., those who are currently or were previously engaged in dirty work, incarcerated, or homelessness). Sometimes this may mean walking away from professional opportunities when the goals for a particular collaboration are not aligned with these principles or our professional and personal values. This can be difficult, but we need to remember that positive change can begin (and end) with one person and as OHP professionals, we simply cannot turn a blind eye toward intentional disregard of workers' rights.

Second, researching and practicing outside the boundaries of typical work organizations will also challenge our methodological skills; our challenge is to develop and use methods that fit the population and context in which we are working. As a couple of examples, even though internet-based surveying is convenient and effective for many research efforts, in some work and community settings, limited access to technology, different communication preferences (e.g., in person vs. phone vs. email), and population-specific language and reading level needs may require us to use a different approach. When engaging in research, we may also need alternative sampling methods like *purposive sampling* to target specific demographic groups or *snowball sampling* to reach members of difficult-to-reach populations. We have learned that it is incredibly valuable to talk with members of your target population before all methodological decisions are finalized. Ultimately, learning about the work and nonwork experiences of workers and other community members can help us design and implement more effective research studies and interventions. As an example, Cuervo et al., (2020) used a qualitative research approach to understand the community context for low-wage immigrants, along with their work experiences and work-related values. Another impressive undertaking along these lines is presented by Saxena (2015), who used qualitative ecological momentary assessment methods to understand and prevent the spread of a communicable disease among rural farmers in northeastern India (see also Saxena & Burke, 2020 for a discussion of policy-related implications associated with this work).

Applying our OHP knowledge and methods to the types of issues outlined in this chapter may also require us to be creative and clever in the design and implementation of our research and intervention efforts. For example, Reeves et al. (2017) demonstrated positive effects of a national minimum wage increase in the United Kingdom on workers' mental health using data gathered through a natural experiment, showing an effect equivalent to that of antidepressant medications. A second example comes from Backman et al. (2011), who demonstrated (with a quasi-experimental design) how providing fresh fruit at the worksite (not just education about nutrition) significantly increased workers' consumption of fruit and vegetable consumption in *and* outside of the workplace. This study is a great example of how an intervention at work can have positive effects that extend to the health of workers, families, and communities.

Third and finally, OHP professionals can broaden their impact by expanding their collaborative networks. We will not be able to create effective change initiatives that transcend the work environment without the help of those who have expertise in these broader domains. We can work with health psychologists, economists, community psychologists, anthropologists, public health specialists, epidemiologists, and others (as noted in Chapter 1) to develop more effective solutions to the broad health, safety, and well-being challenges faced by workers and other members of society. While such collaborations will often open our minds to new perspectives, there will also be certain common principles that are applicable in a wide variety of research and practice situations. As one example, the theory of planned behavior has been applied to encourage adherence to safe work practices (Johnson & Hall, 2005), but also as a framework for understanding and intervening to affect a wide range of health behaviors in larger communities (e.g., smoking, diet, safe sex practices, teeth brushing; Godin & Kok, 1996). Our main point here is that we do not necessarily have to create anything new or complex to extend the impact of OHP to the broader societal challenges outlined in this chapter. However, we are likely to improve our chances of making a difference if we collaborate with professionals from other disciplines.

Concluding Thoughts and Reality Check

Our world is unfortunately characterized by a great deal of inequality and injustice linked to people's differential access to material, psychological, and social forms of resources needed for maintaining health, safety, and well-being. There is a real opportunity to address the concerns outlined in this chapter through the reach and influence of work organizations and governing institutions who can collaboratively ensure that workers and communities have access to the resources needed to survive and thrive. It is our hope that all OHP professionals will openly consider and seek out opportunities to protect human dignity and rights, consider social determinants of health, safety, and well-being, and contribute to social justice and peace in whatever area of research or practice they are primarily engaged.

Protecting WHSWB is a direct way of strengthening organizations, but also families, communities, and society. Work environments are often the most controlled environments that people regularly occupy and the reservoir from which we all draw some of our most significant resources (i.e., income and benefits, meaning and purpose, affiliation with others). If we keep this in mind and broaden our perspectives even just a little bit, the potential impacts of OHP research and practice on society are tremendous. Although the effects of work are complex, the work we do as OHP professionals can be simply described: We support strength and resilience in workers, organizations, families, communities, and societies by protecting and promoting worker health, safety, and well-being. In all seriousness, your work as an OHP professional can literally improve the world.

Media Resources

- Business news article summarizing ways in which organizations are taking action against various forms of social injustice: https://www.businessinsider.com/george-floyd-protests-companies-responses-actions-apple-target-mcdonalds-nike-2020-6
- News story discussing challenges for determining a true living wage: https://www.npr.org/local/305/2019/10/03/766727610/paying-a-living-wage-in-d-c-isn-t-easy-it-s-also-not-enough-workers-say
- News story regarding barriers faced by those with a criminal record https://www.usatoday.com/story/money/2019/03/19/business-commitment-hiring-those-with-criminal-record/3091463002/

Discussion Questions

1) What are some ways in which WHSWB concerns transcend the work environment?

2) How can you apply your knowledge and skills to address one or more of the broad health, safety, and well-being issues discussed in this chapter?

3) Are there other ways in which OHP research can be translated into society-level impacts?

4) What is the connection between the topics discussed in this chapter and the broader issue of corporate social responsibility?

5) What are implications of the topics discussed in this chapter and the overall book for how organizations treat individuals who are marginalized by or somehow seen as "undesirable" within society?

Professional Profile: Robert R. Sinclair, Ph.D.

Country/region: USA

Current position title: Professor of Psychology

Background: I have been working to improve the health, safety, and/or well-being of workers for 24 years. In 1995, I earned my Ph.D. and M.A. in Industrial-Organizational Psychology from Wayne State University. I also have a B.A. in Psychology from the University of Maine at Farmington. After completing my Ph.D., my first academic job was the University of Tulsa where I worked from 1995–1999. I then moved to Portland State University (2000–2008) and then to Clemson University (2008–present). I am a traditional academic engaged in research, teaching, and professional/university service. I also engage in applied research consultation on an occasional basis, mostly with the United States Army. I currently belong to the Society for Occupational Health Psychology, Society for Industrial-Organizational Psychology, and the American Psychological Association.

How my work impacts WHSWB: My current research program addresses three broad issues: (1) how economic stressors affect workers' occupational safety, health, and well-being, (2) how organizations can build organizational climates that enhance worker safety, health, and well-being, and (3) applications of occupational health science to special populations such as health care workers and military personnel. One recent example of an intervention I was involved with comes from a dissertation project in which a student in Clemson's engineering program sought to create safer work environments for university staff working in laboratories where workers were exposed to radiation. The project involved a safety training program, creation of a safety-focused newsletter, and safety-related signage that could be posted around the campus offices. The research program involved assessment of safety climate before and after the intervention as well as direct behavioral observation of staff members with regards to issues such as wearing personal protective equipment and proper disposal of hazardous material. The intervention showed a variety of positive effects on worker safety behavior as well as their perceptions of the extent to which their employer valued safety.

My motivation: Although I did not know anything about OHP at the time, in some of my earliest work experiences, I could see how managers' actions could create supportive or detrimental environments for workers. I was initially interested in the role of labor unions in creating safer and healthier work environments and as time passed, that evolved into a broad interest in workers' safety, health, and well-being. These days, what sustains my interest is my increased awareness over time of the many ways (both good and bad) in which work affects health. Given the importance of work to most people's lives and the fact that many of the threats to worker health lie beyond workers' direct control, occupational health seems to me to be a vitally important social concern where many interesting research questions remain.

Professional Profile: Jennifer Cullen, Ph.D.

Country/region: USA
Current position title: Director of Global Culture and Engagement
Background: I have been working to improve the health, safety, and/or well-being of workers for 18 years. I received a B.S. in Psychology from Virginia Tech, then a Masters and Ph.D. in Industrial-Organizational (I-O) Psychology from Portland State University. I was one of the recipients of the first year OHP awards granted by NIOSH for graduate students doing OHP research with specific OHP professors (Dr. Leslie Hammer in my case). I belong to the Society of OHP, Society of I I-O Psychology, the American Psychological Association, People Analytics & Future of Work, the Bay Area People Analytics Meetup group, and the People Geek Community.

I thought I was going to become an academic . . . However, I was recruited out of graduate school to become an applied Research Psychologist working for the Safety & Health Assessment Research Program within the Washington State Department of Labor & Industries. I worked with an incredible group of interdisciplinary researchers, led by Dr. Barbara Silverstein, focused on various aspects of worker health and well-being, and the environmental and organizational drivers of those. Doing that work, I observed that for some workers, there was no amount of intervention that could lead them to a place of well-being because they were simply working in ill-suited roles. That lesson, considered together with a desire to relocate to sunny California, led me to take on even more applied role at Evolv, Inc. (eventually acquired by Cornerstone) developing pre-hire assessments. During that time, I moved from an internal consulting role as a Selection Scientist into a strategic role as Chief Scientist, where I oversaw all selection science and talent management deliverables. I led a team responsible for building, deploying, and validating selection products for 25 different job families used by global clients spanning across nine countries.

The land of talent acquisition and placement is less of a traditional OHP focus, but no less deserving as a starting point for employee and organizational health. From there, I decided to focus more on workplace culture and environment, together with employee engagement and experience. I joined the team at Culture Amp, designing better ways to collect and receive employee feedback, facilitate open and candid discussions around ideal culture and the employee experience, and creating joint ownership and responsibility between employer and employee about the world of work. Learning a ton from the Culture Amp team, the Culture Amp community of People Geeks, and the organizations I got to work with served as a jumping off point for me to move to an internal role at Visa, Inc.

At Visa, I am amplifying and strategically moving how they think about and approach culture and engagement across the 20K employee members. At Visa, we believe it is everyone's responsibility to develop the best talent and that by acting as one we can create a compelling experience that shifts mindsets and behaviors to drive long-term impact for everyone, everywhere. I am responsible for nurturing culture and, through employee feedback, improving the experience and conversation between managers and employees. I also serve as an internal adviser, initiating ideas across the organization that influence strategy, engagement, and a collaborative culture that drive pride, passion, inclusion, and advocacy.

How my work impacts WHSWB: Overseeing Visa's employee listening strategy, I am having a direct impact on the way that 20K global employees experience their day-to-day work, and thus their emotional, intellectual, and behavioral well-being. In addition, I lead the employee survey strategy, so when it comes to asking employees about their health, safety, and well-being, my team and I decide what to ask, how to ask it, and what we and other leaders can do with the important feedback (data) collected.

One simple example of my work with an impact on employee well-being was when I was working closely with a large telecommunications company on designing their Employee Listening Strategy. As part of a survey to collect employee feedback on their revamped learning and development (L&D) program, the company observed lower than acceptable ratings (by their account) on reports around adequate training opportunities. After unpacking the data, we detected it had less to do with "access" and more to do with not having adequate time to take advantage of the existing L&D opportunities. Following employee insights from participatory focus groups, they optimized the way the company socialized L&D opportunities internally and provided managers and employees with guiding principles and normative expectations for how much time all organizational members should be able to carve out for professional and personal development. It doesn't take a giant leap to envision the positive impact this employee feedback ended up having on the ultimate career progression of folks and the perceptions employees and managers developed around their employer's investment in them.

My motivation: I grew up a latchkey kid. As a witness to the consequences and impact of stressful work in my impressionable years, I became very interested in the research and the conversation about work-life blend at a young age. While that interest and dialogue has since extended to cover the whole employee and the entirety of the employee lifecycle, I remain mission-driven around doing what I can to move the needle on how we design our workplaces and approach our work lives to make them as meaningful, beneficial, and fulfilling as possible, regardless of location, function, level, or task.

Chapter References

Alimujiang, A., Wiensch, A., Boss, J., Fleischer, N. L., Mondul, A. M., McLean, K., Mukherjee, B., & Pearce, C. L. (2019). Association between life purpose and mortality among US adults older than 50 years. *JAMA Network Open, 2*(5), e194270. https://doi.org/10.1001/jamanetworkopen.2019.4270

Ashforth, B. E., & Kreiner, G. E. (1999). "How can you do it?": Dirty work and the challenge of constructing a positive identity. *The Academy of Management Review, 24*(3), 413–434. https://doi.org/10.5465/amr.1999.2202129

Ashforth, B. E., & Kreiner, G. E. (2015). Dirty work and dirtier work: Differences in countering physical, social, and moral stigma. *Management and Organization Review, 10*(1), 81–108. https://doi.org/10.1111/more.12044

Ashforth, B. E., Kreiner, G. E., Clark, M. A., & Fugate, M. (2007). Normalizing dirty work: Managerial tactics for countering occupational taint. *The Academy of Management Journal, 50*(1), 149–174. https://doi.org/10.5465/amj.2007.24162092

Backman, D., Gonzaga, G., Sugerman, S., Francis, D., & Cook, S. (2011). Effect of fresh fruit availability at worksites on the fruit and vegetable consumption of low-wage employees. *Journal of Nutrition Education and Behavior, 43*(4 Suppl 2), S113–S121. https://doi.org/10.1016/j.jneb.2011.04.003

Berner, M., Ozer, T., & Paynter, S. (2008). A portrait of hunger, the social safety net, and the working poor. *The Policy Studies Journal, 36*(3), 403–420. https://doi.org/10.1111/j.1541-0072.2008.00274.x

BLS. (2020, July). *A profile of the working poor, 2018.* https://www.bls.gov/opub/reports/working-poor/2018/home.htm

Bosmans, K., Mousaid, S., De Cuyper, N., Hardonk, S., Louckx, F., & Vanroelen, C. (2016). Dirty work, dirty worker? Stigmatisation and coping strategies among domestic workers. *Journal of Vocational Behavior, 92*, 54–67. https://doi.org/10.1016/j.jvb.2015.11.008

Burton, J. (2010). *WHO healthy workplace framework and model: Background and supporting literature and practices.* https://apps.who.int/iris/bitstream/handle/10665/113144/9789241500241_eng.pdf?sequence=1

CDC. (2020, August 19). *About social determinants of health.* https://www.cdc.gov/socialdeterminants/about.html

Cortina, L. M., Kabat-Farr, D., Magley, V. J., & Nelson, K. (2017). Researching rudeness: The past, present, and future of the science of incivility. *Journal of Occupational Health Psychology, 22*(3), 299–313. https://doi.org/10.1037/ocp0000089

Creed, P. A., & Klisch, J. (2005). Future outlook and financial strain: Testing the personal agency and latent deprivation models of unemployment and well-being. *Journal of Occupational Health Psychology, 10*(3), 251–260. https://doi.org/10.1037/1076-8998.10.3.251

Crowl, J. N. (2017). The effect of community policing on fear and crime reduction, police legitimacy and job satisfaction: An empirical review of the evidence. *Police Practice and Research, 18*(5), 449–462. https://doi.org/10.1080/15614263.2017.1303771

Cuervo, I., Tsui, E. K., Islam, N. S., Harari, H., & Baron, S. (2020). Exploring the link between the hazards and value of work, and overcoming risk for community-based health interventions for immigrant latinx low-wage workers. *Qualitative Health Research.* https://doi.org/10.1177/1049732320964262

de Boer, J., Lok, A., Van't Verlaat, E., Duivenvoorden, H. J., Bakker, A. B., & Smit, B. J. (2011). Work-related critical incidents in hospital-based health care providers and the risk of post-traumatic stress symptoms, anxiety, and depression: A meta-analysis. *Social Science & Medicine, 73*(2), 316–326. https://doi.org/10.1016/j.socscimed.2011.05.009

Donkin, A., Allen, M., Allen, J., Bell, R., & Marmot, M. (2014). Social determinants of health and the working-age population: Global challenges and priorities for action. In S. Leka & R. R. Sinclair (Eds.), *Contemporary occupational health psychology: Global perspectives on research and practice* (Vol. 3, pp. 1–17). John Wiley & Sons, Ltd.

Dyrbye, L., Herrin, J., West, C. P., Wittlin, N. M., Dovidio, J. F., Hardeman, R., Burke, S. E., Phelan, S., Onyeador, I. N., Cunningham, B., & van Ryn, M. (2019). Association of Racial Bias With Burnout Among Resident Physicians. *JAMA Network Open, 2*(7), e197457. https://doi.org/10.1001/jamanetworkopen.2019.7457

Eurofound. (2020). *European Working Conditions Surveys (EWCS).* https://www.eurofound.europa.eu/surveys/european-working-conditions-surveys-ewcs

Friedland, D. S., & Price, R. H. (2003). Underemployment: Consequences for the health and well-being of workers. *American Journal of Community Psychology, 32,* 33–45. https://doi.org/10.1023/A:1025638705649

Froidevaux, A., Hirschi, A., & Wang, M. (2016). The role of mattering as an overlooked key challenge in retirement planning and adjustment. *Journal of Vocational Behavior, 94,* 57–69. https://doi.org/10.1016/j.jvb.2016.02.016

Gloss, A., Carr, S. C., Reichman, W., Abdul-Nasiru, I., & Oestereich, W. T. (2017). From handmaidens to POSH humanitarians: The case for making human capabilities the business of I-O psychology. *Industrial and Organizational Psychology, 10*(3), 329–369. https://doi.org/10.1017/iop.2017.27

Godin, G., & Kok, G. (1996). The theory of planned behavior: A review of its applications to health-related behaviors. *American Journal of Health Promotion, 11*(2), 87–98. https://doi.org/10.4278/0890-1171-11.2.87

Hobfoll, S. E. (1989). Conservation of resources: A new attempt at conceptualizing stress. *American Psychologist, 44*(3), 513–524. https://doi.org/10.1037/0003-066X.44.3.513

ILO. (2019a). *Work for a Brighter Future.* https://www.ilo.org/infostories/en-GB/Campaigns/future-work/global-commission#intro

ILO. (2019b). *The working poor or how a job is no guarantee of decent living conditions.* https://ilo.org/wcmsp5/groups/public/---dgreports/---stat/documents/publication/wcms_696387.pdf

ILO. (2019c). *World of Work: The ILO at 100, working for peace and social justice.* https://www.ilo.org/wcmsp5/groups/public/---dgreports/---dcomm/documents/publication/wcms_710860.pdf

Johnson, S. E., & Hall, A. (2005). The prediction of safe lifting behavior: An application of the theory of planned behavior. *Journal of Safety Research, 36*(1), 63–73. https://doi.org/10.1016/j.jsr.2004.12.004

Klein, B. W., & Rones, P. L. (1989). A profile of the working poor. *Monthly Labor Review,* 3–13.

Kopasker, D., Montagna, C., & Bender, K. A. (2018). Economic insecurity: A socioeconomic determinant of mental health. *SSM – Population Health, 6,* 184–194. https://doi.org/10.1016/j.ssmph.2018.09.006

Landsbergis, P. A., Grzywacz, J. G., & LaMontagne, A. D. (2014). Work organization, job insecurity, and occupational health disparities. *American Journal of Industrial Medicine*, 57(5), 495–515. https://doi.org/10.1002/ajim.22126

Lee, Y. K., Kim, S. B., Chung, J., Jung, M. J., & Kim, M. H. (2011). The "Chair Campaign" in Korea: An alternative approach in occupational health and safety for service workers. *New Solutions*, 21(2), 269–282. https://doi.org/10.2190/NS.21.2.i

Lipscomb, H. J., Loomis, D., McDonald, M. A., Argue, R. A., & Wing, S. (2006). A conceptual model of *work and health disparities in the United States*. *International Journal of Health Services*, 36(1), 25–50. https://doi.org/10.2190/BRED-NRJ7-3LV7-2QCG

Macik-Frey, M., Quick, J. C., & Nelson, D. L. (2016). Advances in occupational health: From a stressful beginning to a positive future. *Journal of Management*, 33(6), 809–840. https://doi.org/10.1177/0149206307307634

NIOSH. (2019, December 18). *Occupational health equity*. https://www.cdc.gov/niosh/programs/ohe/default.html

NIOSH. (2020, June 29). *What is Total Worker Health?* https://www.cdc.gov/niosh/twh/totalhealth.html

O'Connor, A., Peckham, T., & Seixas, N. (2020). Considering work arrangement as an "exposure" in occupational health research and practice. *Frontiers in Public Health*, 8, 363. https://doi.org/10.3389/fpubh.2020.00363

Our World in Data. (2019). *Distribution of population between different poverty thresholds, World, 1981 to 2015*. https://ourworldindata.org/grapher/distribution-of-population-poverty-thresholds?stackMode=relative

Park, W., & Baek, J. (2019). The impact of employment protection on health: Evidence from fixed-term contract workers in South Korea. *Social Science & Medicine*, 233, 158–170. https://doi.org/10.1016/j.socscimed.2019.05.002

Porter, B., Hoge, C. W., Tobin, L. E., Donoho, C. J., Castro, C. A., Luxton, D. D., & Faix, D. (2018). Measuring aggregated and specific combat exposures: Associations between combat exposure measures and posttraumatic stress disorder, depression, and alcohol-related problems. *Journal of Traumatic Stress*, 31(2), 296–306. https://doi.org/10.1002/jts.22273

Probst, T. M., Sinclair, R. R., Sears, L. E., Gailey, N. J., Black, K. J., & Cheung, J. H. (2018). Economic stress and well-being: Does population health context matter? *Journal of Applied Psychology*, 103(9), 959–979. https://doi.org/10.1037/apl0000309

Ravenelle, A. J. (2019, December 8). The gig economy's sexual misconduct problem, and how to fix it. *New York Times*. https://www.nytimes.com/2019/12/08/opinion/uber-sexual-misconduct.html

Reeves, A., McKee, M., Mackenbach, J., Whitehead, M., & Stuckler, D. (2017). Introduction of a national minimum wage reduced depressive symptoms in low-wage workers: A quasi-natural experiment in the UK. *Health Economics*, 26(5), 639–655. https://doi.org/10.1002/hec.3336

Saxena, M. (2015). Communicable disease control in South Asia. In I. McWha-Hermann, D. C. Maynard, & M. Berry (Eds.), *Humanitarian work psychology and the global development agenda* (pp. 69–83). New York, NY: Routledge. https://doi.org/10.4324/9781315682419-6

Saxena, M. (2017). Workers in poverty: An insight into informal workers around the world. *Industrial and Organizational Psychology*, 10(3), 376–379. https://doi.org/10.1017/iop.2017.29

Saxena, M., & Burke, M. M. (2020). Communicable diseases as occupational hazards for agricultural workers: Using experience sampling methods for promoting public health. *International Perspectives in Psychology: Research, Practice, Consultation*, *9*(2), 127–130. https://doi.org/10.1037/ipp0000129

Schill, A. L., & Chosewood, L. C. (2013). The NIOSH Total Worker Health program: an overview. *Journal of Occupational and Environmental Medicine*, *55*(12 Suppl), S8–11. https://doi.org/10.1097/JOM.0000000000000037

Sears, L. E. (2008). *Work-related outcomes of financial stress: Relating perceived income adequacy and financial strain to job performance and well-being.* [Unpublished Master's Thesis, Portland State University]. Portland, OR.

Shoss, M. K. (2017). Job insecurity: an integrative review and agenda for future research. *Journal of Management*, *43*(6), 1911–1939. https://doi.org/10.1177/0149206317691574

Sinclair, R. R., & Cheung, J. H. (2016). Money matters: Recommendations for financial stress research in occupational health psychology. *Stress & Health*, *32*(3), 181–193. https://doi.org/10.1002/smi.2688

Siqueira, C. E., Gaydos, M., Monforton, C., Slatin, C., Borkowski, L., Dooley, P., Liebman, A., Rosenberg, E., Shor, G., & Keifer, M. (2014). Effects of social, economic, and labor policies on occupational health disparities. *American Journal of Industrial Medicine*, *57*(5), 557–572. https://doi.org/10.1002/ajim.22186

Vandoros, S., Avendano, M., & Kawachi, I. (2019). The association between economic uncertainty and suicide in the short-run. *Social Science & Medicine*, *220*, 403–410. https://doi.org/10.1016/j.socscimed.2018.11.035

Watson, B., & Osberg, L. (2017). Healing and/or breaking? The mental health implications of repeated economic insecurity. *Social Science & Medicine*, *188*, 119–127. https://doi.org/10.1016/j.socscimed.2017.06.042

WHO. (n.d.). *Healthy workplaces: A WHO global model for action.* https://www.who.int/occupational_health/healthy_workplaces/en/

World Bank. (2020, October 7). *Poverty: Overview.* https://www.worldbank.org/en/topic/poverty/overview

INDEX

abusive supervision: effects of 187, 191; measures of 192; *see also* bullying; interpersonal mistreatment

accidents: and associated costs 257; factors affecting 260–262; and microaccidents or near misses 264, 268–269; prevalence and impact of 262; as safety-related metrics 264; and stress 143; *see also* safety interventions in the workplace

Adler, A. 98–99

affect 83; as a resource 82–84; states *vs.* traits 83; work implications of 83; *see also* psychological states and conditions

allostatic load model 133; *see also* stress

attention restoration theory 136; and involuntary attention 136; *see also* resource recovery

boredom 84; *see also* psychological states and conditions

breaks from work, for rest 120, 163, 239, 244, 260; *see also* physical and environmental demand and resource interventions; resource recovery

bullying: cyberbullying 187; definition of 186–187; prevalence and impact of 190; prevention resources for 193; *see also* interpersonal mistreatment; interpersonal mistreatment interventions

burnout 84–85; associated with job demands and resources 85; as a disease 85; *see also* psychological states and conditions

bystander training 194–195; *see also* interpersonal mistreatment interventions

cardiovascular health 110–111; cardiovascular disease 110; connections to stress 110–111

Carroll-Garrison, M. 198–199

chronic physical health conditions 110, 112

Civility, Respect, and Engagement in the Workplace (CREW) intervention 194; *see also* interpersonal mistreatment interventions

cognitive demands and resources 162–164; *see also* psychosocial demands and resources

collaboration in OHP: community partnerships as examples of 287; importance of 289; interdisciplinary and multidisciplinary forms of 2, 4, 8, 10, 14–16, 26

conservation of resources (COR) theory 83, 138–139, 208; applied to physical health 112; applied to work and nonwork role dynamics 208; and resource caravans 139; and resource loss spiral 139

constraints 162–163; *see also* cognitive demands and resources

contingent labor and working situations 281, 284, 285

control 163; *see also* cognitive demands and resources

coping strategies 145–146; *see also* stress

counterproductive workplace behavior (CWB) 184; *see also* interpersonal mistreatment

coworker/peer support 94; *see also* psychosocial demands and resources

critical incident debriefing 266; *see also* safety interventions in the workplace

Croghan, S. 248–249
Cullen, J. 293–294

Day, A. 20–21
de-escalation training 32, 198; *see also* interpersonal mistreatment interventions
demands 130–148; *see also* resources; stressors
dirty and 3d work 233, 282, 288
diversity and inclusion as important to OHP research and practice 65, 197, 216, 287

EAP (employee assistance program) 89, 95; *see also* psychosocial demand and resource interventions
effort-recovery model 135; *see also* recovery
effort-reward imbalance theory 132; *see also* stress
emotion 83; *see also* psychological states and conditions
emotional demands and resources 164–165; *see also* psychosocial demands and resources
emotional labor 164–165; deep *versus* surface acting 164; emotional display rules 164; *see also* emotional demands and resources
emotion regulation 169; *see also* psychosocial demand and resource interventions
empathy 165; *see also* emotional demands and resources
engagement 84–85; *see also* job demands and resources (JD-R) theory; psychological states and conditions
ergonomic design and training 242, 245; *see also* physical and environmental demand and resource interventions
essential workers as an OHP concern 168, 282
ethical practice of OHP: importance of 288
exposure-related conditions affecting physical health 110–111; extreme temperatures 110; hearing loss 110; respiratory conditions 109; *see also* physical health

family supportive organizational perceptions (FSOP) 208; *see also* role-related demands and resources;

work and nonwork role dynamics interventions
family supportive supervisor behaviors (FSSB) 208, 219, 222–223; *see also* work and nonwork role dynamics interventions
fatigue 84; chronic emotional fatigue 84; *see also* psychological states and conditions
field theory 35
flourishing 86; *see also* psychological states and conditions
future of work 280–281

gig workers 281
Gonzalez-Morales, M. G. 150–151
gratitude interventions 93; *see also* psychological health interventions
growth mindset 171; *see also* psychosocial demand and resource interventions

Hammer, L. 222–223
happy-productive worker hypothesis 83
harassment: defined 186; effects of 190–191; prevalence and impact of 186, 190–191; *see also* counterproductive workplace behavior (CWB); interpersonal mistreatment; interpersonal mistreatment interventions
hazard identification and/or incident monitoring 267–268; *see also* safety interventions in the workplace
health 1
health disparities 111
health-related behaviors 111; *see also* physical health
Healthy People 2080 campaign 283
help-seeking for psychological health disorders 89, 92; *see also* psychological health interventions
human rights 190, 264; *see also* ethical practice of OHP

identity prominence and salience 210, 217; *see also* role boundaries
IGLO (individuals, groups, leaders, organizations) multilevel intervention framework 27–30; *see also* intervention design elements
incentives and incentive programs for safety 259, 267; *see also* safety interventions in the workplace

incivility: defined 185; incivility spiral 185; measures of 192; prevalence and impact of 190; selective incivility 192, 197; *see also* interpersonal mistreatment
inclusion 197, 216, 287
income insecurity and economic stress 285
individual differences: impact and influence of 56–59, 65–67; measurement of 67–69; as more than covariates 69; work and life experience as main source of 58–59
individual differences in cognitive functioning 63; general cognitive ability 63; mindsets and orientations 63
individual differences in demographic characteristics 58–61; age and life stage 60; income insecurity 61; living. wage 62; perceived income adequacy 61; race and ethnicity 61; sex and gender identity 59–60
individual differences in emotion 64; emotional intelligence 64; positive and negative affectivity 64
individual differences in physical and behavioral characteristics 64–65; behavioral tendencies 64; physical abilities 64–65
individual differences in religious and spiritual beliefs and practices 62
injuries, work-related: prevalence of 260, 262; as safety-related metrics 264, 268, 269; *see also* physical health interventions; physical and environmental demand and resource interventions; safety interventions in the workplace
(in)justice 163–164; *see also* cognitive demands and resources; interpersonal mistreatment
(in)securities 163–164; *see also* cognitive demands and resources
institutions supporting OHP 9–15; government-supported organizations 11–13; knowledge dissemination outlets 14–15; non-governmental organizations 13–14; professional societies and associations 11
intentional change theory 144; *see also* intervention design elements; intervention mechanisms
International Coordinating Group for OHP (ICG-OHP) 14; *see also* institutions supporting OHP

interpersonal conflict: definition and examples of 186; measurement of 192; prevalence of 189; *see also* interpersonal mistreatment
interpersonal mistreatment: effects of 190–192; measurement of 192–193; perpetrators of 188; person-level predictors of 188–189; prevalence of 190; situation predictors of 189–190; targets or victims of 188–189; types of 185–188; *see also* abusive supervision; bullying; harassment; incivility; interpersonal conflict
interpersonal mistreatment interventions 194–196; evaluation of 196; for groups 194–195; for individuals 193–194; for leaders 194–195; for organizations 195–196
interrole dynamics and transitions 208, 210–214; as balance or balancing 213–214; as conflict 211–213; as enhancement and/or facilitation 213; role interface models 212; *see also* work and nonwork role dynamics
intersectionality 189
intervention design elements 24–37; how the intervention will work 31–35; importance of context 35–37; primary, secondary, and tertiary forms of 32–33; what to change 26–27; who will be impacted 27–30; why the intervention is important 29–31; *see also* sustaining interventions over time
intervention evaluation 37–45; alpha, beta, and gamma change 44; confounding variables 45; CONSORT (Consolidated Standards of Reporting Trials) 39; data analysis and interpretation 43–46; double blind and single blind studies 42; evaluation over time 40–41; formative and process evaluation 38; measurement forms and sources 43; mixed methods 43; pilot testing 43; RCT (randomized control trial) 32, 41–42; response-shift bias 45; ROI (return on investment) 30, 40; single item measures 43; summative and outcome evaluation 37; timing considerations 39–40; what to measure 38–39
intervention mechanisms 32–35; attitude and intention modification 34; behavior change 34; cognitive processing adjustment 34–35;

environmental force rebalancing 36; knowledge transfer 34–35; social influence 35–36; stimulus-response control 32–33; *see also* intervention design elements

intrarole dynamics 208; *see also* work and nonwork role dynamics

isolation 86; *see also* psychological states and conditions

Jex, S. 71–72

job characteristics model 86; *see also* meaningful work; psychological states and conditions

job crafting 147, 163, 169–170; *see also* psychosocial demand and resource interventions

job demands and control (D-C) model 138, 14, 171

job demands and resources (JD-R) theory 83, 85, 95, 139, 208; applied to work and nonwork role dynamics 208

job enrichment 170–171; *see also* psychosocial demand and resource interventions

job rotation 239, 244; *see also* physical and environmental demand and resource interventions

living wage 62, 285

logic model 38; *see also* intervention design elements; intervention evaluation

loneliness 85–86; as epidemic 85–86; at work 86; *see also* psychological states and conditions

match hypothesis 146–147; *see also* demands and resources in work environments

meaningful work 86; and job characteristics model 86; *see also* psychological states and conditions

methodological skills needed in OHP: alternative sampling approaches 288; qualitative inquiry 288–289; quasi-experimental designs 289; *see also* intervention design elements; intervention evaluation

MHAT (mental health awareness training) 94; *see also* psychological health interventions

mixed method data collection 168; *see also* psychosocial demand and resource interventions; intervention evaluation

mobbing 187; *see also* bullying; interpersonal mistreatment

mood 82–83; *see also* individual differences in emotion; psychological states and conditions

multilevel modeling, applied to safety climate 265

needs assessment 25; *see also* intervention design elements

Nielsen, K. 48

Nigam, J. 174–176

NIOSH (National Institute for Occupational Safety and Health) . 5, 12; *see also* institutions supporting OHP

NORA (National Occupational Research Agenda) 12; *see also* NIOSH

norms and expectations 170; *see also* social intervention mechanisms; demands and resources

occupational health inequities 284; *see also* individual differences

occupational health psychology (OHP): as amplifier of efforts to influence worker health, safety, and well-being 4; and associated disciplines 4, 6–7; defined 1–2; education and training 8–10; history and development 5, 8; as lens or hub 4–5; present challenges 15–16; and scientist-practitioner reality 2–3

Olson, R. 121–122

opportunities for broader impact in OHP: addressing social determinants of health, safety, and well-being 283–286; contributing to social justice and peace 287–288; protecting human dignity and rights 281–283

organizational role theory 207; *see also* work and nonwork role dynamics

OSHA (American Occupational Safety and Health Administration) 13; *see also* institutions supporting OHP

person-environment fit: applied to physical and environmental demands and resources 233; and stress 132; *see also* match hypothesis; stress physical and environmental demand and resource interventions 241–246; evaluation of 245–246; for groups, leaders, and organizations 243–245; for individuals 242–243

physical and environmental demands and resources 233–240; aesthetics of workplace 237; air quality 236; connections to psychological and physical health 238–239; importance of understanding 239–240; individual differences in exposure and responses to 238–239; lighting 236–237; measurement of 240–241; noise 235; prolonged sitting/standing 234–235; temperature 236; *see also* demands; resources

physical health: connections to work 110, 109, 111; definition of 109–110; effects of 112–114; influence of social forces and norms on 111, 117; measuring and monitoring 114–115

physical health interventions 116–119; evaluation of 118–119; for individuals 116–117; for groups and/or leaders 117; for organizations 118

policies and procedures pertaining to worker physical health 118; *see also* physical health interventions

popcorn metaphor, for interpersonal mistreatment 189

population health 114

PPE (personal protective equipment) 243, 245, 267; *see also* physical and environmental demand and resource interventions; safety interventions in the workplace

presenteeism 90, 113, 118; *see also* psychological health; physical health

Probst, T. 271–272

process mapping 240–241; *see also* physical and environmental demands and resources

psychological health: costs associated with 90; effects of 86, 90; leader support for 91–93; measurement of 91–92; stigma and 91, 93; work connection 81–82, 86

psychological health disorders 87; common symptoms and implications of 87; *see also* psychological health

psychological health interventions 92–96; evaluation of 96; for groups

93–94; for individuals 91–92; for leaders 94–95; for organizations 95–96

psychological safety climate 91; *see also* psychological health; safety in the workplace

psychological states and conditions 81–86; affect, mood, and emotion 82–85; connectedness and purpose 84–85; energy-related 84–85

psychosocial demand and resource interventions 169–172; evaluation of 172; for individuals 169–170; for groups, leaders, and/or organizations 170–172

psychosocial demands and resources: defined 161–162; forms of 167–172; importance of 167–168; measurement of 169; *see also* cognitive demands and resources; demands; emotional demands and resources; resources; social demands and resources

psychosomatic symptoms 110; *see also* work-related stress affecting physical health

public health model 38; *see also* intervention design elements; intervention evaluation

reciprocity motives 185, 188, 239; *see also* interpersonal mistreatment; physical and environmental demands and resource

resilience training 93; *see also* psychological health interventions; resource recovery

resource recovery: control and 136; elements of 134–136; mastery experiences and 135; as part of stress process 134–136; psychological detachment and 135; relaxation and 136; *see also* self-care, stressor-detachment model

resources 136–140; resource alignment 137–138; types of resources 136–140; *see also* demands; match hypothesis; physical and environmental demands and resources; psychosocial demands and resources; resource recovery

retaliation 186, 193, 195, 259; *see also* interpersonal mistreatment

retirement transition and identification with work 282

return-to-work programs 118; *see also* physical health interventions

role: definition of 207; influence of 207; *see also* organizational role theory

role accumulation 213; *see also* interrole dynamics and transitions

role boundaries 208–210; clarity of 209; consistency and symmetry of 210; strength of 209–210; *see also* interrole dynamics and transitions; work and nonwork role dynamics

role conflict 207, 211, 212; *see also* interrole transitions and dynamics; work and nonwork role dynamics

role interference 212; *see also* role conflict; work and nonwork role dynamics

role overload 212; *see also* role conflict; work and nonwork role dynamics

role-related demands and resources 207–208; *see also* conservation of resources (COR) theory; demands; job demands and resources (JD-R) theory; resources

safety behavior, predictors of: deviance proneness 261; environmental factors 259–260; fatigue 260, 261; leader behaviors 260; organization safety climate 257, 258–259; policy enforcement 259; safety knowledge 257, 261; work task factors 259–262; worker characteristics 260–261

safety climate 257, 258–259; benefits of 259; measurement of 265; *see also* safety behavior, predictors of; safety interventions in the workplace

safety compliance 257

safety in the workplace: general goals and motivations for 2, 257–258; governing agencies that monitor 257–258; importance of 262–263; measurement of 264–265; *see also* safety behavior, predictors of

safety interventions in the workplace 265–269; evaluation of 269; for groups and leaders 266–267; for individuals 265–266; for organizations 267–268

safety knowledge 257, 261, 265; *see also* safety behavior, predictors of

safety motivation 257, 261, 265; *see also* safety behavior, predictors of

safety participation 257; *see also* safety in the workplace

safety silence 259; *see also* safety in the workplace

safety voice 261, 266 *see also* safety in the workplace

Sauter, S. 18–19

scheduling of work: and physical and environmental demands and resources 244; and physical health 118, 120; and psychological health 94; and safety 260, 267; and work and nonwork role dynamics 208, 215, 218–219; *see also* physical health interventions

sedentary work, adjustments to 234–235; *see also* physical and environmental demand and resource interventions

self-care 142; *see also* resource recovery

self-efficacy 163; *see also* cognitive demands and resources

shiftwork: as physical and environmental demand 237–238, 244; shiftwork sleep disorder 110; *see also* cardiovascular health; scheduling of work

Sinclair, R. R. 291–292

social and normative forces affecting health behaviors 117; *see also* physical health interventions

social demands and resources 165–167; *see also* psychosocial demands and resources

social support 166–167; *see also* social demands and resources

socialization 170; *see also* psychosocial demand and resource interventions

SOHP (Society for Occupational Health Psychology) 8, 11; *see also* institutions supporting OHP

Support-Transform-Achieve-Results (STAR) intervention 219; *see also* work and nonwork role dynamics interventions

sustaining interventions over time 46; *see also* intervention evaluation

stigma: associated with psychological health 91, 93; associated with physical health conditions 112, 117; defined 166; *see also* social demands and resources

stimulus-response model 27–28; *see also* stress

strain 133–134; *see also* stress
stress: affecting cardiovascular health
110–111; business costs of 142–143;
cumulative nature of 131; defined
130, 131–132; effects of chronic
experiences of 133–134; general
effects of 132; as major OHP focal
area 129; measurement of 143–144;
normalization of 140; psychosomatic
symptoms of 110; sleep problems
and 110; *versus* anxiety 131; *see also*
physical health; recovery, stressors,
strain
stress and recovery interventions
142–148; evaluation of 144; for
individuals 144–148; for groups,
leaders, and/or organizations
146–148
stressors: appraisal of 139–140; demands
as 130; examples of 130–131; *see also*
demands; transactional theory of stress
stressor-detachment model 135; *see also*
recovery

theory of planned behavior: applied to
safe lifting 242–243; applied to safe
work practices 289; *see also* physical
and environmental demand and
resource interventions
thriving 86, 90; *see also* psychological
states and conditions
Total Worker Health® 12, 283
transactional theory of stress 139; *see also*
stress; stressors

violence at work: as a form of
interpersonal mistreatment 184,

186–188; influence of past experiences
with 59; interventions against 38;
measurement and monitoring of
192–193; prevalence and impact of
190; as a safety concern 268–269;
see also interpersonal mistreatment
interventions; safety interventions in
the workplace

wellness programs 114, 116–117; *see also*
physical health interventions
work and nonwork balance and
balancing 213–214; *see also* interrole
transitions and dynamics
work and nonwork role dynamics
206–216; importance of 214–216;
measurement of 216; *see also* interrole
transitions and dynamics
work and nonwork role dynamics
interventions 217–220; evaluation
of 220; for groups, leaders, and
organizations 218–220; for individuals
217–218
work and nonwork supportive benefits,
culture, and policies 215, 218; *see
also* family supportive organizational
perceptions (FSOP); family supportive
supervisor behaviors (FSSB); work and
nonwork role dynamics interventions
workforce and succession planning 172;
see also psychosocial demand and
resource interventions
working poor 285
workload 162; *see also* cognitive demands
and resources; demands
work-related musculoskeletal disorders
(WMSD) 109; *see also* physical health

Made in United States
North Haven, CT
23 October 2025

81191824R00176